Surface Area and Porosity Determinations by Physisorption

Surface Area and Porosity Determinations by Physisorption

Measurement, Classical Theories and Quantum Theory

Second Edition

James B. Condon

ELSEVIER

Elsevier
Radarweg 29, PO Box 211, 1000 AE Amsterdam, Netherlands
The Boulevard, Langford Lane, Kidlington, Oxford OX5 1GB, United Kingdom
50 Hampshire Street, 5th Floor, Cambridge, MA 02139, United States

© 2020 Elsevier B.V. All rights reserved.

No part of this publication may be reproduced or transmitted in any form or by any means, electronic or mechanical, including photocopying, recording, or any information storage and retrieval system, without permission in writing from the publisher. Details on how to seek permission, further information about the Publisher's permissions policies and our arrangements with organizations such as the Copyright Clearance Center and the Copyright Licensing Agency, can be found at our website: www.elsevier.com/permissions.

This book and the individual contributions contained in it are protected under copyright by the Publisher (other than as may be noted herein).

Notices

Knowledge and best practice in this field are constantly changing. As new research and experience broaden our understanding, changes in research methods, professional practices, or medical treatment may become necessary.

Practitioners and researchers must always rely on their own experience and knowledge in evaluating and using any information, methods, compounds, or experiments described herein. In using such information or methods they should be mindful of their own safety and the safety of others, including parties for whom they have a professional responsibility.

To the fullest extent of the law, neither the Publisher nor the authors, contributors, or editors, assume any liability for any injury and/or damage to persons or property as a matter of products liability, negligence or otherwise, or from any use or operation of any methods, products, instructions, or ideas contained in the material herein.

Library of Congress Cataloging-in-Publication Data
A catalog record for this book is available from the Library of Congress

British Library Cataloguing-in-Publication Data
A catalogue record for this book is available from the British Library

ISBN: 978-0-12-818785-2

For information on all Elsevier publications
visit our website at https://www.elsevier.com/books-and-journals

Publisher: Susan Dennis
Acquisition Editor: Susan Dennis
Editorial Project Manager: Theresa Yanetty
Production Project Manager: Omer Mukthar
Cover Designer: Miles Hitchen

Typeset by SPi Global, India

Contents

README first—Organization of this book xi
List of figures xiii
List of tables xxiii
Foreword to the 1st edition xxvii
Foreword to the 2nd edition xxxi

CHAPTER 1 An overview and some uninteresting history of physisorption 1

Introduction 1
   What do you want? 1
   Scope and terminology 2
General description of physisorption 6
Measuring the surface area by physisorption 8
Preliminary analysis 10
   Classical classifications of adsorption isotherm types 10
   Single point methods 14
   The inflection point method 16
   Quantum mechanical classifications of isotherms using χ plots 22
Measuring the surface area from the isotherm 27
   What is the "Henry's law" assumption? 28
   Implications of the Henry's law disproof 30
   The BET theory (Brunaur, Emmett, and Teller) 32
   The "absolute method" by Harkins and Jura 36
   χ and ESW theories that do yield monolayer equivalence n_m 36
   Relating a and A_s to n_m 38
   The advantages of the χ theory formulation 41
   Summary for introduction to Chi and ESW theories 42
   Using standard curves to obtain the n_m 42
Determining porosity by physical adsorption 43
   Micropore analysis 43
   Ultramicropore analysis 46
   Mesopores 47
   Macropores 52
Statistical treatment of isotherms—Error analysis 53
Summary of this chapter 54

What's next? 55
References 56

CHAPTER 2 Measuring the physisorption isotherm **59**

Introduction—Equipment requirements 59
The volumetric method 60
    Equipment description 60
    Determination method 62
    Error analysis for the volumetric method 65
    Advantages and disadvantages of the volumetric
    technique 68
The gravimetric method 69
    Equipment description 69
    Determination method 71
    Error analysis for the gravimetric technique 73
    Advantages and disadvantages of the gravimetric
    technique 73
General error analysis—Common to both volumetric
and gravimetric 74
    Pressure and temperature measurements 74
    Kinetic problems 76
    Sample density problems 77
Calorimetric techniques 77
    Adiabatic calorimetry 77
    Measuring the isosteric heat of adsorption 79
    The thermal "absolute" method 79
    Differential scanning calorimetry 82
Arnold's adsorptive flow system 83
Carrier gas flow method 86
    Engineering aspects: Carrier gas flow method 87
Summary of methods 91
References 92

CHAPTER 3 Interpreting the physisorption isotherm **93**

Interpreting the isotherm 93
Objectives in interpreting isotherms 95
Surface area determination from isotherms 100
    The BET analysis 100
    χ plot analysis 102
    Strategy of the non-local density functional theory 104
    Strategy of the χ/ESW theory 105

- Dubinin et al. method of determining surface area 106
- Tóth T-equation isotherm 107
- The Harkins-Jura absolute/relative method 107
- Porosity determinations from the isotherm 108
- Approximate pore analysis method 109
  - Micropore analysis 109
  - Mesoporosity analysis 112
- Isotherm fits which yield relative numbers for the surface area 118
  - Langmuir isotherm 118
  - Freundlich isotherm 120
  - Polanyi formulations 121
  - deBoer-Zwikker formulation 122
  - The Frenkel, Halsey, Hill (FHH) isotherm 122
  - Analysis using standard isotherms 123
  - Principal problem with standard curves 124
  - Standard isotherms 124
  - Tables of standard isotherms 128
- What about those Type VI isotherms? 138
- Summary of interpreting the physisorption isotherm 140
- References 143

CHAPTER 4 Theories behind the χ plot 145

- Introduction—A short historical background 146
- Theory behind χ plots 147
  - The quantum mechanical derivation of the "simple" χ equation 147
  - A Particle in a box with a "Tooth"—Perturbation theory for adsorption on a surface with a molecule present 150
  - Continuing the derivation 156
  - Converting χ_S to different conditions 160
  - Summary 162
  - What is a? ... and is the hard sphere approximation appropriate? 162
  - The hard sphere geometry and Lennard-Jones correction 167
  - Summary 168
  - In summary of χ theory so far 169
  - The disjoining pressure derivation (ESW) 169
  - The meaning of Γ_m in the hard sphere model 174
  - Relationship to χ-theory 175
- Heterogeneous surfaces 177

The additive principle of χ plots 177
Insensitivity for $\chi \geq \max \chi_c$ 178
Reformulation for a distribution of E_a values 179
Heats of adsorption 180
Isosteric heat of adsorption, q_{st} 180
The integral heats of adsorption 181
Heats of adsorption from BET plot 183
Depth profiles and n-layer calculations with χ theory 184
Assumptions used to calculate the Normal Direction Profile 185
#3 The QMHO approximation for the LJ 6–12 potential 186
#4 The packing problem uncertainty 188
Combining the normal direction distributions with the χ-n layer results 189
Simulations of the isotherm using n-layer 194
Adsorption of binary adsorptives—More than one adsorbate 195
Method 1: Adjusting the grand canonical ensemble 196
Method 2: Binary adsorbate using the n-layer analysis 205
Is the χ plot compatible with the Freundlich and Dubinin isotherms? 208
The thermodynamics and spreading pressure 210
Gibbs' phase rule in systems with surfaces 211
The Olivier monolayer criterion 213
Summary of this chapter 213
References 215

CHAPTER 5 Comparison of the χ and ESW equations to measurements 217

Comparison to standard isotherms 217
Cranston and Inkley standard t-curve 218
deBoer's standard t-plots 218
The α-s standard plots 221
Standard thoria plots 223
Standard curves for lunar soils 224
Isotherms by Nicolan and Teichner 227
Isotherms by Bradley 228
McGavack and Patrick 232
Data by Bray and Draper 233
Conclusion and some comments about carbon 233
Summary with regards to standard curves 234
The observation of χ_c 235

Observations of the energy implications of χ_c 236
Direct observation of χ_c (threshold pressure) 237
Summary and conclusion concerning χ_c 246
Multiplane adsorption..................246
    Examples of two plane adsorption..................246
    The Freundlich, Dubinin-Polanyi, and Tóth isotherms 247
    Summary of comparisons to some other isotherms..............251
    Do isotherms for multiplane adsorption have anomalies? 251
    Conclusion concerning multiple energies..................255
Heats of adsorption..................255
    Data by Harkins and Jura—The absolute method..................256
    Data by T. Berg, Kr on Anatase—Simultaneous measurements..................258
    Data by Kington, Beebe, Polley, and Smith on anatase........260
    Summary of heats of adsorption 261
Adsorption of more than one adsorbate..................261
    Adsorption on non-porous surfaces..................262
    Binary adsorption in micropores..................267
    Conclusions regarding binary adsorption..................274
Statistical comparisons of other isotherms to the χ-plot 275
General conclusions..................277
References..................277
Further reading 279

CHAPTER 6   Porosity calculations..................281

Introduction to porosity calculations..................281
Ultramicropore analysis..................284
Micropore analysis 291
    The BDDT equation 291
    The DR and DA equations..................292
    Standard curve analysis using distributions..................293
    χ theory interpretation of the distribution fit..................298
Analysis of mesoporosity—Classical..................308
    Comments about the standard plot of determining mesoporosity..................309
    What χ theory says about prefilling..................318
Is it microporous or mesoporous and does it matter? 319
    Combined mesopore/micropore equation χ 319
    Using an arbitrary function to determine porosity 320
    Interpretation of mesopore equation using standard curve..................321
    The boundary between mesopores and micropores..............322

Should one use a micropore or a mesopore analysis?...........323
Real data examples..324
What does χ theory say about hysteresis?.......................328
χ theory n-layer simulation 329
Conclusions and summary..330
References..333

CHAPTER 7 Density functional theory (DFT)..............................335

Introduction..335
What is a functional?..337
The functional derivative 339
Correlation functions 340
A quick trip through some partition functions 341
Direct correlation functions..344
The hard rod approximations 345
Hard rods between two walls...347
Percus-Yevick solution expansion for hard spheres.................349
Thiele analytical approximation...351
The Carnahan-Starling approximation 351
Helmholtz Free Energy from the CS Approximation...............352
Non-local density functional theory (NLDFT).........................353
Modeling with the presence of a surface 355
A note about the Monte Carlo technique.................................357
χ theory versus NLDFT, what are the practical differences? ...358
References..360

Appendixes 361
Acknowledgments 401
In Memory of Dr. E. Loren Fuller...403
Author Index...405
Subject Index 409

README first—Organization of this book

This book is organized with the most rudimentary material at the beginning and continues with the more sophisticated material at the end. In this sense it could be considered a textbook, and in fact it was originally designed to be such. As with all textbooks, it is not recommended to start reading in the middle, or to use material at the beginning to do complex analysis that is described later in the book. For example, it would be ill-advised to try to do pore size distribution calculations by simply skipping to Chapter 6 without the background information about the quantum mechanically (QM) derived χ (ESW) theory.[1] It would also not be very accurate to use only the standard curve method to determine porosity, although this might be useful. (This would be taking advantage of the fact that the χ theory breaks the dependence of the standard curve method upon a matching non-porous standard, thus getting around the problem that it is nearly impossible to find a matching non-porous sample.) Thus, if you only read enough of this book to use the χ theory in the standard curve method, then this will not be sufficient today to publish your analysis in the open literature.

My advice, if you only want to use the QM derived χ theory, is to skip the portions of the book that deal with the older theories and just study the relevant χ theory sections. Furthermore, if you are uneasy regarding QM, then simply trust the final χ equation that is derived and skip the QM portion. (When you get to this point, there is a note telling you to do this.)

---

[1] The χ theory (in a few reports listed as "CFS"), the Auto Shielding Physisorption (ASP) theory, the Disjoining Pressure Theory, and the Excess Surface Work (ESW) theory are all fundamentally the same theory, and will usually be referred to as simply χ theory. Each has approached the problem from a different viewpoint but ends up with the same isotherm equation. This makes it problematic if one searches using keywords, and one therefore needs to use all of these terms.

List of figures

Fig. 1.1	A model of adsorption of the "in-register" type—e.g. chemisorption, epitaxy.	6
Fig. 1.2	Models of two types of physisorption: (A) gas-like and (B) liquid-like. ● before encounter, ○ after encounter, hatched is during encounter.	7
Fig. 1.3	Type I isotherm (a sharper and b).	11
Fig. 1.4	Type II isotherm (a →, b offset ←).	11
Fig. 1.5	Type III isotherm.	12
Fig. 1.6	Type IV isotherm (a no loop and b).	12
Fig. 1.7	Type V isotherm with desorption nearly vertical.	13
Fig. 1.8	Type VI(a) isotherm.	13
Fig. 1.9	An illustration of why the "B" point method is not reliable. The position of the "B" point depends upon the abscissa scale used.	15
Fig. 1.10	A simulation illustrating the inflection point method.	17
Fig. 1.11	Type H1 hysteresis loop.	19
Fig. 1.12	Type H2 hysteresis loop.	19
Fig. 1.13	Type H3 hysteresis loop.	20
Fig. 1.14	Type H4 hysteresis loop.	20
Fig. 1.15	Type I isotherm expressed as a standard plot or a χ plot. A simulation of an E_a distribution has been added here.	22
Fig. 1.16	Type II isotherm expressed as a standard plot of a χ plot. No energy distribution is assumed.	22
Fig. 1.17	Type III isotherm expressed as a standard plot or χ plot. A simulation of an E_a distribution has been added here.	23
Fig. 1.18	Type IV isotherm expressed as a standard plot or χ plot.	23
Fig. 1.19	Type V isotherm expressed as a standard plot or χ plot.	24
Fig. 1.20	A standard plot of an alternate VIA isotherm. This is the result of two surfaces with differing E_as.	24
Fig. 1.21	Normal Type VI isotherm expressed as a standard plot or χ plot.	26
Fig. 1.22	The normal isotherm for the alternate Type VI isotherm (VIA). Steps due to different E_as.	27
Fig. 1.23	A χ plot or standard plot of a type I isotherm.	44

List of figures

Fig. 1.24	Data by (S-A,S-A,L R-R) illustrating the "Log-Law," which is often observed.	47
Fig. 1.25	A χ plot or a standard plot of either a Type IV or V isotherm.	48
Fig. 1.26	Illustration of the definitions or A_o area of pore opening, A_w outside "wall" area excluding A_o, r_p pore radius, l pore length, and A_p surface area of the sides of the pores.	50
Fig. 2.1	Schematic of the volumetric system.	61
Fig. 2.2	A drawing of the gravimetric method sample area showing the baffle arrangement.	70
Fig. 2.3	An overview of the system used for the gravimetric method.	71
Fig. 2.4	Consequences of errors in temperature measurement/control in the isotherm.	76
Fig. 2.5	A schematic of a liquid nitrogen cooled adiabatic calorimeter.	78
Fig. 2.6	A schematic of the Harkin and Jura calorimeter to measure the surface area of a powder.	80
Fig. 2.7	A schematic of how the adsorbed film is destroyed when the powder is immersed in the liquid phase thus releasing its surface energy.	80
Fig. 2.8	Schematic of the differential calorimeter by Rouquerol et al.	82
Fig. 2.9	The vacuum flow system developed by Arnold.	83
Fig. 2.10	Suggested arrangement for a sample chamber.	85
Fig. 2.11	Schematic of the signal observed for the flow system.	86
Fig. 2.12	The diffusion-sink situation that one expects from the isotherm type shown on the right. Slabs 1, 2, 3, etc. are filled completely up to n_f. before the front moves on.	88
Fig. 2.13	The contrast of a Type III isotherm with the Type II. The unit step function is not allowed but rather small increments of increase happens at the same time that diffusion is present.	89
Fig. 3.1	Some typical adsorption isotherm for non-porous materials illustrating the problem of identifying the "knee" due to scaling.	96
Fig. 3.2	The transformed BET plot to determine surface area typical of silica material. Linear range is assumed to be 0.05 to 0.35 of P_{vap}.	98

List of figures

Fig. 3.3	The transformed BET plot for an organic material. The 0.05–0.35 range yields a very poor linear fit, thus a high range should be selected.	98
Fig. 3.4	A linear fit to the DRK representation of the adsorption isotherm for a non-porous surface. The fit covers about 2/3 of the $\ln^2$ range.	99
Fig. 3.5	Simulated 2 energy χ-plot and the meaning of the Ss and Is.	102
Fig. 3.6	A schematic of the method used for DFT (NLDFT and QSDFT).	104
Fig. 3.7	Model of the χ method of analysis for physical adsorption isotherm.	105
Fig. 3.8	A typical type I adsorption of possibly microporosity isotherm.	110
Fig. 3.9	The transformed plot using a standard curve to change the x axis.	110
Fig. 3.10	Preliminary parameters obtained from the isotherm to analyze surface area and porosity.	113
Fig. 3.11	Examples of Langmuir isotherms and the position of the "knee" as it varies with adsorption energy.	119
Fig. 3.12	Data by Yahia et al. of CH_4 on MgO at 87.4 K as a χ-plot.	139
Fig. 4.1	Illustration of the error made by assuming an infinite potential well depth versus a finite potential well depth.	151
Fig. 4.2	Diagram of the surface potential well and the energy definitions.	151
Fig. 4.3	A Schematic of the layering of the adsorbate molecule and the vacancies, v, demonstrating how the adsorbed liquid state dimensions are expanded in the plane of the adsorbent but not in the normal direction.	166
Fig. 4.4	The arrangement of an adsorbate molecule "rolling over" another and the distances defined for the treatment of the energy correction.	167
Fig. 4.5	The functionality of surface excess energy, $\Phi(\Gamma)$, with coverage, Γ.	172
Fig. 4.6	Adsorbate molecules between two plates to account for the size of the force between them.	175
Fig. 4.7	Adsorbate molecules between two plates to account for the size of the force between them.	175

List of figures

Fig. 4.8	χ plot of N_2 adsorbed on Vulcan C with two $\overline{E}_a$ straight line fits.	178
Fig. 4.9	χ plot N_2 adsorbed on Sterling FT C with two $\overline{E}_a$ straight line fits.	179
Fig. 4.10	Chi plot of nitrogen adsorbed on high fired thoria indicating two energies of adsorption and another feature by the multiple straight line fits.	179
Fig. 4.11	Dependence of calculated isosteric heat from the BET equation for various input energies.	184
Fig. 4.12	The individual monolayer coverages for layers 1 through 5.	191
Fig. 4.13	Argon depth profiles against a hard-wall of an unlimited surface film.	192
Fig. 4.14	The profiles of the adsorbate "layers" as a function of $\Delta\chi$.	192
Fig. 4.15	Simulated profile fitted to an equation (inside line).	193
Fig. 4.16	Density profile in a slit pore walls separated by 10 nm at $3\Delta\chi s$.	194
Fig. 4.17	Density profile in a cylindrical pore with $r=5$ nm at $3\Delta\chi s$.	195
Fig. 4.18	Denisty profile in a cylindrical pore with $r=2$ nm at $3\Delta\chi s$.	195
Fig. 4.19	Amplitude of the first peak as a function of the calculated.	196
Fig. 4.20	Arnold's adsoptive mixture on anatase of O_2: $N_2=15\% : 85\%$.	202
Fig. 4.21	Arnold's adsoptive mixture on anatase of O_2: $N_2=30\% : 70\%$.	202
Fig. 4.22	Arnold's adsoptive mixture on anatase of O_2: $N_2=50\% : 50\%$.	203
Fig. 4.23	Arnold's adsoptive mixture on anatase of O_2: $N_2=30\% : 70\%$.	203
Fig. 4.24	Arnold's adsoptive mixture on anatase of O_2: $N_2=85\% : 15\%$.	204
Fig. 4.25	χ-plots for the adsorption of pure N_2, ■, and O_2, ×, on anatase by Arnold.	204
Fig. 4.26	A comparison of the χ theory energy distribution and Dubinin-Polanyi (DP) distribution. DP k values used were, starting from the outside, 1, 1.5, and 2.	210
Fig. 5.1	Standard t curve constructed by Cranston and Inkley. The data point by the authors. The line is the least squares χ fit.	218

Fig. 5.2	Standard t-curve data by deBoer et al. The data are the circles and the line is the χ plot least squares fit.	219
Fig. 5.3	Adsorption of I_2 on CaF_2 showing the χ plot according to deBoer.	219
Fig. 5.4	Adsorption of Ar on SnO according to deBoer and Zwikker. The roundoff in the upper portion is probably due to some porosity.	220
Fig. 5.5	N_2 on MgO aerosil according to deBoer et al.	220
Fig. 5.6	N_2 on Ni antigorite according to deBoer et al.	221
Fig. 5.7	Ar on SiO_2 data α-s plot as a χ plot. — is the least squares fit to χ plot.	222
Fig. 5.8	N_2 on SiO_2 data α-s plot as a χ plot. — is the least squares fit to χ plot.	222
Fig. 5.9	N_2 adsorption on thoria normalized to 0.4 P/P_{vap}.	224
Fig. 5.10	Ar adsorption on thoria normalized to 0.4 P/P_{vap}.	224
Fig. 5.11	Adsorption of water at 25°C after several prior adsorption cycles. Data is normalized to 0.4 P/P_{vap}.	225
Fig. 5.12	N_2 adsorption on lunar soil.	226
Fig. 5.13	Ar adsorption on lunar soil.	226
Fig. 5.14	CO adsorption on lunar soil.	226
Fig. 5.15	O_2 adsorption on lunar soil.	227
Fig. 5.16	N_2 adsorption on SiO_2 by Nicolan and Teichner.	228
Fig. 5.17	Ar adsorption on SiO_2 by Nicolan and Teichner.	228
Fig. 5.18	N_2 adsorption on NiO by Nicolan and Teichner.	229
Fig. 5.19	Adsorption of Ar on $CuSO_4$ according to Bradley. Inset is data set #1.	229
Fig. 5.20	Adsorption of Ar on $Al_2(SO_4)_3$ according to Bradley. Inset is data set #1	230
Fig. 5.21	Graph of Bradley-derived van-Hoff plot to determine q_{st}.	231
Fig. 5.22	The χ plot representation of the adsorption of SO_2 on SiO_2 gel according to McGavack and Patrick. Three successive runs with identical conditions but with different samples.	232
Fig. 5.23	The effect water contamination of the SO_2 at three levels. It appears that the higher the H_2O level the higher is the apparent E_a.	233
Fig. 5.24	The RMBM standard α-s carbon curve.	234
Fig. 5.25	$\overline{E}_a$ versus the enthalpy of formation of various oxides.	237
Fig. 5.26	Pc observation from data by GPG with microporous carbon.	238

List of figures

Fig. 5.27	DR plot of the GPG data showing the difference between either the DR fit or the log-law fit to the data.	238
Fig. 5.28	N_2 on Takeda ACF carbon by Nguyen and Do. The "dog-leg" is not evident before the threshold pressure.	239
Fig. 5.29	Ar adsorption on polytetrafluoroethylene (Teflon®) with the normal P/P_{vap} axis by Thompson. In this experiment the threshold pressure is evident even without transforming to a χ plot.	241
Fig. 5.30	χ plot of Ar adsorption on polytetrafluoroethylene (Teflon®) by Thompson. The χ plot makes the threshold pressure more obvious.	242
Fig. 5.31	Various adsorption isotherms on polytetrafluoroethylene (Teflon®) with N_2, Ar, and O_2.	242
Fig. 5.32	χ plot of argon adsorption on H_2 cleaned diamond by Thompson.	243
Fig. 5.33	Nitrogen adsorption on H_2 cleaned alumina.	244
Fig. 5.34	High-resolution N_2 adsorption isotherms at 77.4 K an activated carbon LMA233 on a logarithmic scale.	245
Fig. 5.35	High-resolution N_2 adsorption isotherms at 77.4 K an activated carbon DD52 on a logarithmic scale.	245
Fig. 5.36	χ plot of IUPAC lab "H" adsorption of N_2 on Sterling FT carbon.	247
Fig. 5.37	Comparing DP isotherm with χ-plot with $\sigma = 1$ and $r_F = 1$ (Henry's Law).	248
Fig. 5.38	Comparing DP isotherm with χ-plot with $\sigma = 1$ and $r_F = 0.5$ (Freundlich).	249
Fig. 5.39	Comparing DR isotherm with χ-plot.	249
Fig. 5.40	Comparing Tóth T-equation with Ar χ-plot.	249
Fig. 5.41	Comparing Tóth T-equation with N_2 χ-plot.	250
Fig. 5.42	The energy of adsorption with amount adsorbed.	257
Fig. 5.43	χ-plot of the H_2O adsorption isotherm by HJ to Fig. 5.42.	257
Fig. 5.44	χ plot of data by Berg. Parameters obtain here are constants to calculate the lines in Fig. 5.45.	259
Fig. 5.45	The E of adsorption with amount adsorbed from the data by Berg. Lines are from the χ-plot analysis.	259
Fig. 5.46	χ plot and heat of adsorption Kr adsorbed on rutile at 126K by Dennis.	260
Fig. 5.47	Heats of adsorption by Kington et al. The lines are the least squares fits.	261
Fig. 5.48	χ plot of the pure nitrogen and pure oxygen adsorption on anatase by Arnold.	262

List of figures   xix

Fig. 5.49	The moles of nitrogen adsorbed versus moles of oxygen adsorbed for a 50% mix of gasses.	263
Fig. 5.50	a plot of **exp**$(-\theta)$ for both θs illustrating the linearity expected at low pressure. 50% O_2+50% N_2 data on anatase by Arnold.	263
Fig. 5.51	Adsorption of a gas mix of 15% O_2 75% N_2 on anatase by Arnold.	264
Fig. 5.52	Adsorption of a gas mix of 30% O_2 70% N_2 on anatase by Arnold.	264
Fig. 5.53	Adsorption of a gas mix of 50% O_2 50% N_2 on anatase by Arnold.	265
Fig. 5.54	Adsorption of a gas mix of 70% O_2 30% N_2 on anatase by Arnold.	265
Fig. 5.55	Adsorption of a gas mix of 85% O_2 15% N_2 on anatase by Arnold.	266
Fig. 5.56	Relative molar volume as a function of coverage in terms of monolayer equivalence on a flat surface or $\Delta\chi$.	269
Fig. 5.57	Adsorption of CO-N_2 mix on 5A zeolite at 1 atm.	271
Fig. 5.58	Isotherm of CO on 5A zeolite to extract the constants χ_c and $\chi(1$ bar$)$.	271
Fig. 5.59	Adsorption of CO-O_2 mix on 5A zeolite at 1 atm.	273
Fig. 5.60	Phase diagram of N_2-CO in 5A zeolite.	274
Fig. 5.61	Phase diagram of CO-O_2 in 5A zeolite.	274
Fig. 5.62	A graphical comparison of the BET, χ, and DPtheories respectively from top to bottom. Data points are those of the silica αs standard.	275
Fig. 6.1	N_2 isotherm on LMA233 activated carbon by SSLR.	286
Fig. 6.2	N_2 isotherm on DD52 activated carbon by SSLR.	286
Fig. 6.3	N_2 isotherm on SBA-15 silica by SSLR.	287
Fig. 6.4	The amount adsorbed as monolayer equivalents versus the "log law" for various slit sizes.	289
Fig. 6.5	The amount adsorbed as monolayer equivalents versus the "log law" for various cylindrical pores.	290
Fig. 6.6	The amount adsorbed as monolayer equivalents versus $\Delta\chi$ for various slit sizes.	290
Fig. 6.7	The amount adsorbed as monolayer equivalents versus $\Delta\chi$ for various cylindrical pores.	291
Fig. 6.8	The BDDT equation for various values of N_{BET}. The C constant used for this was 20.	292

Fig. 6.9	An example of a DA plot illustrating the straight line fit. The data is N_2 adsorption in 5A zeolite by Danner and Wenzel.	294
Fig. 6.10	Adsorption of CO, N_2, and O_2 on 10X zeolite by Danner and Wenzel.	297
Fig. 6.11	Adsorption of CO, N_2, and O_2 on 5A zeolite by Danner and Wenzel.	298
Fig. 6.12	Adsorption of ethyl ether on carbon by GP.	302
Fig. 6.13	Adsorption of ethylene chloride on carbon by GP.	302
Fig. 6.14	Adsorption of n-pentane on by Goldmann and Polayni.	303
Fig. 6.15	Adsorption of CS_2 on carbon carbon by Goldmann and Polayni.	303
Fig. 6.16	χ plot of water adsorption on Y-zeolites at 298 K according to WW.	307
Fig. 6.17	χ plot of water adsorption on Y-zeolites at 298 K according to WW.	307
Fig. 6.18	The Broekhoff-deBoer model for adsorption in a cylindrical pore. The adsorbate-gas interface creates a hydrostatic pressure which changes the chemical potential of the adsorptive.	313
Fig. 6.19	The BdB model of pore filling and pore emptying. Notice there is no form "E" in the adsorption branch.	315
Fig. 6.20	The isotherms for adsorption on porous SiO_2 according to the Broekhoff-deBoer theory.	315
Fig. 6.21	The relationship between the pore radius in SiO_2 as the critical thickness and the critical relative pressure according to the Broekhoff-deBoer theory. The lines correspond to the **X**s in Fig. 6.20.	316
Fig. 6.22	Comparison of the Broekhoff-deBoer theory with the Kelvin-Cohan calculation for the switch to capillary filling. $F(t)$ uses the α-s for nitrogen on SiO_2 as the model isotherm.	317
Fig. 6.23	Generated standard plot using the modeling that includes mesopores to illustration the transition from "micropores" to "mesopores."	322
Fig. 6.24	Data and fit for N_2 adsorption on MCM-41 by QBZ (offset by n_{ads}).	325
Fig. 6.25	Data and fit for N_2 adsorption on MCM-41 by QBZ (offset by χs).	325
Fig. 6.26	Details of the adsorption on C-10 by QBZ.	326

Fig. 6.27	Methods to calculate the pore "radius" versus to X-ray measurements.	326
Fig. 6.28	χ plot and pore distribution of Ar adsorbed on MCM-41 and the mesopore fit.	327
Fig. 6.29	The hysteresis loop for the data by Qiao et al. showing the original data and the postulated energy correction for the adsorption data.	329
Fig. 6.30	n-layer χ simulation which show the isotherm "jump" as a function of diameter.	330
Fig. 7.1	NLDFT attempt to simulate the non-porous portion of the isotherm. Calculation by LGN.	337
Fig. 7.2	Demonstration of the mismatch between a standard curve and the NLDFT. NLDFT by LGN.	337
Fig. 7.3	Probabilities for the number of layers to be 1, 2, 3, etc. for the hard rod calculation as a function of layer thickness.	347
Fig. 7.4	Number density (total) for the hard rod case as a function of thickness.	347
Fig. 7.5	One-dimensional pressure for the hard rods as a function of thickness.	348
Fig. 7.6	Number density as a function of slit width for one type of rod with a length of a.	349
Fig. 7.7	The effect of an external field on the number distribution.	350
Fig. 7.8	A comparison of the Carnahan-Starling approximation with the Ree and Hoover hard sphere calculation.	352
Fig. 7.9	Results of the NLDFT calculation by Tarazona (solid line) and results of harmonic oscillator approximation from χ theory (dashed).	356
Fig. A.1	The error in precision of the BET equation inside the BET range (0.5–3.5).	379
Fig. A.2	Errors that the BET equation make with respect to the monolayer equivalent determined by BET.	379
Fig. A.3	Absolute error of the BET method in terms of the ratio $\theta_{BET}/\theta_{actual}$.	379
Fig. A.4	Detailed data table from the article by Gregg and Jacob.	381

List of tables

Table 1.1	Some definitions needed to comprehend the first parts of this book	4
Table 1.2	Classifications of physical adsorption isotherms	14
Table 1.3	Correction factor for n_m obtained by least squares	18
Table 1.4	Characteristics and interpretation of hysteresis loop types	21
Table 1.5	Non-linear features of the χ plot	25
Table 1.6	Different adsorption isotherm equations	30
Table 1.7	Linear correlation of the data by VTMOP—reactivity versus BET surface area	35
Table 1.8	Area parameters for some potential adsorbates	40
Table 2.1	Values for λ, and minimum D required at 25°C	67
Table 3.1	Specific or molar parameters obtainable from the isotherm	93
Table 3.2	Parameters from simple mesopore analysis	113
Table 3.3	Geometric analysis comparing the simple and advanced mesopore analyses	114
Table 3.4	Data analysis of N_2 α-s curves on SiO_2, Fransil-I, by BCST	128
Table 3.5	The α-s curve for the data by BCST referenced to $P/P_{vap}=0.4$	129
Table 3.6	Data for α-s curves by Payne et al.	130
Table 3.7	Smoothed α-s curve on silica normalized to $V_{0.4}$ as listed by Gregg and Sing	131
Table 3.8	Data and smooth t-curve-N_2 on alumina by Lippens et al.	132
Table 3.9	IUPAC silica isotherms	133
Table 3.10	IUPAC carbon samples	133
Table 3.11	Standard isotherm for activated charcoal by RMBM	134
Table 3.12	KFG coefficients for a standard curve extracted from carbons	134
Table 3.13	α-s curve using coefficients from Table 3.12	134
Table 3.14	Standard Nitrogen isotherms of low temperature out-gassed thoria	135
Table 3.15	Standard curve to water adsorption of thoria	135
Table 3.16	Argon adsorption on 25°C out-gassed thoria	136

List of tables

Table 3.17	N_2 adsorption of non-porous lunar soil	136
Table 3.18	Argon adsorption on non-porous lunar soil	137
Table 3.19	Adsorption of O_2 on non-porous lunar soil	137
Table 3.20	CO adsorption on non-porous lunar soil	138
Table 4.1	Error expected from the QM for the χ equation	155
Table 4.2	Densities of the liquid and solid state for some potential adsorbates	164
Table 4.3	IUPAC vs LJ 6–12 calculated molecular area	165
Table 4.4	Parameters to determine spacing and distribution of "layers"	188
Table 4.5	Comparison of O_2 vs N_2 slope for adsorption versus bulk liquid	204
Table 5.1	N_2 on MgO and Ni antigorite—A 3rd power χ fit	221
Table 5.2	Data from the α-s curves by Sing et al. of Ar and N_2 on SiO_2	223
Table 5.3	The statistics for the adsorption of gasses on 25°C out-gassed thoria	225
Table 5.4	The statistics for the adsorption of gasses on lunar soil	227
Table 5.5	Results of the χ plot fit for data by Nicolan and Teichner	229
Table 5.6	The analysis of the data by Bradley of Ar on crystalline $CuSO_4$	231
Table 5.7	The analysis of the data by Bradley of Ar on crystalline $Al_2(SO_4)_3$	231
Table 5.8	Data for the adsorption of Ar on polytetrafluoroethylene (Teflon®)	241
Table 5.9	Parameters extracted from data by Harkins and Jura	258
Table 5.10	Analysis of the parameters for binary adsorption versus the pure adsorbates	272
Table 5.11	Statistics comparing the BET, DP, and χ theories	276
Table 5.12	Statistics comparing the BET, DP and χ theories	276
Table 6.1	DA analysis: N_2 on 5A and 10X	294
Table 6.2	Fit to Eq. (6.12) for the data by Danner and Wenzel—uninterpreted	299
Table 6.3	Fit to Eq. (6.12) for the data by Danner and Wenzel—interpreted	299
Table 6.4	Data from Goldmann and Polanyi	304

Table 6.5	Microporosity analysis: water on Y-zeolites by Wisniewcki and Wojsz	306
Table 6.6	Some δ values for the Tolman equation to correct γ for surface curvature	317
Table 6.7	r_p analysis of modeled data by the two techniques and by data	323
Table 6.8	The dependence of γ_{gl} on r_p. r_p given as a percent of the original	324
Table 6.9	Mesopore analysis of the data by Qiao et al.	325
Table 7.1	The values of the functional, F, from Eq. (7.2) given the function y	338

Foreword to the 1st edition

The objective of this book is to present the practice of measuring and interpreting physical adsorption. It is intended to be a practical guide and not an extensive review of either the literature or the theories involved with physical adsorption. Extensive reviews are available, and the book by Gregg and Sing [1], though about 20 years old, is still highly recommended for specific details about a variety of adsorption experiments. Another recent book is the 2nd edition of the book by Rouquerol et al. [2]. A couple of more recent theoretical aspects are not covered in these books. These are density functional theory (DFT) and χ theory, for which there are no comprehensive reviews. A review by Evans [3] and additional articles by Tarazona [4–6] et al., would be a good start for DFT. χ theory [7–9] is rather simple and will be explained in detail throughout this book. (The one adsorbate-one energy of adsorption is easy, but needless to say, the more complicated the system, the more complicated are the resulting calculations.)

As with all scientific writing, there are various levels that can be presented. For example, infrared spectroscopy could be used on simply the pattern recognition level or at the more sophisticated level of quantum mechanics. The same is the case with physical adsorption. One can use the data from physical adsorption measurements as a simple control device, i.e., "Does this powder have the right adsorption isotherm to meet production requirements?," or, on a different level, "What is the meaning of the isotherm in terms of surface and pore structure and chemical attractions?" For most applications, the level of sophistication is somewhat intermediate.

In this book, the simple interpretations of the physisorption experiments are presented in Chapter 1. Chapter 2 presents the important details on how to make the measurements usually associated with physical adsorption. If one already has a commercial instrument, this chapter may be irrelevant. Chapter 3 is designed to present step-by-step analysis of the isotherms by a few methods and to present other isotherm interpretations. It is generally *not* a good idea to rely upon manufacturers' software supplied with the instruments. Although the programmers are quite knowledgeable about physisorption, it is still best to examine the data carefully. Chapter 4 presents extensive derivations of some theories of adsorption, starting with the disjoining pressure approach. The derivations of most isotherms have been extensively reviewed in other books (e.g., see Gregg and Sing). After all, most have been used for more than 50 years. However, the more recent χ theory and DFT have not been reviewed. Therefore, more detailed descriptions of χ theory and DFT are presented, along with some results. The analysis of one of the more promising techniques for studying adsorption, calorimetry, is not presented. A variety of others that are useful for porosity measurements, such as X-ray, NM, FAIR, etc., are also not presented. There is a vast body of literature on these latter subjects which have been used extensively, especially for the zeolites.

For most practical applications using commercial instruments, and given that one is accustomed to analysis that physical chemists use, Chapter 1 could suffice. The results of the theories formulated will be used in a "cookbook" fashion in Chapter 1 with little explanation. The caveat to the simplified treatments is that occasionally a simple explanation for the behavior of the adsorption is not appropriate. It is hoped that by recognizing patterns in the original or transformed isotherms, most misinterpretations can be avoided. The pattern recognition utilized the set of isotherm "types" as originally presented by deBoer and modified by Brunaur et al. [10,11] and later expanded by Gregg and Sing [1] and by χ plot features. Recognition of the possibilities of the complicating features beyond the simple isotherm is important for physical adsorption to be of value. Such features may be interpreted in terms of multiple surface areas, pore sizes and volume, energies of adsorption, and the distribution of pore sizes or of adsorption energies. The isotherms are generally interpreted in terms of these features, which have associated with them physical quantities that in many cases would be useful to know.

Unfortunately, the physical quantities associated with the physical features listed above must be extracted using some theoretical assumptions and the associated mathematical manipulations. It is not at all certain at this time that any generalized theory is capable of this. The theories available yield quite different values for these quantities, and at the time of writing there has not been resolution as to which interpretation, if any, is correct. Most theories of adsorption do not even yield values for these physical quantities, and some that claim to do so in reality do not. For example, the only theories that have a theoretical basis for calculating surface area of unknown samples are the BET (Brunauer et al. [12]) and the χ theory. Both of these will be explained in the theoretical portions of this book. The BET is unquestionably the most widely used theory to calculate surface area, but it has some very serious flaws. The χ theory is a recent development that has not been tested thoroughly. Another possibility is the continuing development of DFT, which has so far not been successful in calculating the surface area independent of the BET results or from assumed equations of state. There are numerous theories and methods for determining pore volume, both microporosity and mesoporosity. However, to determine the pore radius, most methods rely upon the BET. Furthermore, the BET is used as a correction in these methods as well. Most theories yield approximately the same answer, within a factor of 10, due to an obvious feature in the isotherm that would allow an educated guess to be correct.

It is this author's hope that in the future, some of these questions will be resolved, but for the moment there is a need for some answer, even if only approximate. It is unlikely that any theory will yield answers with the precision which chemists or physicists are used to, say better than 1%, due not just to the uncertainties of the theories and the associated calculations, but also to the defining questions regarding the physical quantities. For example, what is the pore size for pores in the range of 2 nm diameters? Where is the inner boundary for these pores? Atomic sizes begin to have meaning in this range. How does one account for surface roughness on a nearly atomic scale (a classic fractal problem)? Again, there is the same uncertainty in definition. Luckily, these questions may not be of practical importance in many applications. If a pore is large enough to allow, say, methanol to adsorb but not ethanol, there is a parameter that one could possibly extract to yield the distinction. If a catalyst's activity is proportional to the surface area, whatever that means, there is probably a parameter that is proportional to the surface area to make a relative distinction. So, in spite of the theoretical uncertainties, the measurement of physical adsorption is a very useful tool and promises to be more so in the future.

References

[1] S.J. Gregg, K.S.W. Sing, Adsorption, Surface Area and Porosity, Academic Press, London and New York, 1982. ISBN 0-12-300956-1.
[2] F. Rouquerol, J. Rouquerol, K.S.W. Sing, P. Llewellyn, G. Maurin, Adsorption by Powder and Porous Solids, Principles, Methodology and Application, Academic Press, Oxford, 2014. ISBN 9788-0-08-097035-6.
[3] R. Evans, in: D. Henderson (Ed.), Fundamentals of Inhomogeneous Fluids, Marcel Dekker, Inc., NY, p. 85, 1992. ISBN 0-8247-8711-0.
[4] P. Tarazona, Phys. Rev. A 31 (1985) 2672.
[5] P. Tarazona, Phys. Rev. A 32 (1985) 3148.
[6] P. Tarazona, U.M.B. Marconi, R. Evans, Mol. Phys. 60 (1987) 573.
[7] E.L. Fuller Jr., J.B. Condon, Colloid Surf. 37 (1989) 171–181.
[8] J. B. Condon, The Derivation of a Simple, Practical Equation for the Analysis of the Entire Surface Physical Adsorption Isotherm. U.S. D.O.E. Declassified Report Y-2406, 1988.
[9] J.B. Condon, Microporous Mesoporous Mater. 38 (2000) 359–383.
[10] S. Brunaur, L.S. Deming, W.S. Deming, E. Teller, J. Am. Chem. Soc. 60 (1938) 309.
[11] S. Brunaur, L.S. Deming, W.S. Deming, E. Teller, J. Am. Chem. Soc. 62 (1940) 1723.
[12] S. Brunaur, P.H. Emmett, E.J. Teller, J. Am. Chem. Soc. 60 (1938) 309.

Foreword to the 2nd edition

It has been more than a decade since the first edition of the book. A promise was made to update the information presented, and this edition is intended to fulfill that promise. Not much has changed theoretically, but of course there have been changes in the availability and capabilities of the instrumentation.

The first foreword ended with an implied question as to whether the situation described would change and become clearer in the future. The implication was somewhat hopeful. Sad to say, not many advancements have been made. Much of this is due to attempts to calculate from the adsorption isotherm the physical properties of the adsorbent. It is my belief that this is missing the major points, theoretically and practically. It seems that much of the tweaking being performed with the DFT-type theories is in response to this drive to "get it right." It is because of these questions that this book underplays the importance of "surface area" and "pore dimensions," and emphasizes the "monolayer equivalence" and uses the word "layer" with a new meaning.

An opinion concerning BET and other similar models

In my opinion, it is time to discard the BET theory and other isotherms analysis based on the "Henry's Law" assumption. Firstly, these Henry's Law isotherms do not yield the correct value for the surface area or the energy of adsorption, nor do they have any basis for measuring porosity. This latter point has become obvious to almost every investigator in the field, and is the driving force for the continued development of DFT theories. Most of the Henry's Law methods do not yield values for surface area or porosity at all, and are essentially empirical fits to the isotherm. There are probably more than a hundred isotherms available, mostly just empirical. The BET plot is an exception in that it does yield a value—albeit wrong—for the surface area and energy of adsorption. Several other isotherms are based on modifying the Langmuir isotherm. The assumptions made to derive the BET, as well as these other isotherms analysis, are very questionable. This was pointed out early in the development by Halsey [1] and others. Much of the literature is devoted to finding an alternative to the BET, but if the BET were valid, this question would be settled and the search for alternatives would have ended long ago. Furthermore, the BET is not reliable to within a factor of at least 4 for the surface area. It essentially yields meaningless values for energy, which are in stark contradiction to calorimetric values. Indeed, the value for the BET energy of adsorption is often a (mathematically) complex number. Both of these facts have long been known, in addition to seven other weaknesses as pointed out by Gregg and Jacobs [2]. However, until the 1980s, there was a lack of something better, and therefore the BET continued to be used. It is the surface area basis for most standard curves and DFT calculations. An extensive list of problems with the BET, both theoretical and practical, is available in a section authored by Sing in a book on adsorption [3]. Such an extensive discussion is not presented here, but some additional problems are pointed out in this book. Today there are better theories and techniques.

Another problem with most of the older theories including the BET, is that they cannot be used for analyzing isotherms for adsorbents that have more than one energy of adsorption present, referred to here as "multiplane adsorption." (This is sometimes referred to as "heterogeneous." This designation seems to indicate a surface of variable composition rather than multiple crystallographic planes. However, Multiple crystallographic planes are more common than variable composition.) The reason for this is given in the section entitled "Do isotherms for multiplane adsorption have anomalies?", which demonstrates that the BET theory creates a mathematical anomaly as normally used. The phrase "normally used" indicates the practice of using statistics on a transform of the original equation. This anomaly problem is also true for any isotherm analysis that uses a transform which mixes the pressure and the amount adsorbed.

An opinion on the way around the BET problems

The exceptions to the criticism in the previous paragraphs arise by use of the standard curve methods and the χ theory, along with the disjoining pressure theory. Presumably this applies to the NLDFT (*non-local density functional theory*) and the Monte Carlo technique. (Although there is usually agreement between NLDFT and Monte Carlo, this should not be surprising since they use the same classical assumptions. The real test is, Do NLDFT and Monte Carlo agree with experimental data?) It would be challenging to imagine that either one of these latter techniques would produce an anomaly, although the use of the BET precondition would invalidate this statement. In terms of normal practice, it is doubtful that any instrumentation available today has multiplane analysis as a possibility.

In terms of theoretical developments, the analysis referred to as NLDFT analysis is now included in several of the instruments. This is quite handy since the calculations are, for most users of these instruments, confusing at best. The imbedded NLDFT along with the value for the surface area might yield the porosity correctly, especially in a comparison mode. However, the question for such "black box" analysis is: "Specifically, how does it work and what problems might crop up that yield incorrect answers, for a variety of systems?" An analogy that is relevant today is: "Would you be willing to drive a self-driving car if you had no idea what or where the brake pedal or the steering wheel were?" For example, does NLDFT account for multiple energies of adsorption or a distribution of energies? Does it really use the disjoining pressure theory (or the χ theory, but using χ makes the DFT irrelevant) to calibrate? What combination of pore analysis is used, and does it take into account the thermodynamic limits that are now well known?

It is puzzling why ESW and the χ theory are not included as an option in these instruments, especially since the ESW has been used to support the NLDFT calculations to determine a monolayer. The ESW and χ theories, which can be demonstrated as identical, are much simpler and more precise and versatile than NLDFT. Energy distributions and pore size distributions, such as they are, are easy to analyze. The transformed χ plot is intuitive in interpretation once one knows what to look for. In most cases, a simple χ plot and a calculator with ln function are all that are needed for a complete analysis. For more sophisticated analysis, simply good data (necessary for all methods) and a computer program to take the first and second derivative of the χ plot are all that are needed.

As the reader will discover, the χ theory and disjoining pressure theory have been demonstrated as identical. This one theory has been derived from three different approaches:

- both classical mechanical and classical statistical mechanics (ASP theory);
- both perturbation and WKB approximations quantum mechanics (χ); and
- continuum mechanics of the liquid-solid surface and classical thermodynamics (disjoining pressure theory or ESW) with an assumed one monolayer potential decay constant.

From this point of view, χ/ESW theory is well grounded theoretically. It neither resorts to "sites" or equilibrium between "sites," Nor does it depend upon "sites" that are on top of other adsorbate molecules (the primary Halsey criticism).

Given the solid theoretical grounding and the simplicity of analysis, it seems strange that the χ-ESW method is not used more often. It may be that researchers are not familiar with the derivations of either one, because if they were, they would realize how powerful a few reasonable assumptions can be. For example, the disjoining pressure theory relies upon the concept of disjoining pressure, Π, which is not widely known by adsorption investigators, even though it is standard in many courses on surface chemistry. For the χ theory, there may be some confusion about the separation of variables that are used. The direction normal to the surface is treated differently than the plane of the surface. The normal direction uses the stand-by Lennard-Jones (6-12) potential similar to most DFT, whereas the plane of the surface uses a modified version of the "particle-in-the-box." The populations of the various "layers" are calculated from the result obtained from the plane-of-surface calculation, which then allows one to calculate the distributions normal to the surface. The method predicts that for most temperatures of adsorption and adsorbates, the error introduced by some geometrical factors should be only ∼2.5%. In this revision of the book, the philosophy behind this method will be explored in more detail, and hopefully in a convincing fashion. It will also be pointed out that the theories resolve some problems that Brunauer continued to puzzle over, such as where the transition to the liquid phase is and why so few isotherms follow not Henry's Law[1] but rather

---

[1] I must make a comment here. Henry's Law can*not* be derived from the laws of thermodynamic as some authors and books claim. For that matter, neither can the ideal gas law. For solution chemistry, it is an extremely reasonable assumption and observed almost all of the time, but even for this there are some rare exceptions. Although there are reasons for this, they are still exceptions.

the Freundlich isotherm at low pressures. (Just because the isotherm *appears* to intersect at the origin does not mean it is following Henry's Law!)

Comments about problems with many publications

Given the discussion above, one would be justified in asking, "Why are you presenting the older theories? Why not just stick to NLDFT, ESW, and χ theories? Why do I have to wade through all these other mathematical presentations?" The answer is that unfortunately, most of the literature is analyzed based on these theories. Furthermore, many articles do not present the original data in a form suitable for re-analysis. So what can one do? There are several places in this book where these older theories are compared to the χ theory. This is done for three reasons:

- as an explanation as to why the older theories almost fit the data (but never with an $r^2 > 0.990$);
- as an argument to convince the reader that the χ theory applies universally to the conditions specified (the liquid-like conditions); and
- to demonstrate that one can obtain the parameters of the χ theory by refitting the old data, even if it is only available in the form of the fitting parameters of the older theories.

The latter point might not work well, since the fitting data of the older theories may cover only a portion of the isotherm. For example, the BET is usually fitted over the pressure range from 0.05 to 0.35 of P/P_{vap}. This is much too short a range for one to have any confidence in the answers. Unfortunately, some of the "standard curves" have been biased toward the BET fit, with the original data not available.

This latter statement brings up another problem with archived data and even data that are taken today by most investigators. It is very unusual for isotherm data be collected below 0.01 P/P_{vap}. As will become obvious while reading this book, this limit is insufficient to obtain accurate results in terms of adsorption energies and distributions, surface areas and their distribution and pore sizes and type. Experimentally, this problem is easy to solve. It is relatively easy to use equipment that has a high vacuum capability or even an ultrahigh vacuum to measure in the low pressure range. It is also important to obtain accurate data as one approaches P_{vap} ($P > 0.9 P_{vap}$ at least). This problem seems to have been ignored in early data, as is evident from the deviations between various investigators for the high pressure data. It is harder, however, to extend the pressure range to approach P_{vap}. It requires very careful temperature control of the sample, something that most instruments commercially available do not have. Proper baffling and attention to the kinetic/laminar flow transition are also usually lacking. One should attempt to push the experimental limit ever closer to P_{vap}. This requires better vacuum system design, better temperature controls, and particular attention to the areas of the instrument that affect temperature control using baffles and awareness of the laminar to turbulent transition zone. It is naive to assume that the sample has the same temperature as the refrigerant, and that the pressure is the same in the cold zone as it is in the rest of the vacuum system.

What's new in this edition?

In this book, there is one change on the theoretical side for the χ theory. There are enough data available to make a decision about how the adsorbed molecules are organized on the surface, whether as individual molecules or as patches of molecules. It seems obvious now, from both theoretic considerations and experimental data, that the molecules must cluster, at least in patches. If this is the case, then the previously questionable geometric factor, f, listed as in the first edition, has a value of nearly 1. This makes sense from the classical point of view, which considers that there are lateral forces. It is also consistent with the assumption used to derive the ESW. However, this has not yet been *specifically* addressed and confirmed with experimental data.

There is general uneasiness for those dealing with the theory of adsorption and assigning a value to the molar area, $\overline{A}$, or monomolecular, a. This uneasiness is not new; it is evident from the earliest days of the BET development. Along with this comes the question, "What is meant by surface area?" It is no secret that with adsorption experiments, one obtains a variety of answers depending on the conditions of the adsorbent and especially on the adsorbate used. The concept and visualization of surface area are quite simple, but can also be misleading. In this book, I have put forward the concept of a specific monolayer equivalent, n_m, as the physical property that is important to tabulate. It is acceptable to list the surface area, provided that the specific monolayer equivalent, which is more fundamental, is supplied. Therefore, for now, the important quantity that is not in question—that is, the monolayer equivalence n_m—will be tabulated. The conversion to surface area with $\overline{A}$ or a, as preferred by the reader, is always possible.

On a more practical note, the symbolism in this edition has been changed to be more in alignment with the 3rd edition of the IUPAC's book, *Quantities, Units and Symbols in Physical Chemistry*. There have been many changes since the 2nd edition that apply here. However, superscript notations have been avoided as much as possible, since these may easily be confused with exponents. The IUPAC convention that numbers and quantities be *italic*, whereas distinguishing letters be roman type, does not seem to be sufficient to keep things straight. Therefore, the quantities that use superscript for distinction are not used here. Other methods, usually distinct subscripts, are used to make the differentiations. (The only distinguishing superscript used is the standard symbols, "θ" or "*", which are so unique that there is little chance for confusion.) This is done as some of the advanced mathematical treatments are difficult enough without such a notation confusion. Even for those who are reasonably familiar with the subject, it would be wise to make a quick survey of the symbol list before starting, in order to check what equivalent symbols and names apply in this book. You may want to go to www.genchem.net/symbollist.pdf and print out the symbol list and keep it handy as a reference. There are several subtle distinctions for some quantities, such as energies and heats, that are important to recognize. A further modification in this book is to use the molar mass that chemists normally use—that is, with units of g mol^{-1}. This is given the symbol (capital bold) "**M**" versus "$\overline{M}$" which is with units of kg mol^{-1}, as explained in the symbol list. (The IUPAC alternative bar-over designation is used for molar quantities.) Furthermore, the word "molar," meaning per mole of material, is used throughout and not the word "enplethic," which the IUPAC recommends. Most scientists today recognize the word "molar" and not "enplethic."[2] In my opinion it seems unlikely that the word "enplethic" will catch on any time soon (if ever), since it does not appear in any dictionary. (It only appears in the 3rd edition of the IUPAC's Green Book, or in texts discussing terminology that quote the Green Book.)

Where are all the thermodynamics?

Many books present very long explanations about the thermodynamics of adsorption. Examples include the 2nd edition of the Elsevier book *Adsorption by Powders and Porous Solids: Principles, Methodology and Applications* [3], starting with Section 2.3 on page 33, or even throughout A. Adamson's book *Physical Chemistry of Surfaces* [4], in particular section XVI-12, starting on page 633. It might be interesting reading for some, but most of it is unnecessary. There are some thermodynamic derivations included when needed, but other than heats of adsorption and disjoining pressure, thermodynamics is a minor part of this book.

You may wonder why this is. The reason is simple. One can*not* derive any isotherm or any other phenomena in this book using only the four laws of thermodynamics. What one really needs is the equation of state for the material with which one is working. Given the equation of state, the phenomena are usually explainable without using thermodynamics. Furthermore, without equations of states, thermodynamics is not very useful. The assumption that the thermodynamics alone yield answers for physical adsorption is a dilution. Take, for example, the integral often used to justify the postulate of Henry's Law-like behavior of:

$$\Pi = \frac{RT}{A} \int_0^P \frac{n_{abs}(P)}{P} dP \qquad (1)$$

The implied assumption is that the number of moles of adsorbate material $n(P) > 0$ when $P > 0$. If this were the case (and it is obvious that without a pressure of adsorbent, the adsorbate goes to 0 as P goes to 0), then the lower limit of P for the integral

---

[2] It seems that the IUPAC committee has a problem with the words "molar" and "molarity." I believe they want to eliminate any confusion for molar quantities and molar solution. I do not see how anyone with a chemistry background would be confused between, for example, molar volume of 18 mL and a 10 molar solution. Their attempt to ban the words "molar," "molarity," and "percent" have been unsuccessful because these terms are so widely used and understood.

would drop out using L'Hospital's rule. This may not be (and generally is not) the case.

The most important point, as the reader will discover, is to obtain the equation of state correctly. The thermodynamics that can be used to make relationships have been known for a very long time and it is assumed that the reader is aware of this very old and pervasive tool. To obtain the equations of state theoretically correctly and not by measurements, the gold standard is quantum mechanics (QM) linked with statistical mechanics, for which there is a long history of success. For the χ theory, very few assumptions are made going into the QM derivation, especially for the determination of the amount of adsorption.

Some final advice if you're in a hurry

To understand the full implications and the theory of χ, you need to read all the sections on χ theory in Chapters 1–6. In other words, you can skip the sections of all the more established theories. If you would like the background to DFT, then Chapter 7 is appropriate.

A tribute to Loren Fuller

This is a final history note. Dr. Loren E. Fuller, Jr., worked very diligently on the ASP (χ) theory for many years, working against the odds, but was able to publish several papers. Indeed, some very prominent scientists[3] were furious whenever the theory was mentioned. Loren finally promised one of these scientists, out of great and deserved respect, not to publish the theory until that scientist's passing. The basis and use of the ASP theory was usually downplayed in order to make the publications acceptable. I was resolved to be open about the theory and therefore was not very successful in my publications record. However, I have worked in situations where publications did not matter, which is a considerable advantage in being straightforward.

Shortly before Loren's death, I noticed that the χ-plot always passed through a common point, and I mentioned this to him. His immediate reaction was, "Yes, the point of the relative pressure of 0.368 (1/e) is the magical point from which one can extract a lot of information!" Indeed, this is a jaw-dropping point along with the early data points in the isotherm. By itself, this point yields the energy of adsorption and surface area for a flat surface or, as Loren would indicate, for a fractal factor of 1. Together with the early part of the isotherm, it yields the fractal factor other than 1, usually porosity but maybe very low particle sizes. This will become clear when you read about the inflection point technique in Chapter 1, with details in Appendix IIJ.

---

[3] I will not name these scientists here, but I assure you that most scientists would immediately recognize them.

References

[1] G. Halsey, J. Chem. Phys. 16 (1948) 931.
[2] S.J. Gregg, J. Jacobs, Trans. Faraday Soc. 44 (1948) 574.
[3] F. Rouquerol, J. Rouquerol, K.S.W. Sing, P. Llewellyn, G. Maurin, Adsorption by Powders and Porous Solids: Principles, Methodology and Applications, second ed., Elsevier/Academic Press, Oxford, UK, 2014, p. 239. ISBN 9788-0-08-097035-6.
[4] A.W. Adamson, Physical Chemistry of Surfaces, fifth ed., John Wiley and Sons, Inc., New York, NY, 1990. ISBN 0-471-6101804.

CHAPTER 1

An overview and some uninteresting history of physisorption

Introduction
What do you want?

Here is an important question for you: "What do you expect to get out of this book?" It may be that you are measuring the physisorption isotherm[1] as a quality control method. The machine that you are using yields two parameters called the BET surface area and the BET "C" constant. You then assume that as long as these parameters are constant from batch to batch that everything is OK. This may or may not work for you. You are more likely to have a too high rate of rejection than a too low, but, as a cautionary note, the too low is also possible. So, what is going wrong? Can the instrument be telling you the wrong information? Do you need a better, and probably more expensive, instrument? Or is it that the analysis for the physical isotherm incorrect? This latter point is very likely.

It may be that you are a graduate student and want to get some fundamental physical quantities from the physisorption isotherm. You discover that the BET method of analysis is highly disputed in the literature with endless caveats. You're wondering if there is some other method of analysis available, or at least some other ways of interpreting the isotherm. Well, do we have the book for you!

Are you an engineer, for example, working on columns for separations? Well, this book is devoted to the equilibrium portion and may be a little more than you need. There are some parts that are important. These are:

1. the basic χ theory, which yields to correct equations you need for the isotherm along with the energies;
2. the values for the surface area for the pores of porous materials and the external surfaces; and
3. most importantly, how one can do screening experiments for binary mixtures using χ theory as a guide

The latter point significantly cuts down on the measurements that one needs to make to determine promising adsorbents. It may be possible to use the technique without a

---

[1] The phrase often used to describe the method is the "BET method." This is indeed inaccurate. The expression used here is the "physisorption," "physical adsorption," or simply "adsorption" isotherm.

detailed knowledge of the theory; thus in the section on this method, a quick "how to" section is provided.

Maybe you are an investigator and wish to do more than just the isotherm, but also calorimetry, porosimetry, X-ray analysis, and other measurements. You find that cross relating these measurements has many problems. It is hoped that this book will help you in your quest. Indeed, this may get you started on the very important task of cross-correlating these measurements—little of this has been done elsewhere, under the conditions addressed in this book. Or maybe, as an investigator, you just want to solve the problem of the experimental contradictions and puzzles you have encountered. You might be able to find some answers here.

Scope and terminology

This book is arranged with the most elementary and easy to understand material in this chapter. More sophisticated material is introduced in subsequent chapters. It is written from the viewpoint of a physical chemist, and therefore the terminology is that used by chemists. It is also written to minimize the amount of information that the reader needs to wade through. This being the case, if the reader wishes more experimental detail and examples, it is recommended that they refer to one of the other excellent books that are available; we particularly recommend the book by Rouquerol et al. [1]. However, one cannot find information on the most recent methods of analysis, other than NLDFT,[2] in any other book. A review of NLDFT by (See Chapter 7 in this book for a background to start with, or more detail in the book by Henderson, ed. [2], or by Davis [3]. There is a review of the topic by Ravikovitch et al. [4], which is also a good starting point.)

The first and very important viewpoint is the distinction that chemists make between "intermolecular forces" and "chemical bonding." Intermolecular forces include dipole-dipole moment attractions, dipole-induced dipole attractions, ionic-dipole attractions, ionic-induced dipole attractions, London forces, and "hydrogen bonding."[3] These attractions are weaker than chemical bonds and can be operational even when not localized. Chemical bonds include covalent bonds, which includes coordinate covalent bonds or "dative" bonds, and ionic bonds. Chemical bonds are generally localized and changing positions of the molecules requires overcoming a large energy barrier either by shear energy input or by some intermediate chemical species, which is usually referred to as a "catalyst," or possibly a "promoter."

The term "physical adsorption" or "physisorption" refers to the phenomenon of gas molecules adhering to a surface at a pressure less than the vapor pressure. It involves intermolecular forces—the delocalized attractions. Physisorption can also

---

[2]We are not aware of any book that gives a complete description of NLDFT. One needs to piece together information from the literature. A good place to start is Chapter 7. Another starting point is the book edited by Douglas Henderson.

[3]"Hydrogen bonding" is written in quotes to remind the reader that this is not a true chemical bond with exchange of charge, thus not a chemical bond, but rather a very strong directional attraction.

be observed from the liquid state, in which case one normally refers to the physical adsorption of the solute. This could also be treated with the same mathematical constructions as the gas phase adsorption at the saturation pressure of the solvent. Many situations of adsorption from solvents, however, involve ionic attractions or chemical bonds. The physisorption attractions between the molecules being adsorbed and the surface are relatively weak, as are most intermolecular forces, and definitely not covalent or ionic.

In contrast, the term "chemisorption" refers to the phenomenon of molecules adhering to a solid surface by chemical bonding. Chemisorption is often associated with catalysis and other (possibly) surface reactions such as corrosion and electrochemistry. Chemisorption, electrochemistry, and corrosion are very broad subjects and are not topics covered in this book.[4]

To get started, there are some words that you need to become familiar with. To understand the language in this book, first refer now to the vocabulary[5] in Table 1.1. These definitions are used in this book and in most of the literature on physisorption [5].

One definition of particular importance is the word "specific." The word indicates that the results are being reported on the basis of 1 g of (adsorbent) sample. This is the normal default condition in the field. Molar quantities are usually in reference to the adsorbate and it is highly unusual to refer to the adsorbent in this fashion. Thus, for example, the molar enthalpy of adsorption is referring to the energy per mole of adsorbate (the adsorbing gas) and not of adsorbent (the solid upon which adsorption is occurring); whereas, the *s*pecific surface area is referring to the surface area of one gram of adsorbent. Often the word "specific" is left off as well as the understood default unit g^{-1}. Likewise, you will find the designation "monolayer equivalent" moles is a "specific" quantity. This refers to the number of moles of an adsorbate that would be contained in a theoretically evenly spread single layer of molecules (a monolayer) on a gram of sample.

Symbolism is also somewhat confusing, therefore a symbol list is provided in Appendix IV. This symbol list follows most of the conventions specified by the IUPAC "Quantities, Units and Symbols in Physical Chemistry" (the Green book) [6] and the IUPAC Technical Report [7] on "Physisorption of gases, with special reference to the evaluation of surface area and pore size distribution." Exceptions to these conventions are: not using superscripts of something other than physical quantities or a pure number as a power. There are a few exceptions to this exception, but this will be obvious by the symbols used (see symbol list).

---

[4]Not all corrosion reactions are surface reactions or surface limited. For example, there are corrosion reactions that occur by internal nucleation of products. Some electrochemical reactions might also be something other than surface reactions.

[5]The practical distinction between the various pore types depends upon conditions and especially the adsorbate. The above definition is based on nitrogen adsorbate and the IUPAC definition. In this book these distinctions will be made on the basis of the standard curve characteristics and the method of analysis required.

Table 1.1 Some definitions needed to comprehend the first parts of this book

Term	Definition (See also appendix IVA)
Adsorbate	The molecules adsorbed on the surface of the solid material
Adsorbent	The solid material upon which the adsorbate is adsorbed
Adsorption	Addition of adsorbate to the adsorbent by increasing the adsorptive pressure
Adsorptive	The gas in equilibrium with the adsorbate
Chemisorption	Enhancement of the amount of gas molecules on the surface of a solid caused by covalent or ionic bonding
Chi(χ) plot	A plot of the amount adsorbed (usually mmol g^{-1}) versus $-\ln(-\ln(P/P_{vap}))$
Desorption	Removal of adsorbate from the adsorbent by decreasing the adsorptive pressure or increasing the temperature
Hysteresis	The phenomenon of the desorption isotherm over a limited range being greater than the adsorption isotherm
Isotherm	The measurement of the amount adsorbed versus adsorptive pressure at constant temperature
Macropores	Pores with diameters greater than 50 nm[a] (IUPAC definition [5, 6])
Mesopores	Pores with diameters between 2 and 50 nm[a] (IUPAC def. [5, 6])
Micropores	Pores with a diameter of less than 2 nm[a] (IUPAC definition [5, 6])
Monolayer	A theoretical uniform liquid film of adsorbate one molecular layer thick.
Monolayer Equivalent	The amount of adsorbate that has the same number of molecules as the Theoretical monolayer. Symbol for this is n_m
Physical adsorption	Enhancement of the amount of gas molecules on the surface of a solid caused by van der Waal forces (includes dipole-dipole, dipole-induced dipole, London forces, and possibly hydrogen bonding)
Physisorption	Same as physical adsorption
Specific	Means "per gram of sample." Quantities throughout the book are assume to be "specific"—always implied unless otherwise stated or irrelevant
Standard plot	Refers to one of these: {alpha}-s plot, the t-thickness plot, the χ plot, and others that may be specific to an adsorbate-adsorbent pair. A generalize standard plot function will be designated as $\mathbf{F}(P/P_{vap})$ in this book

[a]These definitions are modified for the χ-theory as a function of adsorbent/adsorbate pair.

For most adsorption experiments the temperature at which the measurements are made is at or below the boiling point of the gas being used but above its freezing point. The restriction of being at or below the boiling point is normal instrument limitations. This is no theoretical reason why one cannot to go above the boiling point if the instrument can be safely pressurized. Theoretically, at some point below the freezing point the adsorption begins to look like epitaxy and not physisorption. This being the case, one would normally expect that the adsorbate characteristics resemble the liquid phase rather than the solid phase of the adsorptive. This liquid-like behavior of the adsorbate is the normal assumption used for most adsorption theories. This behavior is also often observed even several degrees below the freezing point.[6]

The principal measurement performed as an adsorption experiment is the measurement of the adsorption isotherm. The adsorption "isotherm" is the measurement of the amount adsorbed versus adsorptive pressure at constant temperature. The pressure is usually plotted on the abscissa as the adsorbate pressure, P, divided by the vapor pressure at the temperature of the adsorbent, P_{vap}. This is the easiest measurement to make.

The isotherm measurement is one type of experimentation to obtain useful information. Another measurement is calorimetry. One form of calorimetry measures the amount of heat evolved as the adsorptive is adsorbed. Another form measures the heat capacity of the adsorbate. There are various forms of calorimetry but the most accurate methods are very difficult to perform and only a few examples are available in the literature.

Another form of calorimetry, which is easier to perform, is scanning calorimetry. This calorimetry method is a good tool to determine qualitative features of the adsorption and to yield a fair indication of the physical quantities. It is not very precise or accurate.

Sorption research is a very broad subject and even physical adsorption is fairly broad. The range of temperatures, pressures, adsorbents and gases, and mixtures is nearly limitless. The number of theoretical approaches, that is interpretation methods, is also very broad; therefore, for practical purposes there are further restrictions in this book. The number of theories has been limited to the most commonly used. Theoretical descriptions of the models are limited except for the BET and χ theories (+ESW=excess surface work) and Density Functional Theories (DFT and NLDFT=non-local DFT). The reason for this will become apparent upon inspection of the material throughout this book. The pressures and temperatures are restricted to low pressure, usually defined as below 10 bar but conditions may dictate otherwise, which sets the upper limit on temperature. The lower limit for the temperature is usually the freezing point (but certainly above about $0.5T$ freezing, which is above the accepted 2 dimensional freezing point.)

The distinction between the low and high pressure and the importance of this has been reviewed by Myers and Monson [8]. They make this distinction along with the concepts of "excess adsorption," usually indicated by the Gibbs' excess, and "absolute adsorption" with a shift in the thermodynamic system definition. This essentially is making a distinction as to what are the boundaries of the thermodynamic system under consideration. The distinction is particularly important for porous adsorbents at high pressures. In both cases the systems are thermodynamic open systems;[7] but in

---

[6]There is a statistical thermodynamic reason for the liquid behavior to extend below the freezing point of the bulk material. The reader need not be concerned about this at this point.
[7]Define as a system for which energy, both work and heat, and matter can enter and leave freely.

the case of the "excess adsorption" the system's boundaries are approximately where the "liquid" phase ends and the gas phase begins,[8] whereas for the "absolute adsorption" the entire adsorbent including the pore volume with the adsorbate and adsorptive in the pores is the system.

General description of physisorption

For the purposes of this book, distinctions will be made between physical adsorption for the liquid-like state and the solid-like state. There could also be a gas-like state, which is difficult to detect due to the low concentrations of adsorbates. Typically in the bulk phases, the density of the liquid phase is about 1000 times that of the gas phase at the temperatures of concern here. Consider, for example water at room temperature. The liquid density is about $1.0\,g\,mL^{-1}$ versus the gas at about $8.0 \times 10^{-4}\,g\,mL^{-1}$. Thus typically, one would expect that the gas adsorbed state would be less than 1% of the liquid state at one monolayer equivalent; far too low for most experiments to measure.

Figs. 1.1 and 1.2 illustrate the atomic scale difference between the different forms of the adsorbed state proposed. For the solid-like state of Fig. 1.1 the adsorbate molecules are located on definite sites in relation to the underlying atoms of the adsorbent. For example, they lie directly over one of the atoms or in between two or three atoms in a defined geometry. One could refer to this as an "in-register" adsorption or even "epitaxy." Chemisorption, where the attraction between the adsorbate and adsorbent is a covalent or ionic bond, would be an example of such adsorption. The chemical bonds would be directed toward one and possibly two or three neighboring surface atoms. For example, in field emission microscopy one can readily

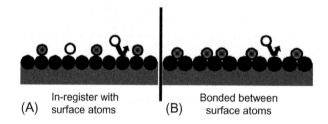

FIG. 1.1

A model of adsorption of the "in-register" type—e.g. chemisorption, epitaxy.

---

[8]Literally, the density of the gas phase far from the adsorbent is subtracted from the adsorbate/adsorptive from the solid boundary outward. This yields a fuzzy system boundary, which is very inconvenient in thermodynamics. With the formation of the liquid-gas interface this might work out, since there would be an identifiable boundary, at least to within a 3 molecular distance.

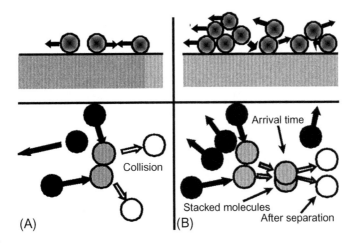

FIG. 1.2

Models of two types of physisorption: (A) gas-like and (B) liquid-like. ● before encounter, ○ after encounter, hatched is during encounter.

discern the difference between the physical adsorption and chemisorption since the physically adsorbed material has a much higher molecular mobility. An example of this would be the switch of the sorption of benzene from physisorption to a chemisorption involving a benzene π-metal bond [9]. Adsorption well below the triple point in temperature would also be expected to follow this pattern. Additional adsorption above the first layer, might also be "in register." Deposition below the freezing point to begin solid growth by epitaxy might also be another example of an "in register" adsorption process.

The other modes of adsorption, which are the subject of this book, are illustrated in Fig. 1.2. These are the gas-like and the liquid-like. These are what is normally referred to as physisorption, in which the adsorbent provides an overall attraction for which no particular site has a strong enough attraction to localize the adsorbate. In other words, the adsorbate molecules are free to skate over the entire surface, at least for a fair distance, even though there might be "bumpy spots."

For this physisorption picture there can be further distinctions: one where the adsorbate is behaving as a gas and there is only adsorption on top of the adsorbent surface, or one where the adsorbate behaves like a liquid, where adsorbate molecules can "roll over" one another and an adsorptive molecule can adsorb upon an adsorbate molecule. Most adsorption isotherms are performed under conditions where the liquid-like condition is assumed to exist. Calculations of the gas-like state indicate that the amount that can be adsorbed in this fashion is very low for most practical experimental conditions. Nevertheless, one would expect some of this to exist even with the presence of the liquid-like adsorbate.

Measuring the surface area by physisorption

There are two principal methods to measure the adsorption isotherm: volumetric and gravimetric. In addition, there is the carrier gas flow method and the high vacuum flow method. Calorimetry is best performed in conjunction with one of the techniques that measures the isotherm. In both principal methods the adsorbent is held at a constant temperature, usually near or at the boiling point of the adsorptive. The adsorptive pressure is increased stepwise and held constant for a period of time. This time allows the adsorption enough time to occur and the temperature of the adsorbent to re-equilibrate. The length of time required depends upon the physical arrangement and the system being studied. Since re-equilibration might take hours in some cases, it is best to monitor the progress of the adsorption to determine when equilibrium is achieved. Most automated instruments do not check the progress of the adsorption with time to determine if equilibrium is achieved. They are more likely to have a set amount of time programed in. When buying an instrument, it is highly recommended that the automation can detect when the equilibrium is approached well enough to switch to the next data point. The amount adsorbed is measured, in the case of the volumetric system, by measuring the pressure change and comparing this to the expected pressure change if the adsorbent were absent. In the case of the gravimetric measurement, the amount adsorbed is indicated by the mass gain. In both cases, some corrections to the raw data must be performed to take into account the experimental set-up. Details of why and how this is done are presented in Chapter 2.

A typical isotherm, then, is a plot of the amount adsorbed versus the adsorptive pressure. Usually the pressure is expressed as a ratio of the adsorptive pressure, P, to the saturated vapor pressure over the bulk liquid, P_{vap} ($P°$ usually seen for the BET notation.) The preferred unit for adsorbate amount is millimoles or micromoles adsorbate per gram of adsorbent ($mmol\,g^{-1}$ or $\mu mol\,g^{-1}$).

The literature has a variety of units for adsorbate amount, often with mL at STP per gram sample, or perhaps mass per mass sample, preferred in most of the older literature. Other units, such as layer thickness, are problematic since they transform the experimental data obtained using theoretical considerations and the original data may be irretrievably lost. This is especially the case when the BET surface area is used for this transform.

Each of the principal methods of measuring the isotherm has advantages and disadvantages. Both isotherm measuring methods normally cool the sample to, or below, the boiling point of the adsorptive. The sample is then exposed to adsorptive gas while the gas pressure is measured. Since the temperature of the sample is known, usually by use of a "gas-liquid thermometer," then the vapor pressure of the adsorptive over its liquid is known and thus the ratio P/P_{vap} can be calculated. This is the most precisely measured physical quantity; although P_{vap} could be significantly off if the temperature of the sample is not carefully checked. This measurement is common to all the techniques. The other measurement differs, depending upon the technique.

The most common measurement of the isotherm is "volumetric method." This method has the advantage that it is the simplest and is relatively inexpensive. It has the disadvantage of a greater uncertainty in the results. In this technique the amount of gas adsorbed is determined by measuring how much gas is used from a reservoir. This is sometimes referred to as a "gas burette." There are several corrections that need to be checked, the principal one being what is referred to as the "dead space." In this technique, temperature measurements, both in the cooled zone and for the gas burette portion, are very important.

Normally, in the volumetric method the gas is admitted to a cul-de-sac sample area; that is, the tube leading to the sample chamber is used for both admission and evacuations of the adsorptive. There is another method for this, which has been referred to (confusingly) as a flow system. This other method will be referred to in this book as the "Arnold flow method," after the developer [10] of the method, or the "vacuum flow method." In this method, the adsorptive is not admitted in a static fashion but is allowed to flow through the adsorbent without any carrier gas before being held static. This speeds up the approach to equilibrium and cuts down on the time needed to obtain a precise isotherm. This method is not normally available commercially since it is more expensive, might have problems with exhaust gas, and requires more sophistication in operation. However, it has a higher through-put speed and an especially huge advantage for the measurements of mixed-gas adsorptives.

A low-cost alternative to the volumetric method is the "carrier gas flow system." This system is entirely different from the Arnold flow system described above. In this system a carrier gas is used to carry the adsorptive. The disadvantage of this method is that the results are not very accurate or precise and normally do not yield the isotherm. Furthermore, the carrier gas can interfere dramatically, even if it is helium. This is especially true for adsorbate-adsorbent systems that have high energy of adsorption, or for porous adsorbents. For some engineering purposes this might be a good alternative, but it has little value as a research instrument. For engineering purposes it simulates many flow systems used in industry. The calculations for this method miss a very important limitation if it is used to purify a gas.

Generally, the gravimetric method is more accurate and precise. However, such instrumentation is more expensive and requires more skill and patience to operate. Normally, one uses a balance that is referred to as a "microbalance." The balance should have at least a relative sensitivity of 1.0×10^{-6}. For example, if the normal load on the balance is about 1.0 g then it would normally be sensitive to, at worst, 0.1 μg. Indeed, it is not unusual that a 1-g microbalance can measure to 1 ng. The higher sensitivity is recommended since the very low pressure portion of the isotherm is needed to get an accurate measurement of the energy of adsorption. Of course, the higher the sensitively the higher the cost. For the most sensitive measurements, one must make buoyancy corrections as described in Chapter 2.

Calorimetric measurements are less common than the measurements mentioned above and yield a different physical quantity. To be effective, the calorimetric method needs to be combined with either the volumetric technique, which is normal, or combined with the gravimetric technique, which is a little more difficult for high quality work. Both methods are used. Calorimetry measures the temperature change as the adsorption occurs. This, along with heat capacity measurements of the resultant adsorbate-adsorbent combination, yields the heat of adsorption as a function of pressure. Other calorimetric measurements that are less precise measure only the heat evolved. Without measuring the heat capacity the actual energy of adsorption cannot be obtained, but a relative measure may be all that one needs. This gives some idea of the various adsorption mechanism involved. Calorimetry is not widely used since accurate calorimetry is an extremely difficult technique to perform and requires a great amount of time and effort. It is especially difficult to measure the adsorption amounts at the same time.

Preliminary analysis

The measurement of the amount of adsorbate present at equilibrium as a function of the adsorptive pressure at constant temperature is referred to as an "adsorption isotherm." This is the most common piece of data used in physisorption research and process control.

There are several forms of the isotherm that have been officially recognized. Identifying the form, although not necessary, can give some hints as to how to proceed with the analysis. It is also useful to have some idea as to at what point in the isotherm is a rough equivalent of a monolayer present. Traditionally speaking, this is referred to as the "B point analysis." It is generally recognized that is method is inaccurate and has largely been abandoned. Another method, referred to in this book as the "inflection point analysis," is accurate for nonporous adsorbents. However, it does not help much for a sample which has porosity, nor for a type of sample that has not previously been checked for porosity.

The usual method is to start with an analysis of the isotherm to determine what is referred to as the "type" of isotherm. The types of isotherms are labeled "I" through "VI," "a" and "b." However, all of these types may be described more simply by using the standard isotherm transformation method. The reason for this will become obvious once one grasps the meaning of the features displayed in the standard curve representation.

Classical classifications of adsorption isotherm types

Some of the forms of the isotherm are shown in Figs. 1.3–1.8. These types are labeled I through VI according to the classification developed by deBoer, codified by Brunauer et al. [11], Brunauer [12], and supplemented by Prof. Gregg and Prof. Sing [13].

These classifications are widely used in the literature on physisorption and normally have the interpretations listed in Table 1.2. Although the numbers are I through VI, there are sub-types added after this classification with labels "a" and "b," which yield nine at this time.

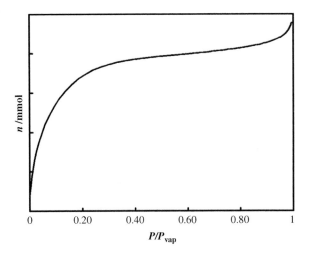

FIG. 1.3

Type I isotherm (a sharper and b).

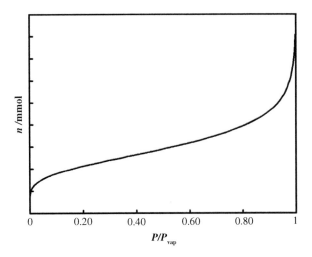

FIG. 1.4

Type II isotherm (a →, b offset ←).

Sometimes Type VI is interpreted to be associated with an initial chemisorption step. This chemisorption occurs along with physisorption, however the chemisorption portion should be somewhat irreversible. If chemisorption is what makes this isotherm unique, then after a mild desorption if the isotherm is measure again, then the follow-on isotherm should differ from the first measured isotherm.

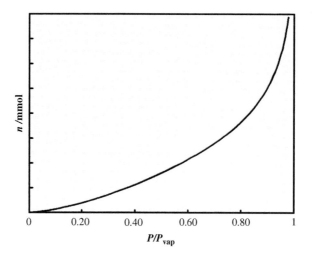

FIG. 1.5

Type III isotherm.

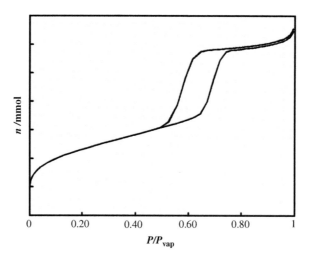

FIG. 1.6

Type IV isotherm (a no loop and b).

The units on the ordinate in these figures are arbitrary. Today the units are usually "mmol g^{-1}." In most of the early literature the units were milliliters of gas (adsorptive) at standard pressure (1 atm) and 0°C. (Read the articles very carefully to obtain the units.)

The first step in analysis of the isotherm is to determine to which classification the isotherm belongs. A further recommendation is to determine the classification of the

Preliminary analysis 13

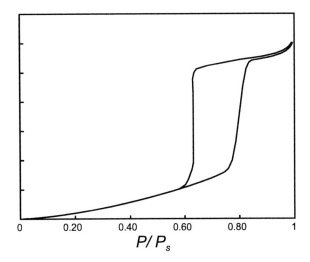

FIG. 1.7

Type V isotherm with desorption nearly vertical.

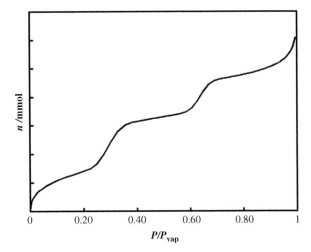

FIG. 1.8

Type VI(a) isotherm.

isotherm according to the standard curve representation or the χ plot representation. This used to be more difficult than at present, since each adsorbent-adsorbate combination had its own standard curve that had to be numerically obtained on special "non-porous" and "smooth" samples of "identical composition." This was always and still is difficult, with a great deal of uncertainty about the criteria given here in quotes.

Table 1.2 Classifications of physical adsorption isotherms

Type	Interpretation
I	This is characteristic of either a chemisorption isotherm (in which case the final upswing at high pressures may not be present) or physisorption on a material that has extremely fine pores (micropores)
II	This is characteristic of a material that is not porous, or possibly macroporous, and has a high energy of adsorption
III	This is characteristic of a material that is not porous, or possibly macroporous, and has a low energy of adsorption
IV	This is characteristic of a material that contains mesoporosity and has a high energy of adsorption. These often contain hysteresis attributed to the mesoporosity
V	This is characteristic of a material that contains mesoporosity and has a low energy of adsorption. These often contain hysteresis attributed to the mesoporosity
VI	This type of isotherm is attributed to several possibilities, the most likely being if the temperature is below the adsorptive triple point, that the adsorbate is more like a solid forming a structured layer—i.e., epitaxial growth. Other possible explanations include multiple pore sizes. If the steps are at the low pressure portion of the isotherm, then the steps may be due to two or more distinct energies of adsorption. If the steps are at the high pressure part of the isotherm, then the steps might be due to sharp steps on the adsorbate surface

This has become easier today with a universal representation [14] of the standard curve based upon a quantum mechanical theory of adsorption. This representation is referred to as the chi, χ, representation. An χ plot is a plot of the amount adsorbed versus the quantity $-\ln(-\ln(P/P_{vap}))$. This drops the number of types, since some look identical with just a shift in the abscissa. The explanation for this will be presented more fully in Chapter 4. In general the χ plot of a nonporous adsorbent, provided there is only one energy of adsorption for a particular adsorbent-adsorbate combination, is a straight line. Thus, deviation from this straight line in either the positive or negative direction indicates deviations from this simple case. In Table 1.5 is a summary of the possible features that the χ plot can have in addition to a straight line.

Single point methods

The following are some ways to make an initial estimate of the monolayer coverage value. The older method, the "B Point" analysis, was used very early and is very inaccurate. It might be useful for a quick comparison. The second method is the "inflection point," which is more accurate but requires a considerable amount of analysis, is relatively easy with present day computers and even smart phones.

A very big note of caution is that one needs to know what a typical isotherm for the type of sample one is working one is. The question is, "does it follow a typical classical isotherm which involves no porosity?" If not, it is highly recommended that these techniques not be used.

The "B point" analysis

The "B point" check was used very early in the research of the physical adsorption isotherm [15]. It is assumed that this method applies to nonporous adsorbents. In analogy to the Langmuir isotherm, the bend in the isotherm should be near the equivalence of a monolayer. The B-point method, however, is not very reliable. Confined to only Type II isotherms, which includes most of the standard curves such as the α-s and t-plot, the shapes of the curves are very similar on a plot of amount versus P/P_{vap}. The problem with this method is scaling, and the visual selection of the "B point" shifts. This is illustrated in Fig. 1.9. Depending upon the range selected, the B point yields a different visual answer. Thus, it is best to get the overall isotherm in a consistent manor.

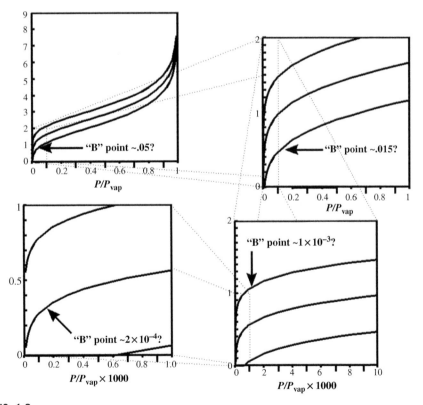

FIG. 1.9

An illustration of why the "B" point method is not reliable. The position of the "B" point depends upon the abscissa scale used.

The "B point" method is good within a factor of 10, and is a very subjective estimation.

There is another mathematical point of minimum curvature that can be derived from χ theory at $P/P_{vap}=0.3679$. However, the value of the monolayer does not correspond to this value and must be calculate with the slope at this point. This method yields the monolayer equivalence and the adsorption energy. This method is discussed in the next section.

There is also another similar "one point" method used in conjunction with the BET theory. This too is very arbitrary and essentially depends upon the reproducibility of the BET plot for samples that are not really alike.

The next "one point method" is more up-to-date and more reliable, but still, without the full isotherm may be misleading. To be safe, it is best to use three points on this method rather than just one. With three points, the answer should be very accurate.

The inflection point method

Another one point method is referred to here as the "inflection point method." The derivation of this method is not presented at this point since it requires the knowledge of the quantum mechanical modeling of the isotherm. The detailed algebra for this method is presented in the Appendix II H. The derivation starts with the results of the simple χ-theory, so one needs to have the knowledge presented at the beginning of Chapter 4 to understand how this works. The same caveat as the "B" point method is assumed, that is it applies to nonporous adsorbents.

Inflection point method is based on quantum mechanics and, therefore, has a much firmer foundation than the "B" method or the BET single point method. The inflection point method requires very little judgment to perform. Since it works only for nonporous adsorbents one should have some evidence that the adsorbent is nonporous. This could be determined by several methods, for example scanning electron microscopy, or previous experience with full linear χ-plots that indicate the material is nonporous.

The method is as follows:

1. Determine the number of moles adsorbed at the relative pressure (P/P_{vap}) of 0.3678 (the value of e^{-1}).
2. Determine the slope of the isotherm at this point. The isotherm should be close to a straight line from a relative pressure of about 0.2 to 0.6. If it is not, then the adsorbent has some porosity and an alternative method must be used. For this one need not do the entire isotherm but simply a few data points about (P/P_{vap}) of 0.3678 ($=1/e$).
3. To obtain the monolayer equivalence, divide the slope by e (2.718…).
4. If the energy of adsorption for the first adsorbate molecules is required, substitute into the equation:

$$\chi_c = -\frac{n_{ads,l}}{n_m} \Rightarrow \overline{E}_a = -RT\exp\left(-\frac{n_{ads,l}}{n_m}\right) \tag{1.1}$$

where $n_{ads,I}$ is the number of moles adsorbed at the inflection point. This molar energy, $\overline{E}_a$, is for the first molecule adsorbed (even though it is measured in mid-isotherm.) One expects it to change with addition adsorption.

From these two pieces of information, one can also obtain the function, $E(n_{ads})$. However, due to the multiple definitions of "heat of adsorption" one needs be knowledgeable of the information contained in the section "Heats of Adsorption" in Chapter 4. (Note: $\overline{E}_a$ is not a *function* of n_{ads}.)

The graph in Fig. 1.10 the extraction of the information from the isotherm by this technique is illustrated. The straight line fit for the tangent line is very good (error of 2%) over the P/P_{vap} range of 0.2 to 0.6. It has an error of 0.1% over the P/P_{vap} range of 0.30 to 0.44. How close one can get the right value is more a matter of how good the data is.

However, there is a systematic error due to the fact the both major deviations biases the error to the high side. This systematic error is dependent upon the range picked. In Table 1.3 are some systematic errors as a function of range with 12 even increments on both sides of $P/P_{vap}=1/e$. To use this, determine the abscissa range that you are using and use the correction factor to compensate. Whatever number the inflection point slope yields according to Eq. (1.1), divide by the values given in this table.

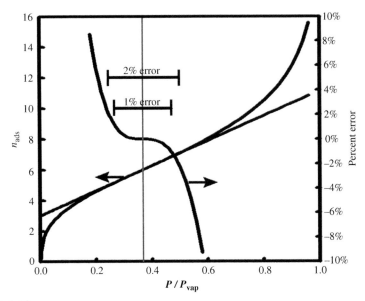

FIG. 1.10

A simulation illustrating the inflection point method.

Table 1.3 Correction factor for n_m obtained by least squares

Interval		
From	**To**	$\frac{n_m \text{ input}}{n_m \text{ output}}$
0.1500	0.5858	0.956
0.2000	0.5358	0.975
0.2500	0.4858	0.988
0.3000	0.4358	0.996

The BET single point method

The BET single point method is as follows:

1. Measure the adsorbent amount, n_{ads}, at a $P/P_{vap} = 0.30$.
2. Divide the value obtained by 0.30 to yield the slope of the line going through the origin.
3. The resultant number is the monolayer equivalence.

Why the value of 0.30 has is used is probably because this usually gives agreement with the full BET treatment.

Summary of one point methods

The "B point" and the inflection point methods are referred to here as single point methods. Both of them, along with the BET single point method, are not as accurate as a full analysis of the isotherm.

Assuming non-porous materials, the "B point" method still suffers from the fact that the selected point is arbitrary and scale dependent.

The BET single point method has some similarity to the inflection point method, in that it selects a recommended data point of $P/P_{vap} = 0.30$. The slope at this point is assumed to be a line though the origin, $P/P_{vap} = 0.0$. Thus, BET single point method assumes "Henry's" law, which on a P/P_{vap} basis is not bad, but assumes that this slope is the monolayer equivalent coverage. Thus, according to the quantum mechanical model, this method is off by a factor of $e^1 \sim 3.$[9] This is a problem along with errors inherent in the BET itself. It is claimed to work for sufficiently large C_{BET} values, but nevertheless, an arbitrary data point is selected.

The inflection point method is discussed. This method is not quite a single point technique but requires at least three accurate data points about the value of 1/e (=0.3678...). A step-by-step procedure is given in the section for the inflection point. This procedure eliminates the problem of using the origin to set the slope and should yield an accurate value for the monolayer equivalence, especially if the corrections given in Table 1.3 are utilized.

---

[9] And it is also a bit off the magic point of 0.3678, mentioned in the forward that Fuller discovered.

Characterization of hysteresis loops

Hysteresis loops are classified into four types. These types were given the designation of H1 through H4 by an IUPAC committee [16]. In Figs. 1.11–1.14 are schematics of these four types.

The characteristics and conventional interpretation of these hysteresis loops are given in Table 1.4. However, there is much work still being performed to understand these forms.

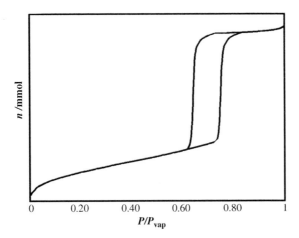

FIG. 1.11

Type H1 hysteresis loop.

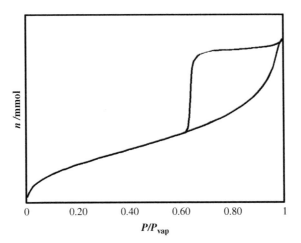

FIG. 1.12

Type H2 hysteresis loop.

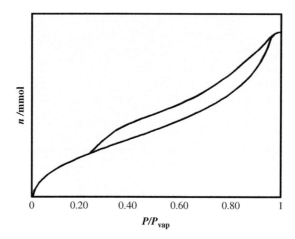

FIG. 1.13

Type H3 hysteresis loop.

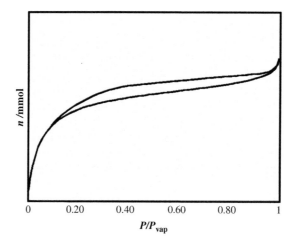

FIG. 1.14

Type H4 hysteresis loop.

Modeling by Roja et al. [17] indicates that the loop shapes, at least Type H1 and H2, depend upon two factors:

1. the size difference between spherical pore chambers and connecting pore passages; and
2. the number of pore passages versus pore chambers.

The shape of these hysteresis loops and their position is still a matter of much discussion. For example, Neimark and Ravikovitch [18] have modeled adsorption in

Table 1.4 Characteristics and interpretation of hysteresis loop types

Type:	Characteristics	Usual interpretation
H1	Nearly vertical and parallel adsorption and desorption branches	Regular even pores without interconnecting channels
H2	Sloping adsorption branch and nearly vertical desorption branch	Pores with narrow and wide sections and possible interconnecting channels
H3	Sloping adsorption and desorption branches covering a large range of P/P_{vap} with underlying Type II isotherm	Slit-like pores for which adsorbent-adsorbate pair which would yield a Type II isotherm without pores
H4	Underlying Type I isotherm with large range for the hysteresis loop	Slit-like pore for the type I dsorbent-adsorbate pair

MCM-41 type zeolite with NLDFT methods. Their conclusion is that the adsorption branch corresponds to the spinodal condensation, i.e., metastable situation, and the desorption branch corresponds to the equilibrium capillary condensation/evaporation situation. Kowalczyk et al. [19], have calculated the hysteresis using a lattice density functional theory. The basis of their work stems from the similar simulations by Arnovitch and Donohue [20]. Their calculations demonstrate the H1 type hysteresis loop due to the curved moving meniscus (this latter publication gives an extensive review of previous work). Although some interpretations are given in Table 1.4, there is probably not any consensus at this time.

There appears to be a relative pressure (P/P_{vap}) below which hysteresis does not occur.

According to Harris [21], the value for this is 0.42 for nitrogen adsorption (an alternate proposal is presented later in the porosity section). Even for samples which demonstrate hysteresis above this value, if the loop extends to this value, then a sudden cutoff will occur. Trens et al. [22] have correlated the intersection of the desorption branch with the adsorption branch at low pressure (referred to as the "reversible pore filling" or "rpf") with thermodynamic properties. Specifically, it seems to follow the Clausius-Clapeyron equation and follows that relationship expected from corresponding states relationship. This indicates that the reversible pore filling is characteristic of a first-order gas-liquid transition. The enthalpy of this transition is somewhat higher than the liquid-gas transition in the bulk, which should not be surprising since the interaction of the solid with the adsorbate should supply extra energy.

Further complicating the comparisons of the various modeling with experimental data is the possibility that the energy of adsorption might shift, and possibly in a reproducible manner, from the adsorption branch to the desorption branch. Although such a shift cannot explain all of the hysteresis, especially the types other than H1, it creates problems in comparing modeled hysteresis with observed hysteresis. Type H4, however, is precisely what one would expect from such a shift. The source of such a shift, if that is the case, is at the moment unknown.

Quantum mechanical classifications of isotherms using χ plots

The quantum mechanical classification of the isotherms is quite different since the abscissa is from the χ transformation of P/P_{vap}. Only the pressure is transformed. This transform is:

$$\chi = -\ln\left(-\ln\left(P/P_{vap}\right)\right) \tag{1.2}$$

Types I to VI isotherms given in Figs. 1.3–1.8 are presented in the χ-plot representation in Figs. 1.15–1.20. In Table 1.5 are some of the distinguishing feature that show up in the χ-plot. Thus, the χ plot is simply plotting the raw adsorption data, usually in millimoles per gram sample, as a function of the pressure transformed

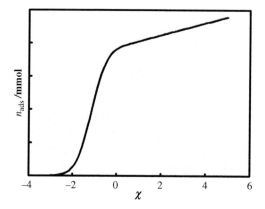

FIG. 1.15

Type I isotherm expressed as a standard plot or a χ plot. A simulation of an E_a distribution has been added here.

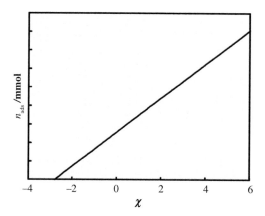

FIG. 1.16

Type II isotherm expressed as a standard plot of a χ plot. No energy distribution is assumed.

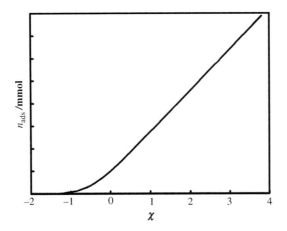

FIG. 1.17

Type III isotherm expressed as a standard plot or χ plot. A simulation of an E_a distribution has been added here.

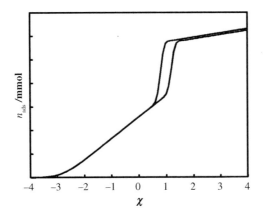

FIG. 1.18

Type IV isotherm expressed as a standard plot or χ plot.

by the function $-\ln(-\ln(P/P_{vap}))$. One of the advantages of the χ plot is that often the entire isotherm when plotted in this fashion is a straight line. The most common isotherm plot, indeed the "standard curve" plot, is Type II, which is a precisely a straight line, as illustrated in Fig. 1.16. When transformed, Types II and III are qualitatively identical and indistinguishable in instruments that do not have high vacuum capability. Types IV and V are also normally not distinguishable.

In addition, a second Type VI plot, VI(b), is presented that differs from the one presented in Fig. 1.8, which has the χ plot feature 5. This characteristic plot is

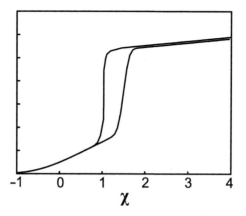

FIG. 1.19

Type V isotherm expressed as a standard plot or χ plot.

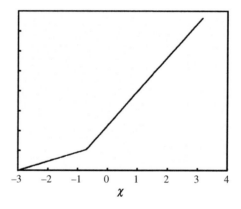

FIG. 1.20

A standard plot of an alternate VIA isotherm. This is the result of two surfaces with differing E_as.

indicative of adsorption on two distinctive surfaces; that is, on two surfaces with different energies of adsorption Thus the χ representation cuts down on the number of isotherms to consider and specifies exactly the physical feature that each χ plot feature corresponds to. One of the possible Type VI isotherms that shows feature 5 in Table 1.5 can be distinguished from the pore filling feature 3 in the χ plot, whereas for the untransformed, i.e., classical, isotherm this discernment is not possible. If one adds a distribution in E_a, such as a Gausian, then a rounding of the plot at low pressures appears and may create the situation where as $P \to 0$ that $n_{ads} \to 0$ (almost). This creates a Henry's law artifact.

Table 1.5 Non-linear features of the χ plot

Item	Feature	Interpretation
1.	Positive curvature at the lowest pressures	A distribution of adsorption energies
2.	Negative curvatures	Decrease in adsorption potential due to filling of pores
3.	Large positive curvature followed by negative curvature to yield a slope of the χ plot that is less than at lower pressures	Mesopore filling due to capillary action
4.	Hysteresis associated with item 3	Several possibilities: A shift in adsorption energy, odd shaped pores, major and minor pores, distortion of the adsorbent
5.	A break in the straight line at moderately low pressures	Similar to 1, except the distribution may be 2 or 3 distinct values

There are two possibilities for a Type IV (a) isotherm for which the classical appearance is in Fig. 1.8:

1. Normally this is found below the freezing point of the adsorbent and the some steps may be phase changes. Specifically, one could be the phase change from the adsorbed liquid state to the adsorbed solid state.
2. This is indicative of multiple distinct E_as along with microporosity and mesoporosity.

There are some characteristics to look at that might also show a distinction between these two possibilities. For example, if case 2 is correct, then the successive slopes must decrease.

Summary of adsorption types

The Type I through VI isotherms of Figs. 1.3–1.8 are shown in Figs. 1.15–1.20 as the transform to the standard curve form. These have been generated by transforming the abscissa according to the quantum mechanical model. Later the reader will be able to determine that this is proper, especially for determining the qualitative features, which is being discussed here (Fig. 1.21).

Type I isotherm is usually described as being a Langmuir isotherm, but this may not be the case. However, looking at Fig. 1.3, the characteristic leveling-off of the curve is not really present and indeed, as P/P_{vap} approaches 1.0 the curve increases rapidly. This does not occur with the Langmuir isotherm. In the Type I transform to standard curve (Fig. 1.15) as a standard curve is the curve characteristic of a sample with microporosity. It is important when one speculates that the Langmuir isotherm is operative that the ultrahigh vacuum data be obtained. This is the most certain way to make the distinction since as $n_{ads} \to 0$ the $P \to 0$ for the Langmuir isotherm. If

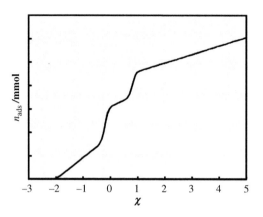

FIG. 1.21

Normal Type VI isotherm expressed as a standard plot or χ plot.

there is some low pressure at which $n_{ads}=0$ and $P>0$, then the Langmuir model does not hold and this isotherm is indicative of microporosity. Unfortunately, this low pressure range is usually not observed experimentally due to equipment limitations. In most cases when it is stated that microporosity is operative, the material has been compared to the nonporous standard sample stand-in. This is probably safe, but there may be some areas in the pores that promote chemisorption (indeed many commercial zeolites have promoters imbedded; after all, that's the whole point).

Type II and Type III are basically identical except for the value for E_a. A distribution of E_a was added to the Type III here for illustration purposes and is not a necessary feature. Types IV and V are also identical except for the differences in E_as.

There is a possibility of obtaining a false hysteresis loop if the energy of adsorption (E_a) shifts to a lower value for the desorption isotherm. This type of shift would not have nearly vertical boundaries for the hysteresis loop. This occurrence is very likely, so a plot of n_{ads} versus $\Delta\chi$ for each branch might be more appropriate to yield the difference in the E_as.

There is another plot that one can observe in the χ representation that might appear as a Type VI plot. This is characteristic of samples with two or more types of surfaces with differing energies (E_a) of adsorption. The feature, however, might be obscured in the normal isotherm representation. The standard or χ plot is represented in Fig. 1.20; whereas, Fig. 1.22 is the normal isotherm one would obtain with the two different E_a. The χ plot of Fig. 1.20 clearly reveals the two E_as that are present. The abscissa intercept yields one E_a and the abscissa value at the break yields the other E_a. Thus, one infers from Fig. 1.20 that there are two energies of adsorption due to the break. This feature may not be obvious in the normal isotherm representation and depends upon the various values and the data scatter. It is however very obvious in the standard or χ plot representation. This plot is very characteristic of nonporous carbon as an adsorbent.

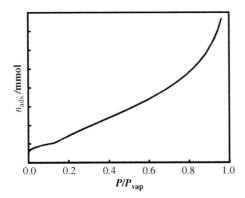

FIG. 1.22

The normal isotherm for the alternate Type VI isotherm (VIA). Steps due to different E_as.

Measuring the surface area from the isotherm

As hinted at in the previous section, if one can determine the amount of material in one monolayer of an adsorbate, then one might be able to calculate the surface area from this. One simply needs to know what the average cross-sectional area of the adsorbate molecule is. If this is possible then the calculation is rather simple. However, there has been considerable discussion in the past as to what this relationship should be. Furthermore, the important parameters obtained, for practical purposes, are the monolayer equivalence, n_m, and the energy of adsorption. One might say it's the esthetics of obtaining the surface area that is sought, regardless of the usefulness. Therefore, the following will proceed with how to make this calculation.

If n_m were [subj.] the number of moles of an adsorbate in a monolayer and a was the cross-sectional area of the adsorbate molecule, then the surface area, A_s, would be given by:

$$A_s = n_m N_a a \qquad (1.3)$$

where N_A is Avogadro's number ($6.022 \times 10^{23}\,\text{mol}^{-1}$).

Two problems are involved with this. Firstly, "how does one arrive at n_m?" Secondly, "what is the value for a?" Both of these questions were, and still are, problematic in the early mathematical hypotheses that attempted to solve these. In the first instance, some theoretical equations should yield n_m. Once the theoretical equation is in place, the second question might be answerable. There appear to be only two equations that yield an answer for the n_m. These two are:

1. the theory by Brunauer et al. [23], the BET equation, and
2. The χ theory or disjoining pressure theory (or Excess Surface Work or ESW) which yield the same transform for isotherm and the same value for surface area or monolayer equivalence. ESW is derivable from χ theory, but independently proposed, which supports the validity both theories. These theories are very different from the BET and other Henry's law based theories.

Still today, almost all the other methods depend on the surface area value determined by the BET. This includes the standard curves (unless based on χ theory or disjoining pressure theory) and possibly the various forms of DFT or "Monte Carlo" calculations. Before explaining these, the notion of "Henry's law" needs to be examined.

What is the "Henry's law" assumption?[10]

"Henry's law," or more precisely, the Freundlich isotherm with a coefficient of 1 as applied to physical adsorption states that:

1. As the adsorptive pressure approaches zero, $P \rightarrow 0$, the amount of adsorbate approaches zero, $n_{ads} \rightarrow 0$; and
2. so long as $P > 0$, $n_{ads} > 0$. In other words, $n_{ads} = 0$ if and only if $P = 0$.

The first part of this makes sense and must be so according to thermodynamics. However the second part, that n_{abs} is finite so long as $P > 0$, is not required by thermodynamics in spite of what experts and intuition say. An argument has been put forth that it must be true since the adsorbate can be described by the virial equations. Firstly, the virial equations assume the validity of "Henry's law" and as such can be viewed as a correction to Henry's law, thus revealing circular reasoning. Secondly, the virial equations are not derived from thermodynamics alone but include other assumptions.

Some background about this assumption is in order. A more thorough thermodynamic discussion will be presented later in this book, but for edification a quote from the book by Rouquerol et al. [24] will be presented here. In the attempt to relate the spreading pressure, π, to the surface excess, Γ or n_{ads}, the following equation is derived that must be evaluated:

$$\pi = RT \int_0^P \frac{\Gamma}{P} dP \quad (1.4)$$

Eq. (1.4) is often referred to as the "Gibbs' adsorption equation" where the interdependence of Γ and P is given by the adsorption isotherm. The value of the spreading pressure, for any surface excess concentration, may be calculated from the adsorption isotherm drawn with the coordinates n_{ads}/P and P, by integrating between the initial state ($n_{ads} = 0$, $P = 0$), and an equilibrium state represented by one point on the adsorption isotherm.

A mathematical difficulty arises in the evaluation of the integral in the region below the first experimental point ($P = P_1$). First, the exact form of the adsorption isotherm is unknown in this range, and second, the ratio is indeterminate as P tends to zero. **A possible solution is to assume a linear variation** of n_{ads} versus P in this

---

[10]What is being addressed here is physical adsorption and not solution chemistry. The following discussion has been criticized on the basis that this author called "Henry's law" a myth. Firstly, this author never did call Henry's law, per se, a myth and continues to use it for solution chemistry. Secondly, read on.

lowest part of the adsorption isotherm. **By analogy** with solutions at infinite dilution, for which Henry proposed his linear law, it is the custom to refer to this linear part of the isotherm as the Henry's law region. Thus:

$$n_{ads} = k_H P \qquad (1.5)$$

(The emphasis here is inserted along with the symbolism used in this book.)

The problem with this is that any monotonic increasing function, or even with Γ (or n_{ads}) = 0 for a distance, is quite easily integrated so long as $n_{ads}=0$ when $P=0$; in other words, that assumption 1 above holds true, but not necessarily assumption 2. The other problem is that it creats an obsession in the field of physical adsorption that Eq. (1.5) must be followed. Thus, a popular corollary, which violates all mathematical logic, to the definition of Henry's law is:

> "Any proposed theory of physical adsorption must include in its resulting mathematics the Henry's law limit or it is invalid."

This is obviously and absolutely false.

Finally, the assumption made it seen unnecessary to collect lower pressure data when there was a large portion of a monolayer, 0.1 equivalent monolayer or more—sometimes more than one equivalent monolayer—already adsorbed by the first data point. This is very often an important range in the isotherm. Thus, for many experiments, important low pressure information that is required for correct analysis has been left off or not collected. Indeed, in some literature there are suggestions to take data only around a $P/P_{vap}=0.30$. As a consequence, this Henry's law assumption has crippled the theoretical advance in the field for at least the last 35 years, if not 90 years, and almost all of the data has been incorrectly interpreted. This is a very upsetting revelation, and only in recent years have there been efforts to correct this mistake.

It is a well-known principle in science that any theory that is proven incorrect with only one type of measurement must be either invalid or a poor approximation to reality. When an old theory is disproved, it might still be useful as an approximation. There are many examples of this in science; for example, General Relativity versus Newtonian mechanics or valence bond theory versus various quantum mechanics approximations. However, if a theoretical prediction is reliably found to be false, then in actuality the entire theory is false. Such has been the case for the Henry's law criterion since at least 1955 with the work of Lopez-Gonzalez et al. [25] at the National Bureau of Standards. Since that time, others have also observed this phenomena without comment. If the Henry's law criterion is shown incorrect for any isotherm, then by extension, all theories that require or predict Henry's law are proven incorrect (warning: physical adsorption is being addressed here not solution chemistry. In Chapter 4, the thermodynamics associated with disjoining pressure theory also disproves all Henry's law theories). This is a very severe test. In order to recover any semblance of the disproved theory, there must be some practical reason to continue using it, especially if the new theory is simple to use.

Table 1.6 Different adsorption isotherm equations

Names of isotherm	Isotherm equation[a]	Type of adsorption
Henry	$bP = \theta$	Any
Langmuir isotherm	$bP = \theta/1 - \theta$	Localized, without lateral
Fowler-Gugenheim isotherm	$bP = [\theta/(1-\theta)] \exp(-c\theta)$	Localized, with lateral interactions
Volmer isotherm	$bP = \frac{\theta}{1-\theta} \exp\left(\frac{\theta}{1-\theta}\right)$	Distributed, without lateral interactions
Hill-deBoer isotherm	$bP = \frac{\theta}{1-\theta} \exp\left(\frac{\theta}{1-\theta}\right) \exp(-c\theta)$	Distributed, with lateral interactions
Freundlick isotherm	$\theta = K_F P^{r_F - 1}$	Derived from ideal gas law
BET theory	$\theta = \dfrac{CP}{(P_{vap} - P)\left[1 + (C-1)(P/P_{vap})\right]}$	Localized, without lateral interactions

[a] $\theta = n_{ads}/n_m$ where n_m is the monolayer equivalent.

Implications of the Henry's law disproof

There are many isotherms, both theoretical and practical, that use "Henry's law" for physisorption to the surface liquid state and by the above criterion may all be tested together. In Table 1.6 are five commonly referred to isotherms that fit into this category summarized by Alfonso et al. [26] (sources are: Langmuir [27], Fowler and Gugenheim [28], Volmer [29] and Hill-deBoer [30–32]). Added to this list are the Freundlich isotherm and the BET. It is easy to demonstrate for each of these that as $\theta \rightarrow 0$ that $P \rightarrow 0$ absolutely also. It is not surprising that all of these isotherms and several others fulfill the "Henry's law" criterion, since this has been used as an "acid test" for the possibility that any proposed equation is correct. It is today clear that it is quite the opposite. It is the "acid test" that the theory is incorrect.[11]

As will be pointed out later in this book, there are several instances where Henry's law is not followed, but rather n_{ads}, or θ, drops to zero before P does. This has now been demonstrated by several experiments and investigators. This leaves very few adsorption theories as potentially totally valid.

The phenomenon of n_{ads} going to zero at a finite adsorptive pressure has been referred to as the "threshold pressure." Thus, the status of the BET theory is proven invalid in an absolute sense, since the threshold pressure for physical adsorption has now been confirmed by multiple investigators. There were many theoretical and practical problems with the BET theory before the observation of the threshold pressure, but the BET has continued to be used, since traditionally there was no substitute and a very large database has been created based upon it.

---

[11] This comment applies only to the physisorption in the liquid range as described earlier in this chapter. Do not apply this to experimental conditions different from those described in Chapter 2, i.e., liquid conditions.

This may seem to be a rather dark situation; however, there are two hopeful ways around the problem:

- Firstly, there are many engineering applications which need only the isotherm and any good fit to it will suffice. One example is the carrier gas flow systems that are not used to purify the carrier gas, the carrier flow technique, which deals only with relatively high pressures. It really needs only an empirical equation that does a decent job of fitting the data above $\sim 0.01 P/P_{vap}$. Even there, however, it is possible to conceive of some applications and conditions where one would go very far astray.
- Secondly, there are now theories that fit the isotherm better and account for the threshold pressure. If the original data is available then everything may be recalculated. With only the BET surface area and C_{BET} (the BET "C" constant) the information is not useful. An exception to this is if C_{BET} remains constant and the Sing qualifications are followed, then the BET surface area is monotonic with the n_m but not proportional.

Furthermore, the BET or some similar theory may be useful for the adsorption in the solid state, since the theory was originally based on surface sites, similar to the Langmuir equation. Thus, chemisorption and epitaxy are excluded from the above test.

As outdated as the BET is, one still needs to know the details of this theory to be proficient in the area of physisorption. The reasons for this are twofold:

1. Almost all previous data has been analyzed using the BET, so one needs to be familiar with this theory.
2. Given the BET transformed data and the BET surface area and the C_{BET}, it might be possible to recover the original data.

Therefore the BET equation and its use are presented here in sufficient detail to take advantage of these reasons. If one is not concerned with prior data information, it is not necessary to be proficient in this topic.

Summary about Henry's law and BET

The reason for learning about BET theory is to be able to interpret old data. Sometimes, the paper or report that has utilized the BET cannot be salvaged. The reader is cautioned about using the BET without providing some means for others to recover the original data, since the BET theory is clearly incorrect. The BET has been a subject of much criticism in the literature and in this section one of the most important problems is addressed: that is the Henry's law requirement that is assumed for the validity of an adsorption theory. It is therefore not surprising that there has been not much advance theoretically for at least 40 years. Those advancement that have occurred have essentially been ignored for at least another 30 years.

The Henry's law assumption, which was proposed on shaky grounds to begin with, has been disproved for the type of adsorption addressed in this book (more mathematical disproof is presented in later chapters). Along with the BET, there is a group of isotherms, referred to here as Henry's law isotherms, that also need to be abandoned.

Nevertheless, the BET is too important at this point not to be explained. Anyone doing physical adsorption experiments or measurements will quickly come across it. Therefore, it will be explained in the following sections.

The BET theory (Brunaur, Emmett, and Teller)
Original assumptions

The BET theory assumes that there exist, much like the Langmuir theory, sites on the adsorbent surface that bond with adsorbent molecules and are in equilibrium with the adsorbent. This is similar to a chemical reaction but with lower energies of bonding. Additional adsorbent molecules then bond to the first layer adsorbate molecules and are also in equilibrium with the adsorbent gas but with a much smaller bonding energy. Subsequent layers of molecules bond in a similar fashion to successive layers. It is then obvious by normal thermodynamics that the second, third, etc., layers are also in equilibrium with the adsorbent first layer species, i.e., multiple equilibria.

The derivation was originally made using normal equilibrium reasoning and was later codified with a statistical mechanical derivation. Since the model was the same for both methods it should and does yield the same equations. The BET theory equation is:

$$\frac{n_{ads}}{n_m} = \frac{C_{BET} P}{(P_{vap} - P)\left[1 + (C_{BET} - 1)(P/P_{vap})\right]} \tag{1.6}$$

where P_{vap} stands for the vapor pressure of the adsorptive at the temperature of the sample. In most of the literature, the symbol $P°$ is used for this. In this equation, n_{ads} is the amount of the adsorbate, n_m is a monolayer amount, and C_{BET} is the (so-called) BET constant. C_{BET} is related to the energy of adsorption by the equation:

$$C_{BET} = B_{BET} e^{-\frac{E_{BET}}{RT}} \tag{1.7}$$

where E_{BET} is an energy of adsorption and B_{BET} is a constant.

C_{BET} is an equilibrium constant in the original BET derivation and Eq. (1.7) expresses this. Since, in general, $B > 0$ for the equilibrium constants, then C_{BET} can never be negative. However, in measurements it is often observed to be negative and various corrections to using this equation have been proposed to compensate for this problem. It should be noticed that if:

$$\lim P \to 0 \; n_{ads} \to 0 \tag{1.8}$$

and thus it fulfills the "Henry's Law," or more precisely the Freundlich isotherm. This is not surprising since the starting assumption is that the first monolayer is assumed to follow the Langmuir isotherm, which in turn follows Henry's law.

The BET Linearization for analysis

Eq. (1.6) is normally rearranged to yield a linear equation with variables:

$$x = \frac{P}{P_{vap}} \text{ and } y = \frac{P}{n_{ads}(P_{vap} - P)} \tag{1.9}$$

so that the rearranged equation:

$$\frac{P}{n_{ads}(P_{vap} - P)} = \frac{1}{n_m C_{BET}} + \frac{C_{BET} - 1}{n_m C_{BET}} \frac{P}{P_{vap}} \text{ becomes : } y = mx + b \tag{1.10}$$

with the constants $b = 1/(n_m C_{BET})$ and $m = (C_{BET} - 1)/(n_m C_{BET})$

By making a plot of the defined y as a function of the defined x one can determine the slope, m, and the intercept, b. One then uses these parameters in the equations:

$$C_{BET} = \frac{m+b}{b} \text{ and } n_m = \frac{1}{m+b} \tag{1.11}$$

in order to obtain n_m, which is interpreted as the monolayer coverage of the surface, and the BET constant C_{BET}. Notice, from Eq. (1.6) as $C_{BET} \to \infty$ that:

$$n_m = n_{ads}\left(1 - \frac{P}{P_{vap}}\right) \text{ or } n_{ads} = \frac{n_m P_{vap}}{P_{vap} - P} \tag{1.12}$$

two of the ways the single point calculation can be written.

The plot should be usually taken over the 0.05 to $0.35 P/P_{vap}$ range. Beyond these values the linear characteristic of the plot breaks down. However, this range is variable depending upon conditions and materials, as explained by Kupgan et al. [33].

The normal sequence recommended in the derivation C_{BET} and n_{ads} to divide Eq. (1.10) is to invert both sides of Eq. (1.6). The resulting equation is then multiplied both sides by P/P_{vap}. One might wonder why the latter operation is performed. The reason becomes obvious to anyone who has checked the plot without multiplying by P/P_{vap}. When this is done, the data does not yield a linear plot, if one were to plot P_{vap}/P versus $P_{vap}/[n_{ads}(P_{vap} - P)]$. Thus, this rearrangement is simply to get a partial linear fit, which is not a statistically valid method without an appropriate weighting factor, and maybe not even then.

BET problems and restrictions

The problems with the BET hypothesis were apparent very early. However, as time lapsed, no other satisfactory solution for determining the surface area appeared. Therefore, it became the sole measure of surface area. Initially there were four types of isotherms identified, which were expanded later to at least nine. These are usually referred to as: Ia, Ib, IIa, IIb, III, IVa, IVb, V, and VI. The first problem is that the BET theoretic equation did not fit any one of these isotherms well over the full range of pressures used. The pressures were usually expressed in terms of P/P_{vap}, and had a range $1 > P/P_{vap} > 0$. To partly solve this problem it was stated that the fit should be only over the range $0.05 < P/P_{vap} < 0.35$ and even then there could be exceptions. The latest solution to these inconveniences were to specify four BET criteria for finding the linear portion of the transformed BET equation to yield an answer [13]. These are:

1. It should be used only on Type II and IV isotherms.
2. C_{BET} must be positive and the range should be adjusted to make it positive. Some investigators recommend that adjustments be made to keep C_{BET} between 50 and 150.
3. The equation should be restricted to the range where the term $[n_{ads}(P_{vap} - P)]$ increases with P.
4. At the pressure where $n = n_m$ should be in the range where the calculation is made.
5. The calculated value of $(P/P_{vap})n_m$ should not differ from the (P/P_{vap}) corresponding to the calculated value n_m. If it isn't, move the range until this criterion is fulfilled.

BET forbidden isotherm types

Even if all these criteria are met, Type III and Type V isotherms cannot be analyzed. It is also not recommended to use the BET for a Type I isotherm to yield surface area. BET analysis of Type II and Type IV isotherms are to be regarded as *effective* surface area, good possibly only for relative comparison purposes. However, notice that by criterion 1, one could end up with a range of answers by adjusting the C_{BET} from 50 to 150; a strange result indeed. Finally, the Type VI is found for adsorption experiments below the freezing point of the adsorbate, thus layering epitaxially, or possibly for adsorbents that have two or more energies of adsorption.

Contrast the above restrictions to the χ theory criterion. Use the entire isotherm to obtain information with no restrictions except that the temperature of the adsorbent not be below the freezing point of the adsorbate. Sometimes even this last restriction can be extended. Since it can be demonstrated that the standard curve methods, such as the α-s and others, may be derived from the χ theory, the same criterion applies to them.

Correlation of BET to surface dependent reactions

The relationship between the BET surface area measurement and surface limiting reaction has always assumed to be the following:

1. The reaction rate is surface dependent if and only if it correlates to the BET surface area value
2. The reaction rate is not surface dependent if and only if it does not correlate to the BET surface area.[12] As you will see, in the following example, the "only if" part may not be correct.

In a publication by Volsone et al. [34] (VTMOP) a correlation was proposed between the BET surface area measurement and the reactivity to various gasses on clays. The

---

[12]This is an assumption that I also made in the hydriding reactions. The conclusion was, since there was hardly any correlation to surface area, thus, the rate limiting step was not surface inhibited. As you now see, the reasoning was faulty. However, in the case of the hydriding reactions there were other avenues to prove the independence of the reaction on surface area other than the BET measurement.

Table 1.7 Linear correlation of the data by VTMOP—reactivity versus BET surface area

Molecule	CO_2	N_2	C_2H_2	CO	CH_4
r^2 with	0.81	0.47	0.86	0.43	0.54
Intercept =	0.026	0.016	0.033	0.038	0.010
Max value[a]	0.30	0.049	0.26	0.086	0.075
r^2 thru 0,0	0.79	0.15	0.80	−1.3	0.48

[a]Maximum value is not necessarily the value at the highest surface area.

correlation r^2 has been calculated for the postulated linear fit and is given in Table 1.7.

One definitely expects a linear correlation, but the question is: how good is the correlation? The r^2 values determine this. Furthermore, the linear fit does not go through the origin as it should. These intercepts are given in the table with the maximum value for comparison. Forcing the linear fit through the origin yields the r^2 in the last column. It is always a judgement call as to whether the values are high enough to draw a conclusion, especially when there is no other model to compare with. Given the various r^2 values seen throughout this book, except perhaps for C_2H_2, these numbers do not look supportive.

The bottom line for the BET—Don't use it

As mentioned in the forward and the previous section, it is time to discard the BET analysis for several reasons as provided above. There are multiple others pointed out in the literature as well. But the most serious problem is that it does not yield the surface area within a factor of 3. This problem is very obvious if one uses the single point method. One can get just as good an answer by laying a finger down randomly on the isotherm graph and reading the number. For those that have checked the results against other methods such as electron microscopy[13] and those who have tried to use it to determine porosity, this has become painfully obvious.

There are today other ways to approach the problems of surface area analysis. These include the χ theory as described in this book, the disjoining pressure theory (ESW), and perhaps in the future, the various forms of density functional theory (DFT.)

There is one more method that yields the correct answer for some cases. The method is difficult to do since it is calorimetric, which is always difficult, with additional difficulties. The restrictions on this method become obvious once one considers what materials to use. For example, it is probably the easiest to do the method with water and a ceramic material adsorbent, which does not react with water. This method has been named the "absolute method" and is described in the next section.

---

[13]I know this problem well, I tried for years to get the measurements to match with well prepare samples, good computer techniques for SEM measurements, trying to get consistent results between porosimetry and negative porosimetry on the macroporous rare earth plasters and other well behaved samples with no luck.

Summary of the BET method

The BET equation is presented along with the linearization technique. Some of the restrictions are given on what is assumed to be appropriate caveats. According to the literature, only Type II and Type IV isotherms are appropriate for its use and the restrictions apply. Most important, the BET C_{BET} constant should be between 50 and 150 and the range restriction can be modified to be in anywhere in that range (negative C_{BET} constants are definitely not allowed since these yield an imaginary energies). Other restrictions apply, and only the most important ones are listed here.

The "absolute method" by Harkins and Jura [35]

The "absolute method" by Harkins and Jura is explained in more detail in the section devoted to it. The method does not measure the properties of the adsorbate-adsorbent interface, but rather the properties of the adsorbate-gas phase. It can be used to yield a value for n_{ads} and E_a, the energy of adsorption when $n_{ads} = 0$, with some clever manipulations in energy accounting. It is assumed that the adsorbate is a thin film of liquid, which as far as we know today is nearly correct. By adsorbing an adsorbate that has a high liquid-gas interface tension, and thus energy, and determining the heat released upon quenching in water at the same temperature, one obtains an energy release that is related to the surface area.

Experimentally, this technique is very difficult and the interpretation, which depends upon the film thickness, is somewhat questionable. The question is: at what point of adsorption can one assume the adsorbate-gas interface is the same as a bulk liquid-gas interface? The answer turns out to be at a multiple layer thicknesses. This can be seen from the quantum mechanical calculations of either the χ-formation calculation or from the density functional theory method.

The interpretation that there is a surface film thickness is not quite correct, but the theoretical development is not dependent upon this assumption. In spite of the initial interpretation problems, this method is on a sound theoretical foundation.

Summary of the "absolute method" introduction

The absolute method by Harkins and Jura should yield the surface area and the heat of adsorption as a function of exposure to adsorbates. The method is difficult and has some restrictions, but seems not to be flawed in any way. The method and the results will be addressed later in this book.

χ and ESW theories that do yield monolayer equivalence n_m

Another method to determine the monolayer equivalent comes from quantum mechanical χ theory (and the classical mechanical equivalent named auto-shielding physisorption or "ASP" by Fuller[14] et al. [36]) and the disjoining pressure theory [37]

---

[14] Fuller has published several dozen of articles on this subject. All of them are not cited in this book, but an extensive search for "E. L. Fuller" will bring up many interesting publications.

or Excess Surface Work (ESW) theory. These two theories are compatible and indeed, the disjoining pressure starting equations, which were derived from thermodynamics with an additional assumption, can be derived from the χ theory equations. Thus, they are in reality the same theory. The values obtained by this method, as analyzed by Condon [38], seem to agree with some other methods, such as the "absolute method" of Harkins and Jura [35] and with the conclusions by Kaganer [39, 40]. It is also consistent with X-ray analysis for some porous samples and SEM scans for some nonporous solids. For a non-porous, single-energy surface the following equation holds, according to χ hypothesis [36]:

$$\theta = (\chi - \chi_c) U(\chi - \chi_c) \quad (1.13)$$

The derivation for this equation from quantum mechanics is presented in Chapter 4 for nonporous surface, and Chapter 6, for porous materials. The following clarifies the left side of this equation. The symbol θ designates what has been referred to as the specific "coverage," which in more precise language would be called the "specific monolayer equivalents" (all the quantities here are specific quantities, i.e., the number of moles have been divided by the mass of the adsorbent). This means that if all the molecules were in the first monolayer, then θ would be equal to 1 exactly (notice the subjunctive).

Defining the quantity n_{ads} is the specific (:= per gram of sample) number of moles adsorbed and n_m is the specific number of moles in one monolayer equivalent. θ is related to these quantities by:

$$\theta = \frac{n_{ads}}{n_m} \quad (1.14)$$

Thus for the specific surface area, A_s:

$$A_s = n_m \overline{A} \quad (1.15)$$

where $\overline{A}$ is the area needed for the perfectly complete monolayer of 1 mole of adsorbate, that is, the molar area.

On the right side of Eq. (1.13), the quantity U is the unit step function; that is, negative values of n_{ads} are meaningless, and:

$$\chi = -\ln\left(-\ln\left(\frac{P}{P_{vap}}\right)\right) \text{ and } \chi_c = \ln\left(\frac{-\overline{E}_a}{RT}\right) \quad (\overline{E}_a < 0 \text{ absoption is exothermic}) \quad (1.16)$$

From the plot of n_{ads} versus χ one can obtain n_m as the slope and $\mathbf{E}(\theta)$, the energy of adsorption as a function of coverage. E_a is a constant and $E_a = \mathbf{E}(0)$. These parameters may be calculated from the abscissa (x-axis) intercept, which yields χ_c. For the slope:

$$n_m = \frac{\partial n_{ads}}{\partial \chi} \quad (1.17)$$

and for the energy of adsorption:[15]

$$\mathbf{E}(\theta) = (E_a - \varepsilon)\exp(-\theta) + \varepsilon \tag{1.18}$$

E may or may not be what one thinks of as the "heat of adsorption." This is dependent upon what "heat" one is referring to, since there are several definitions. Some "heats" can be interconverted, whereas others cannot, due to loss of information in the recording process.

Summary of the χ theory introduction—For non-porous single energy physisorption

One makes a χ plot as an *x-y* plot with the abscissa, *x* axis, labeled as χ as given in Eq. (1.16). For the ordinate, *y* axis, the amount adsorbed is given, preferably in moles or millimoles per gram sample. The χ plot should be a straight line for a non-porous single energy adsorbent. An upward bend or curvature is probably due to multiple energy planes, whereas, a downward curvature is an indication of microporosity. A ventricle upward bend, especially if there is hysteresis, is an indication of mesoporosity. These complicating features will be examined in Chapter 6.

The straight line obtained will intersect the abscissa at a particular value of χ from which one can obtain the adsorption energy from the second part of Eq. (1.16). The slope of the line yields the monolayer equivalence according to Eq. (1.17).

In this chapter, the discussion of the depth profile will be summarized with a few equations. This summary is usually all that most investigators will ever need, especially if they do not want to be bothered with the theoretical derivation. For those interested, the theoretical derivation is presented in Chapter 4.

Relating *a* and A_s to n_m

Eqs. (1.14)–(1.17) are used to obtain the specific surface area. However, Eq. (1.15) has a quantity $\overline{A}$, the molar area, which has not been defined and the quantity *a*, the effective area of an adsorbate molecule, which is dependent upon $\overline{A}$ (or visa-versa.) This is the second challenge in calculating the surface area of a sample from physisorption. There is the philosophical question of what is the meaning of surface area, since the geometry is not flat and also different answers are obtained for n_m depending upon adsorbate.

In Eq. (1.21) is a listing of some typical adsorbates for $\overline{A}$ and *a* using the hard sphere assumption. There are different ways to make the calculation depending upon the assumptions used. In the four methods listed, it is assumed that the bulk liquid or solid of the adsorptive is arranged in a close packed arrangement. There are packing factors for the bulk and the two-dimensional (2D) films. The packing factors account for the empty space between the hard spheres. The packing factor

---

[15] There are several definitions for the heat of adsorption, this equation is typical. For more details about these definitions see Chapter 4 where the most important four are explained.

for the close pack solid/liquid, $F_{3D}=0.7405$, and the packing factor for the film, $F_{2D}=0.9094$.

To create a consistent answer, the IUPAC made a decision that the bulk liquid phase density would be used with both F_{2D} and F_{3D} corrections. If the BET IUPAC convention were to be used, then $\overline{A} = 9.76 \times 10^4 \, m^2 mol^{-1}$ for nitrogen, or one could use for the molar area the formula:

$$\overline{A} = 1.0209 \times 10^4 \frac{F_{3D}^{2/3}}{F_{2D}} \left(\frac{\mathbf{M}}{\rho}\right)^{2/3} \quad (\text{in units of } m^2 \, mol^{-1}) \tag{1.19}$$

Here the molar mass, $\mathbf{M}$, has units of $g\,mol^{-1}$ and the density, ρ, has units of $g\,cm^{-3}$. For the moment the adsorbate molecules will be assumed to be perfect hard spheres. To get the value of a, one could use the IUPAC convention. The IUPAC equation then with both packing factors included is:

$$\overline{A} = 9.189 \times 10^3 \left(\frac{\mathbf{M}}{\rho(l)}\right)^{2/3} \quad (\text{in units of } m^2 \, mol^{-1}) \tag{1.20}$$

This formula uses the method of calculation originally used by Brunauer, Emmitt, and Teller.

Often the quantity a, the area taken up by one adsorbate molecule, is listed. This is given by the equation:

$$a = \overline{A} N_A^{-1} \tag{1.21}$$

For the IUPAC convention (which is somewhat arbitrary) this is $0.163\,nm^2$.

Something that is puzzling about Eq. (1.19) is that the liquid density is used to obtain the molecular radius. There has, in the past, been much discussion about how to relate the liquid or solid density to the way the molecules are on the surface. There was discussion about whether to use the packing factors or not. What about geometric irregularity? Logically, to obtain the hard sphere radius, one would use the solid phase. The question of what is the radius or area of a molecule often comes up in chemistry. Is it the van der Waal radius? But is this a hard boundary? Literature for NLDFT seems to leave these questions up to variation. Throughout this book a convention will be followed, that A_s is not the important indicator of what the surface of the solid is, but rather n_m. This, of course, varies with adsorbate, but this might be an advantage especially when dealing with porous materials. For example, if a monolayer of CH_4, but not one of SF_6, can fit in a pore, what would one predict about benzene, C_6H_6 being able to penetrate?

To use the packing factor, one normally needs the hard sphere radius (again variable definitions), which can be obtained from the solid state. Thus, one would use the three dimensional packing factor and the solid state to obtain the hard sphere radius. The sphere radius is used to calculate the 2D hard sphere area, using the 2D packing factor, $F_{2D}=0.9069$. This should yield the 2D molar area. This is option "D" in Table 1.8, which includes various options for calculating $\overline{A}$ and a for a variety of adsorbates.

Notice that in Table 1.8 it makes a very big difference depending upon what convention is used to calculate these values. There are at least four possibilities left off in this table. These would involve using just one of the packing factors, which seems a

Table 1.8 Area parameters for some potential adsorbates

	M_u (g mol^{-1})	ρ (kg m^{-3})	A_m (m^2 mol^{-1})	a (nm^2)	ρ (kg m^{-3})	A_m (m^2 mol^{-1})	a (nm^2)
			—A–(IUPAC)—		—————B—————	
N_2	28.01	0.809	9.77×10^4	0.162	0.868	9.31×10^4	0.155
Ne	20.2	1.297	5.73×10^4	0.0952	1.444	5.33×10^4	0.0886
Ar	49.95	1.394	9.99×10^4	0.166	1.623	9.02×10^4	0.150
Kr	83.80	2.415	9.78×10^4	0.167	2.826	8.80×10^4	0.146
Xe	131.3	3.057	1.12×10^5	0.187	3.540	1.02×10^5	0.170
SF_6	146.1	1.91	1.66×10^5	0.275			
CH_4	16.04	0.423	1.04×10^5	0.172	0.494	9.35×10^4	0.155
C_6H_6	78.11	0.877	1.83×10^5	0.304	1.012	1.67×10^5	0.277
H_2O	18.01	0.998[a]	6.32×10^4	0.105	0.819	7.21×10^4	0.120
CS_2	76.13	1.266	1.41×10^5	0.234	1.539	1.24×10^5	0.234
			—————C—————		—————D—————	
N_2	28.01	0.809	1.035×10^5	0.172	0.868	1.085×10^5	0.180
Ne	20.2	1.297	6.37×10^4	0.106	1.444	5.93×10^4	0.0984
Ar	49.948	1.394	1.11×10^5	0.184	1.623	1.00×10^5	0.166
Kr	83.798	2.415	1.09×10^5	0.180	2.826	9.78×10^4	0.162
Xe	131.29	3.057	1.25×10^5	0.208	3.540	1.14×10^5	0.189
SF_6	146.1	1.91	1.84×10^5	0.305			
CH_4	16.04	0.423	1.15×10^5	0.191	0.494	1.04×10^5	0.173
C_6H_6	78.11	0.877	2.04×10^5	0.338	1.012	1.85×10^5	0.307
H_2O	18.01	0.998[a]	7.02×10^4	0.117	0.819	8.01×10^4	0.133
CS_2	76.13	1.266	1.57×10^5	0.260	1.539	1.38×10^5	0.228

Columns A: Calculations based on the liquid density at boiling point and IUPAC convention that uses both F_{2D} and F_{3D}.
Columns B: Calculations based on the solid density at the triple point and with F_{2D} and F_{3D}. Columns C: Calculations based on liquid density at the boiling point without packing factors.
Columns D: Calculations based on the solid density at the triple point without packing factors.
[a]*At 25°C. Values are obvious approximate for H_2O given its special properties.*

bit inconsistent. For example, to leave off F_{3D} for the bulk liquid based answer, one would multiply column "A" results by 0.8185 (F_{3D} to 2/3 power). This changes the IUPAC convention to 0.148 nm². On the other hand, Neimark et al. [41] specified a hard sphere radius as 0.179 nm, giving a value of a of 0.101 nm² for N_2. This is much lower than any of the possibilities in Table 1.8. It is even lower if the described low value of the classical shape is included, as explained in the next section. Given this discussion, it is not surprising that many different values of a can be found in the literature.

The physical quantities calculated this way assume the hard sphere assumption. However, classically, due to the rounding of the potential energy as one molecule passes over another, the value could be between 91% to 96% of the numbers listed. This phenomenon is reviewed in chapter 4.

The advantages of the χ theory formulation

The χ theory formulation has several advantages over other formulations. This includes, for the calculations dealing with the plane of the surface:

- There are no assumptions about molecular size or potential needed.
- The monolayer equivalent is obtained without assuming molecular size—this can be converted to surface area with assumptions about molecular size.
- The molar energy of adsorption and its dependence on surface coverage is correctly calculated (and demonstrated to be correct) without any energy or size considerations.
- The calorimetric energies of adsorption are consistent with the isosteric heat from the isotherm.
- The theory is very simple to apply. It can easily be calculated using graph paper and a calculator.
- Even heterogenous surfaces with multiple adsorption energies can often be calculated in this way.
- Even simple porosity calculation of both size and volume may also be easily calculated.

More complex samples, including heterogeneous surfaces and samples with micropores and mesopores together are, of course, more difficult to calculate. However, as modern calculations are concerned, they are still fairly simple and a computer spread sheet and least squares routine are enough to make the calculations. These calculations yield:

- The energies of adsorption for heterogeneous surfaces and their distributions obtainable.
- The pore volume in terms of moles of adsorbate can be determined. Given the molecular size, the volume can then be calculated.
- The surface area of the pores in terms of monolayer equivalents can be obtained. Given the molecular volume, the actual surface area can be calculated.

- Similarly, the surface area and volume of macropores in term of monolayer equivalents and molecular volume can be calculated and the actual surface area and volume can be calculated given the correct molecular size.
- The external surface area and the pore surface area can be separated.
- The distributions for both micropores and mesopores can be obtained as well as adsorption energies.

To obtain the information described above is impossible using any of the older physisorption theories, such as BET, Langmuir, etc. The disjoining pressure theory, of course, can do this if expanded upon, since the reasonable assumption that was made about the decay constant was proved to be correct with χ theory.

The DFT (and NLDFT) formulations have the weakness that a realistic answer cannot be obtained without the details of the adsorbate molecule and some input about the surface area. In other words, the information obtained depends upon the assumed radius and attraction potential of the adsorbate molecules and the adsorbent up front. For the χ theory, this is an afterthought. Furthermore, the DFT calculations are very complex, requiring good computer power and programing skill. They cannot be easily interpreted without a full calculation and the validity of a particular calculation is extremely difficult to check. The NLDFT calculation is quite sophisticated and elegant. Using the energies and distributions from χ theory it has the potential to yield a high degree of detailed information. Indeed, in some cases, the disjoining pressure theory has been used to calibrate the calculation. However, these calculations misinterpret the meaning of the decay constant.

Summary for introduction to Chi and ESW theories

The χ theory and ESW have the advantage of being simple and yield the correct value for the monolayer equivalent moles and the adsorption surface energy function. These items will be demonstrated in various sections throughout this book.

Using standard curves to obtain the n_m

In the section on standard curves in Chapter 3 starting on page 138 is a more thorough explanation of how to use standard curves to determine surface area. The introduction of standard curves was a great advance in the understanding of physisorption and the measurement, at least in a relative sense, of surface area and porosity.

In principle, a different standard curve should be used for the different materials, both adsorbate and adsorbent, and thus a different calibration if the BET method is used for the surface area. Thus, the section on standard curve provides standard curves for a variety of materials. Unfortunately, it is likely that the BET yields an incorrect approximation of surface area, thus rendering the standard curves not very useful for their intended purpose. This is due to the fact that the BET has an absolute error of as high as 300%.

In contrast, if χ method is used for the standardization, then one obtains an excellent value for surface area. This, however, seems to be a round-about and unnecessary step in analysis since the χ method yields the correct answer independent of what material the adsorbate is. In the section on standard curves, this analysis will be obvious from the fits of the χ theory/disjoining pressure theory to the standard curves.

Determining porosity by physical adsorption

There have been several methods to determine porosity using physical adsorption. The first step is to determine if the porosity consists of micropores or mesopores. There is another type of pore called ultramicropores. Ultramicropores are defined here as restricting the adsorption to a single monolayer and there is a special form of the χ theory to describe it. By definition, macropores are too large to show up as pores within the experimental data range using most instruments build today.[16] A Type I isotherm is usually interpreted as an indication of micropores. Type IV, V, and possibly VI are characteristic of mesopores. As far as the χ plot or the standard plots features are concerned, whenever the slope of the plot decreases, with or without an intervening positive increase, pores are present. The presence of the intervening positive increase is an indication (χ definition) of mesopores. In the BDDT [11] designation, a Type II or III isotherm does not indicate porosity, however in the χ transform a Type II or III-appearing isotherm might indeed indicate porosity. If a mix of micropores and mesopores are present, then typing might prove difficult but the χ plot might reveal these individual features. If more than one size of micropore is present, the χ plot has proven to be successful in determining this [42, 43]. Although both micropores and mesopores can be handled simultaneously, for clarity they will be separated in this treatment.

Micropore analysis

Langmuir analysis

Classically, micropores have been treated using the Langmuir [44, 45] isotherm with the assumption that, since the micropores were too small for more than one molecular layer to adsorb, the multilayer consideration (as assumed for the BET) was irrelevant. The Langmuir equation is:

$$\frac{n_{ads}}{n_m} = \frac{KP}{1+KP} \quad (1.22)$$

where K is an equilibrium constant.

This may be rearranged to the following linear form:

$$\frac{P}{n_{ads}} = \frac{1}{Kn_m} + \frac{P}{n_m}. \quad (1.23)$$

---

[16] One problem, even if the instruments are capable, is the possibility of "bed porosity"; that is, the porosity that must exist between powder particles. One type of sample which may be amendable to this measurement is a sintered material or a plaster, which have open porosity for one solid piece. Porosimetry is often used for these materials but also physisorption has been used with, for example the rare earth plasters.

From the plot of P/n_{ads} versus P the slope and intercept can be obtained to yield K and n_m. From an analysis of standard curves, this analysis for n_m will be off by as much as a factor of 4 for physical adsorption in micropores.

BET N-layer formulation

Another possibility in analyzing for micropores is to modify the BET equation to allow for only a certain number of monolayers to adsorb. This introduces another parameter, i.e., the number of allowed monolayers. This is usually referred to as the N-layer BET formulation. A further refinement for porosity with additional parameters, referred to as the BDDT [2, 11] (Brunauer, Deming, Deming, Teller), was also developed. These equations are not widely used and have not proven to be very successful. For the equation restricting the number of layers, n_{max} is the number of allowed layers, the modified equation, called the N-layer BET, is:

$$\frac{n_{ads}}{n_m} = \left(\frac{C_{BET}}{1-C_{BET}}\right) \frac{1-(n_{max}+1)(x)^{n_{max}} + n_{max} x^{n_{max}+1}}{1+(C_{BET}+1)x - C_{BET} x^{n_{max}+1}} \tag{1.24}$$

where:

$$x = \frac{P}{P_{vap}} \tag{1.25}$$

To use this equation, a minimum search routine might be necessary with three variables and a non-integer for n_{max} is difficult to interpret. This interpretation is rarely used and does not seem to work correctly.

Standard curve interpretation

In Fig. 1.23 is an illustration of the technique. Plotting the amount adsorbed versus the standard plot value listed in the figure as "$\Delta\chi$ or $F(P/P_{vap})$" one should obtain two (almost) linear regions. In the case of a χ plot, the function is given by:

$$F(P/P_{vap}) = -\ln\left(-\ln\left(P/P_{vap}\right)\right) - \chi_c. \tag{1.26}$$

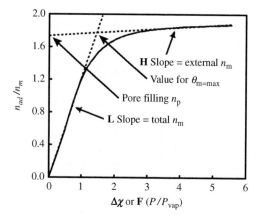

FIG. 1.23

A χ plot or standard plot of a type I isotherm.

The various features have the following meanings:
- The slope of the low linear region (labeled L) is equal to the surface area, including the surfaces of the micropores. This is labeled "total n_m."
- The slope of the upper linear region (labeled H) is equal to the area outside the pores including the pore openings. This is referred to as the external monolayer equivalence, labeled "external n_m."
- The intercept to the ordinate for this line is equal to the number of moles that fill the pores, labeled "n_p." This is, therefore, an indication of the pore volume.
- The monolayer equivalent inside the pores is indicated by the abscissa value.

It has been speculated that the roundoff between these curves is an indication of the geometry of the pore. This is based on the calculation of the monolayer equivalents in the layers, that is the values of $\theta_1, \theta_2, \theta_3$, etc., and how they deviate from the flat surface values explained in the next section. Thus, the maximum sharp transition indicates slit-like pores, whereas a more rounded transition indicates more cylindrical pores. Although reasonable, this has not been confirmed experimentally. However, one must remember that there might be a distribution of pore sizes and, possibly, energies of adsorption. All these factors are considered in Chapter 3 and the method to calculate the physical situation from them.

Layering according the χ theory

In this section, an explanation will be given as to what χ theory says about subsequent "layers" of adsorbate. This information and the QM solution in the normal direction yields the density profile of the adsorbate. This also reveals that there is a simple formula for the case where adsorption is physically restricted to a single monolayer thickness.

In the section: "Depth Profiles and n-layer calculations with χ Theory" in Chapter 4 there is a description of how to determine the amount of adsorbate in each liquid layer according to the χ hypothesis. The resulting equation for the first layer $\theta_{m=1}$ is (m being the index for the "layer" count):

$$\theta_{m=1} = 1 - \exp(-\Delta\chi) \tag{1.27}$$

Notice for the first layer, one can rewrite this as:

$$\theta_{m=1} = 1 - \frac{RT}{\overline{E}_a} \ln\left(\frac{P}{P_{vap}}\right) \tag{1.28}$$

where $\theta_{m=1}$ is defined as the first layer surface coverage in the first layer or $\theta_{m=1} = n_{ads,m=1}/n_m$. (Recall that $\overline{E}_a$ is negative.)

These equations indicate that as $P \rightarrow P_{vap}$, $\theta_{m=1} \rightarrow 1$. This is the often observed "log law" for microporosity. Actually, such pores are extremely small and will be referred to from here-out as *ultramicroporous*. Thus micropores will be defined as pores larger that ultramicropores but not large enough to display pore pre-filling as defined in the next section.

In a later chapter, Eq. (1.28) will be used for some far reaching subtleties that effect how one should think about χ theory and adsorption. It effects how one should calculate porosity and introduces the concept of observed $\chi_{c,obs}$ versus the actual χ_c.

This, in turn, impacts the energy of adsorption calculation and the monolayer equivalence which was originally reported by Fuller.

The remainder of the adsorbate, designated $\theta_{m>1}$ is given by:

$$\theta_{m>1} = \theta - 1 + \exp(-\Delta\chi) \qquad (1.29)$$

For the second layer:

$$\theta_{m=2} = 1 - \exp(-\Delta\chi + \theta_1) \qquad (1.30)$$

and assuming no collapse to pre-filling for subsequent layers:

$$\theta_{m=k} = 1 - \exp\left(-\Delta\chi + \sum_{k=1}^{m-1} \theta_k\right) \qquad (1.31)$$

It is possible to make this a generalized equation, but it is just as messy as using successive calculations.

The question arises of when is it more energetically favorable to form a distinct liquid-gas interface instead of a decay of concentration from the surface. This formation of the liquid-gas interface is referred to as "pore filling" or "pre-filling." Obviously, this cannot happen for the $\theta = 1$. Nor for $\theta = 2$, since there is not enough 2nd layer and above to form the interface. At the very minimum it would be possible if there is enough adsorbate in layers above to the 2nd layer to match the amount in the 2nd layer. This point is at a value of $\theta \geq 2.51\ldots$ (and most likely at 2.92…).

In Chapter 6, a more sophisticated, a little more difficult, N-layer approach to porosity is presented. It takes into account not only the distribution between layers but also the QM distributions of the individual layers. A method using sets of standard plots is proposed similar to the DFT methods, but without a non-porous standard.

Ultramicropore analysis

The standard curve given above applies to an unlimited normal (z) direction. The question is: what happens if the normal direction is restricted? Specifically, what if due to geometrical restrictions only a monolayer can adsorb? Eq. (1.28) above can answer this question. This law applies to the main porosity portion of the isotherm. The up-swing as the isotherm approaches P/P_{vap} is to be expected from the external surface taking up the vapor as the condensation conditions are approached.

This equation can be recast to yield a familiar adsorption equation that is sometimes observed. This is the Log-Law that is fairly regularly observed with ultramicropores. Indeed, for this book, the adsorption that obeys the Log-Law will be the definition of ultramicropores.

In Fig. 1.24 is a demonstration of the Log-Law as recorded by Silvestre-Albero et al. [46] (S-A,S-A,L R-R). The least squares linear fit in this figure is from a $\ln(P/P_{vap})$ of -3 and below (-3.02 to -15.4). The x intercept, as in the χ plot yields the $\overline{E}_a$ and the extrapolated y intercept ($x=0$) yields the monolayer equivalence. For this particular isotherm then:

$$\overline{E}_a = 10.02 \mathrm{kJmol}^{-1}$$
and
$$n_m = 8.94 \mathrm{mmol}$$

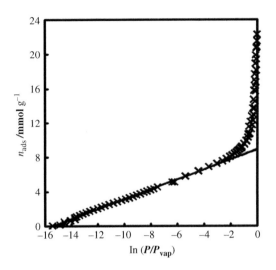

FIG. 1.24

Data by (S-A,S-A,L R-R) illustrating the "Log-Law," which is often observed.

Redrawn here with the permission from the American Chemical Society.

Some of the curves almost fit this Log-Law but might have a bow in the middle. A more sophisticated analysis needs to be performed for these. It may happen that the normal method of using the χ-plot to obtain the porosity measurements due to the lack of low pressure data. In such a case, this analysis could perhaps yield an answer. A more thorough analysis of this type of isotherm will be presented in Chapter 6.

Mesopores

Mesopores generate either a Type IV or V isotherm. In Type IV and V isotherms, a similar strategy as that for micropores can be used, as illustrated in Fig. 1.25. Notice, however in this case that the lower line, L, and the upper line, H, intersect at a higher value of χ than the commencement of the negative change in the slope. This is, of course, due to the "step-like" functionality at the hysteresis loop. The analysis from these lines remains the same as for the micropore case (Fig. 1.23 using the two slopes and the ordinate intercept of H), but there is additional information. One could refer to such pores as "pre-filled" or "capillary filled" since it is normally attributed to capillary action.

From the χ/(ESW) theory point of view, the pre-fill is simply the normal order of molecules specified by Eqs. (1.28)–(1.31) collapsing into the dense liquid, which fills the pores. If all the pores were *exactly* the same, this would be similar to a phase transition with the lines being perfectly vertical. The hysteresis for these isotherms might be indicative of a possible spinodal-like transition.

Notice that this particular part of the analysis is an answer for the moles adsorbate to pre-fill pores, n_p, (relates to the pore volume), total monolayer equivalents, total n_m, (related to the total surface area), and monolayer equivalents external to pores,

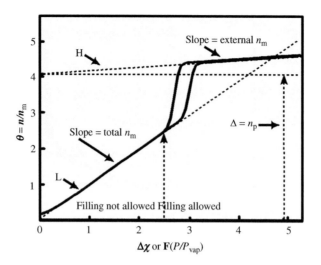

FIG. 1.25

A χ plot or a standard plot of either a Type IV or V isotherm.

external n_m, (related to external surface area) is independent of whether the adsorption or desorption branch of the isotherm is used. In principle, given the related values and some assumptions about the pore geometry, one could calculate a value for d_p the parameter for either pore radius or slit width (or something else).

The caveat for the determination of pore dimension is even stronger than for surface area. If the hard sphere model for the surface area was uncertain, it is even less certain when one starts assuming some sort of packing of molecules within the pores. There are several proposals for this problem. The following is a widely used, but not the only, example.

Another modern example is the non-local density functional theory (NLDFT), which has also been used for these purposes. NLFDT is probably a much better approach than some of the classical approaches, but it has some problems. Although the NLDFT probably correctly uses the interactions between adsorbate molecules (as does χ theory) it uses parameters from similar materials to calculate the adsorbate-adsorbent parameters. For example, for a porous silica material, the parameters used are those determined from measurements on non-porous materials, that is, a standard isotherm material or the α-s plot. As Kupgan et al. [33] indicated, the NLDFT methodology is not suitable for material for which there is no standard plot or that has pores of shapes other than the normal designated shape, such as slit pores. This lack of standard isotherm is more common than otherwise. Furthermore, there is no guarantee the energy of adsorption, $\overline{E}_a$ of the standard plot material is the same as the porous material.

Unfortunately, the literature standard plots are all references to the BET theory, which makes them unreliable. Furthermore, standard plots are usually made up by using some smoothing and possibly an extrapolation method. Unfortunately, these methods are then in many cases biased to be compatible with accepted notions. The standard plots presented in the listings of this book have been reevaluated from the original data. These standard plots are presented later in Chapter 3, in the section "Analysis Using Standard Isotherms."

The interpretation of the hysteresis loop is a matter of some current discussion. The primary explanation is based upon the Kelvin equation as modified by Cohan [47]. The χ theory, however, predicts that before about a $\Delta\chi$ value of 2.5 to 3 monolayer equivalents the adsorbate is forbidden to collapse to form an outer interface. This is the approximate point at which the value of $\mathbf{E}(\theta)$ drops to a level lower than obtained by the formation of the defined liquid-gas interface. Thus, the position of the pore filling is dependent upon the value of χ_c, that is, $\overline{E}_a$.

The Cohan derivation is based on the "pre-filling" of the pore, which corresponds to the increasing pressure branch of the hysteresis loop. This equation (Kelvin-Cohan, "KC," equation) is:

$$RT \ln\left(\frac{P}{P_{\text{vap}}}\right) = \frac{h\gamma_{gl}V}{r_p - t} \qquad (1.32)$$

The following are the meanings of the new symbols:

γ_{gl} = the surface tension of the liquid adsorptive;
V = the liquid adsorptive molar volume;
r_p = the pore radius or half the distance across the pore;
t = the "film thickness" before the prefilling starts; and
h = a constant depending upon the pore geometry.

For h the following is used:

- $h=1$ for slit-shaped pores.
- $h=2$ for cylindrical shaped pores.
- $1<h<2$ for oblate-shaped pores.
- $h<1$ for slits that have many concave sides.
- $h>2$ would be an indication of some fractal arrangement.

The thickness, t, is assumed given by the following equation:

$$t = \frac{n_{\text{ads}} V_m}{A_s} \qquad (1.33)$$

Obtaining pore radius from the two slopes—Using ordinate intercept of H

The following then is the information that one would hope to extract from these plots.

Assuming one can relate the slopes of L and H to areas either by comparison to non-porous standard or through the theoretical χ treatment, one has the areas A_s corresponding to the L slope and A_{ex} corresponding to the H slope (χ treatment would use the analytical expressions of Eqs. (1.14)–(1.16). The pore volume, V_p, is given by the ordinate ($n_{\text{ads}}=0$) intercept for H. These are related to the physical quantities of

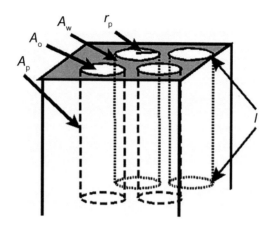

FIG. 1.26

Illustration of the definitions or A_o area of pore opening, A_w outside "wall" area excluding A_o, r_p pore radius, l pore length, and A_p surface area of the sides of the pores.

the total surface, A_s, the surface area inside the pores, A_p, the total area of the pore openings, A_o, and the area of the edge-on walls or the non-porous area of the outer surface, A_w (as illustrated by Fig. 1.26) by:

$$A_s = A_p + A_w \tag{1.34}$$

$$A_{ex} = A_w + A_o \tag{1.35}$$

The total pore volume, V_p, should be well approximated by the ordinate intercept of line H, as mentioned. At this point, a geometry must be assumed to analyze further. If cylindrical pores are assumed, then there will be an average length per pore, $<l>$. Basing the following upon a fixed amount of adsorbent, conventionally exactly 1 g, one can construct the following equations:

$$V_p = \pi r_p^2 \langle l \rangle N_p \tag{1.36}$$

$$A_p = 2\pi r_p \langle l \rangle N_p \tag{1.37}$$

$$A_o = \pi r_p^2 N_p \tag{1.38}$$

where N_p is the number of pore openings per gram and r_p is the pore radius.

For microporous plots, these equations present the problem that there are more physical quantities that need to be extracted than there is information available. For these cases the assumption normally used is that $A_o \ll A_w$, thus making $A_{ex} = A_w$ and $A_p = A_s - A_{ex}$. Other assumptions could be made, for example, if the wall thickness of the zeolite were about the same size as the pore radius, then $A_o \approx A_w$ and therefore $A_w \approx 0.5 A_{ex}$. In general one could insert a factor, $0 < \alpha < 1$ to relate A_w to A_{ex}. The r_p then would be:

$$r_p = \frac{2V_p}{A_s - \alpha A_{ex}} \tag{1.39}$$

Eq. (1.39) is capable, therefore, of yielding a range of values for the pore radius using the slopes and intercepts of the standard plots for both micropores and mesopores.

Using Kelvin-Cohen value of r_p for mesopores

In the case of the analysis of mesopores (Fig. 1.25) a separate determination of r_p may be obtained using Eqs. (1.32) and (1.33). The pressure used in Eq. (1.32) is that at which the sudden increase is observed or the average value in the step. n_{ads} is for the purpose of calculating t extracted from the L portion of the standard plot or its extrapolation. Both of these assumptions are approximations. In general, there is also a distribution of either pore sizes or of adsorption energies, which complicates this simplistic analysis. These complications can be overcome and are addressed in Chapter 6. Nevertheless, the mathematics for a simple treatment follows for mesopore analysis.

The quantities A_s, A_{ex}, and V_p are provided by the isotherm as described above and r_p can be calculated from the Kelvin-Cohen, Eq. (1.32), and the other quantities may be calculated using Eqs. (1.34)–(1.38) to yield Eqs. (1.40)–(1.44) so:

$$A_p = \frac{2V_p}{r_p} \tag{1.40}$$

$$A_w = A_t - \frac{2V_p}{r_p} \tag{1.41}$$

$$A_o = A_{ex} - A_t + \frac{2V_p}{r_p} \tag{1.42}$$

Using the value of A_o obtained:

$$<l> = \frac{V_p}{A_o} \tag{1.43}$$

$$N_p = \frac{A_o}{\pi r_p^2} \tag{1.44}$$

Thus, for an isotherm indicating mesoporosity, one should be able to obtain all of the physical quantities unambiguously. This analysis requires very good data to yield results. Notice that A_p plus A_s should be much larger than A_w. If this is not the case, then there is a high uncertainty in A_w and obtaining a nonsense answer is possible.

What ESW theory says about mesopores

Jürgen Adolphs has examined the thermodynamic side of the χ/ESW theory and came up with some interesting conclusions [48]. The basic χ/ESW equation yields with the absence of curvature in the surface (using symbolism consistent for this book) for slit pores:

$$RT \ln\left(\frac{P}{P_{vap}}\right) \equiv \prod \overline{V} = -\prod_0 \overline{V} \exp\left(\frac{n_{ads}}{n_m}\right) - \frac{2B_1 \overline{V}}{\left(H - 2n_{ads}d_{ads}n_m^{-1}\right)^3} \quad (1.45)$$

where B_1 is a constant from average "film" permeability and d_{ads} is the average center to center distance between layer of adsorbate molecules.

Notice that the right-most term is in addition to the χ form of the theory. The collapse point of the dispersed film is when $d\pi/dn_{ads} = 0$. The film becomes stable when $d\pi/dn_{ads} < 0$. For cylinders, this equation is modified for the surface curvature, r:

$$-RT \ln\left(\frac{P}{P_{vap}}\right) \equiv \prod \overline{V} = \prod_0 \overline{V} \exp\left(\frac{n_{ads}}{n_m}\right) - \frac{2B_1 \overline{V}}{4\left(r - n_{ads}d_{ads}n_m^{-1}\right)^3}$$
$$+ \frac{\gamma_\infty r}{(r - 2d_{ads})(r - n_{ads}d_{ads}n_m^{-1})} \quad (1.46)$$

The break-down of the stable film occurs for cylinders according to the Kelvin equation using decrease in the amount of adsorbate according to:

$$r = -\frac{2\gamma \overline{V}}{RT \ln(P/P_{vap})} + \frac{n_{ads}d_{ads}}{n_m}, \quad \gamma := \frac{\gamma_\infty r}{(r - 2d_{ads})} \quad (1.47)$$

These equations look a bit messy but they work, according to Adophs. This approach looks very promising and it would be appropriate for other researchers to test their validity with more experimental investigations.

One might ask; why do we need to know where the mesopore filling is when we can already obtain the pore volume and the pore radius from the standard plot? That may be the case, but it may also be the case that it would be advantageous to change the mesopore filling or the hysteresis according to Eqs. (1.46) and (1.47). Very often advances in science and engineering lead in unexpected and important paths.

Macropores

Almost by definition, macropores cannot be observed in the isotherm, at least until now. This is because the pore filling would occur at pressures too close to the vapor pressure, P_{vap}, to be reliably measured. This, however, may be changing. By using a differential technique and very good temperature control and handling methods. Denoyel et al. [49] reported being able to extend the reliable pressure range to 0.99985 of P_{vap} and measured porosity up to 12 µm. The analysis method should be identical to that used for mesopores.

Summary of standard curve pore analysis

Analyses of standard curves are the easy methods to determine the porosity of a sample. The biggest problem, which could possibly be solved by NLDFT, is how packing in small pores occurs. A special case for standard curves, specifically χ analysis, is ultramicropores that are distinguishable from other type of pores by a special abscissa transformation. Other pores are analyzed by both the tabulated standard isotherms or the χ plot in the same fashion. The χ method has the advantage of not requiring a nonporous standard.

The easiest way to analyze the porosity is to use the deviations from the linear standard curve. The initial slope of the standard curve yields to total n_m and the final slope yields n_m for the external surface. The total pore volume can be calculated by a back extrapolation of the high linear fit to the start of the standard plot. For micropores, this is the only information available.

For mesopores, the pre-filling phenomenon yields additional information about the volume of the pores which might be useful for obtaining the fractal factor for the pores.

An alternative is the older BET N-layer system. This method has not been very successful in the past and is little used today.

The various DFT methods are not discussed in this section, but the whole of Chapter 7 is devoted to it. One chapter is not sufficient to explain the whole theory, including the more recent versions referred ta NLDFT and QSDFT (Non-Local Density Functional Theory and Quenched Solid Density Functional Theory). References are therefore made to some reviews.

Some preliminary error analysis for the various transformations that are used follows. The best systems for minimizing error are those which do not transform the ordinate (amount adsorbed.) Transforming ordinate can lead to very large errors.

Statistical treatment of isotherms—Error analysis

The principal fact to keep in mind when analyzing an adsorption isotherm is that the pressure reading is invariably much more precise than the measurement of the amount adsorbed. If a transform for P/P_{vap} is used in the analysis, very little error (indeed, usually insignificant error) is introduced in the statistical analysis. Whereas, if a transform is performed upon n_{ads} then a statistical error is introduced unless some compensating weighting factors are introduced (caution: when is a weighting a legitimate method and not just a "fudge factor?").

For example, the linearize BET equation (Eq. 1.10) transforms both P/P_{vap} and n_{ads} but one need only be concerned with the error introduced by the n_{ads} transform. There possibly could be a significant error in the assumed value of P_{vap}, which would create problems at the higher pressures. However, the recommendation is to use only values of P between 0.05 and 0.35 of P_{vap} for a separate reason, which eliminates most of this latter uncertainty. On the other hand, given an absolute[17] error in n_{abs} of ε, and that the function used in n_{ads} is $1/n_{ads}$, then the error in the BET plot is approximated by εn_{ads}^{-2}. Thus the lower the value of n_{ads}, the greater is the introduced error in the transformed plot. In computing a linear least squares on the transformed

---

[17] The average absolute uncertainty is often assumed to be constant regardless of the value of n_{ads}. See def.

plot a weighting factor of n_{ads}^2 should be used. In practice, over the 0.05 to 0.35 P_s range this gives a variation in error of a factor of about 50. A more precise choice would be to use a non-linear least squares routine for the untransformed equation (Eq. 1.6) in which case, the initial estimates of n_m and C_{BET} could be obtained from the transformed plot.

The practical consequence of the not using a weighted regression (least squares) method is that under most practical conditions the error is small when one uses the non-weighted regression. Simulations of adsorption on ceramic materials have indicated that a probable error of about 3% to 4% is introduced by the transform. The answer for the surface area from the non-weighted method is always less than the weighted value. Interpretation of the statistics from the transformed plot, however, is not straightforward. For example, what is the meaning of the statistical outputs such as r^2 and parameter deviations (σ_x and σ_y) indicated in the regression analysis?

The problems of interpretation of statistical parameters and error introduced by data transformation are not present in the standard isotherms, such as the α-s plot, t-plot or others, nor are they problems with the χ plot. In these cases, n_{ads} remains untransformed and the absolute error for the full range is approximately constant. (The relative error, of course varies.) A straight application of a simple linear regression is therefore appropriate. This makes the derived statistics, such as standard deviation, r^2, σ_x, σ_y, etc., easy and straightforward to interpret. These statistics then relate directly to the physical quantities that they are associated with.

Summary of this chapter

This chapter summarizes some of the basic classical concepts of physical adsorption and the vocabulary associated with it. Part of this is the concept of various isotherm types, which may in the future not be as important as in the past due to more modern theories of physisorption.

Some of the classical theories of adsorption were introduced with particular attention to the BET theory. This latter theory dominated the field for nearly 100 years. Even though it gave some comfort to investigators the impression that they were obtaining fundamental physical quantities such as surface area and adsorption energies, its failings were very apparent from the very beginning. The worst failure is that it created anomalies not just between the different measurements but also internally mathematically. It is now known that the theory is both inaccurate and imprecise. By being the recognized theory from three very prominent scientists and accepted without question by almost all the investigators in the field, all other theories were forced to conform to the answers that the BET theory created. However, it is obvious now that the BET theory has been disproved and is misleading.

However, this theory is given in complete detail so that the older literature will make sense. Knowing this theory can allow some of the data in the literature to be back calculated. Some other literature data is not complete enough to recover the data and, unfortunately, is probably lost forever (this is a valuable lesson that the original data should always be presented in a publication for recovery by others, and the investigator, including this one, should always keep in mind that the interpretation put forth may be faulty).

The modern concepts using thermodynamics and quantum mechanics is then presented. These theories include a thermodynamic based theory called "disjoining pressure theory" or "ESW" and the quantum mechanical "χ theory."[18] The disjoining pressure theory has only one assumption. This assumption is that the exponential (mathematical) decaying potential from the surface has a decay constant of one monolayer thickness. The χ theory likewise has one assumption, which is that the overall wave function is amenable to separation of variables; that is, the distance normal to the surface for each atom is separable from the plane of the surface. The disjoining pressure theory can be derived from the χ theory in full. Due to the way the integration and differentiation is arranged to derive the disjoining pressure theory from χ theory, the χ theory cannot be derived from the disjoining pressure theory.

Using the standard plot formulation with the information from χ theory, the measurements of ultramicroporosity, microporosity, and mesoporosity may be determined. Furthermore, the energies of adsorption, the E_a, defined in this case as the energy of the first adsorbed molecule, can be calculated. The energy of adsorption as a function of surface coverage, $\mathbf{E}(\theta)$, then follows, not surprisingly an exponential decay for non-porous adsorbents. This is demonstrated in later chapters by comparing the theoretical χ energies against matched calorimetric data.

What's next?

In the next chapter, the instrumentation for measuring the isotherm and the heats of adsorption by calorimetry are described. If you already have instrumentation, it would still be a good idea to go over this chapter. Why? Because some instruments lack some important features: particularly proper temperature measurement and control, and also proper attention to laminar versus turbulent flow. Also included in this chapter is a general error analysis and special error analysis for each technique. Reading this, you may be surprised to find you have some problems.

Beyond this next chapter can be found additional theoretical derivations and analysis techniques on how to analyze the isotherm completely, including porosity, energies, energy distributions, and porosity distributions. Density functional theory is presented as an alternative to the simpler χ-theory and how the two could possibly be brought together to create a very powerful analysis technique.

---

[18] Some have asked me why I called the theory "χ" theory. Was it after my name? The answer is more boring than that. As I was working through this theory in the late 1970's I kept using the letter X for what is now χ. I tried several names for the theory but reviewers kept criticizing them. I noticed that the BET used the letter X and my written X looked like a χ so that was it. Unbeknown to me, Dr. Loren Fuller finally was able to publish using "ASP" standing for auto-shielding physisorption, so we decided to let that stand for the classical mechanical analogue to the QM χ theory.

The subject of how co-adsorption of two adsorbates, and possibly by extension, more than two, can be calculated is addressed. This leads into the important subject of separations in which the differing energies of adsorption may be used to separate gases. The modification of the binary phase diagrams by physical adsorption is addressed to make these separations. This should save the investigators considerable time and effort in the quest to craft and manipulate gas mixtures.ff

References

[1] F. Rouquerol, J. Rouquerol, K.S.W. Sing, P. Llewellyn, G. Maurin, Adsorption by Powders and Porous Solids, Principles, Methodology and Applications, second ed., Elsevier/Academic Press, Oxford, UK, 2014. ISBN: 9788-0-08-097035-6.
[2] D. Henderson (Ed.), Fundamentals of Inhomogeneous Fluids, M. Dekker, New York, 1992 ISBN: 13:978-0-8247-8711-0, ISBN: 0-8247-8711-0.
[3] H. Ted Davis, Statistical Mechanics of Phases, Interfaces, and Thin Films, VCH Publishers, Inc, New York, NY, ISBN: 1-56081-513-2, 1995.
[4] P.I. Ravikovitch, G.L. Haller, A.V. Neimark, Adv. Colloid Interface Sci. 76-77 (1998) 203.
[5] D.H. Everett, IUPAC manual of symbols and terminology, Appendix 2, part 1, Colloid and surface science, Pure Appl. Chem. 31 (1972) 578.
[6] E.R. Cohen, T. Cvitas., J.G. Frey, B. Holmström, K. Kuchitsu, R. Marquardt, I. Mills, F. Pavese, M. Quack, J. Stohner, H.L. Strauss, M. Takami, A.J. Thor, Quantities, Units and Symbols in Physical Chemistry, IUPAC Green Book, third ed., IUPAC & RSC Publishing, Cambridge, UK, 2008. 2nd Printing.
[7] M. Thommes, K. Kaneko, A.V. Neimark, J.P. Olivier, F. Rodriguez-Reinoso, J. Rouquerol, K.W.S. Sing, *Pure Appl.* Chem. 87 (9-10) (2015) 1051–1069. IUPAC Technical Report.
[8] M.L. Myers, P.A. Monson, Adsorption 20 (2014) 591–622.
[9] R.B. Moyes, P.B. Wells, Adv. Catal. 23 (1973) 121–156.
[10] J.R. Arnold, J. Am. Chem. Soc. 71 (1943) 104.
[11] S. Brunauer, L.S. Deming, W.S. Deming, E. Teller, J. Am. Chem. Soc. 62 (1940) 1723.
[12] S. Brunauer, The Adsorption of Gases and Vapors, vol. 1, Princeton University Press, Princeton, NJ, 1945.
[13] S.J. Gregg, K.S.W. Sing, Adsorption, Surface Area and Porosity, Academic Press, London and New York, ISBN: 0-12-300956-1, 1982.
[14] J.B. Condon, Langmuir 17 (2001) 3423.
[15] S. Brunaur, The Adsorption of Gases and Vapors, Princeton University Press, Princeton, NJ, 1943, p. 293.
[16] K.S.W. Sing, D.H. Everett, R.A.W. Haul, L. Moscou, R. Pierotti, J. Rouquerol, T. Siemienwska, Pure Appl. Chem. 57 (1985) 603.
[17] Roja F., Korhause K., Felipe C., Esparza J. M., Cordero S., Domingues A., Riccardo J. II, (2002) Phys. Chem. Chem. Phys. 4 2346.
[18] A.V. Neimark, P.I. Ravikovitch, Microporous Mesoporous Mater. 44–45 (2001) 697.
[19] P. Kowalczyk, K. Kaneko, L. Solarz, A.P. Terzyk, H. Tanaka, R. Holyst, Langmuir 21 (2005) 6613.
[20] M.D. Donohue, G.L. Aranovich, J. Colloid Interface Sci. 205 (1998) 121.

References

[21] R.M. Harris, Chem. Ind. 1965 (1965) 268.
[22] P. Trens, N. Tanchoux, A. Galareau, D. Brunel, B. Fubini, E. Garrone, F. Fajula, F. Di Renzo, Langmuir 21 (2005) 8560.
[23] S. Bruaner, P.H. Emmett, E. Teller, J. Am. Chem. Soc. 60 (1938) 309.
[24] F. Rouquerol, J. Rouquerol, K.S.W. Sing, P. Llewellyn, G. Maurin, Adsorption by Powder and Porous Solids, Principles, Methodology and Application, Academic Press, Oxford, UK, 0.37 ISBN: 978-0-08-097035-637, 2014.
[25] J. de Dios Lopez-Gonzalez, F.G. Carpenter, V.R. Dietz, J. Res. Natl. Bur. Stds. 55 (1955) 11.
[26] R. Afonso, L. Gales, A. Mendes, Adsorption 22 (2016) 963.
[27] I. Langmuir, J. Am. Chem. Soc. 40 (1918) 1361–1403.
[28] R. Fowler, E.A. Guggenheim, Surface Layers (book section X), Cambridge University Press, Cambridge, UK, 1949, pp. 421–451.
[29] M. Volmer, Z. *Phys. Chem.* 115 (1925) 253–261.
[30] J.H. de Boer, The Adsorption Isotherm in the Case of Two-Dimensional Condensation, first ed., Oxford University Press, London, 1953 (book section VIII).
[31] T.L. Hill, J. Chem. Phys. 14 (1946) 441–453.
[32] T.L. Hill, J. Chem. Phys. 20 (1952) 141–144.
[33] G. Kupgan, T.P. Liyana-Arachi, C.M. Colina, Langmuir 33 (42) (2017) 11138.
[34] C. Volsone, J.G. Thompson, A. Melnitchenko, J. Ortiga, S.R. Palethorpe, Clays Clay Miner. 47 (5) (1999) 647.
[35] W.D. Harkins, G.J. Jura, J. Am. Chem. Soc. 66 (1944) 919.
[36] E.L. Fuller, J.B. Condon, Colloids Surf. 37 (1989) 171–181.
[37] N.V. Churaev, G. Starke, J. Adolphs, J. Colloid Interface Sci. 221 (2000) 246.
[38] J.B. Condon, Microporous Mesoporous Mater. 53 (2002) 21.
[39] M.G. Kaganer, Dokl. Akad. Nauk 122 (1959) 416.
[40] M.G. Kaganer, Zh. Fiz. Khim. 33 (1959) 2202.
[41] N.A.V. Ravikoritch, M. Grün, F. Schüth, K.K. Unger, J. Colloid Interface Sci. 207 (1998) 159–169.
[42] E.L. Fuller, Morphology of carbons deduced from physisorption isotherms: I. Nuclear grade graphite, in: 24th Conference on Carbon, vol. 1, 1999, p. 14.
[43] E.L. Fuller, Morphology of carbons deduced from physisorption isotherms: II. Activated carbon, in: 24th Conference on Carbon, vol. 1, 1999, p. 16.
[44] I. Langmuir, J. Am. Chem. Soc. 38 (1916) 2219.
[45] I. Langmuir, J. Am. Chem. Soc. 40 (1918) 1368.
[46] J. Silvestre-Albero, A.M. Silvestre-Albero, P.L. Llewellyn, F. Rodrígues-Reinoso, J. Phys. Chem. C 117 (2013) 16885.
[47] L.H. Cohan, J. Am. Chem. Soc. 60 (1938) 433.
[48] J. Alolphs, Chem. Ing. Tech. 88 (3) (2016) 274–281.
[49] Denoyel B., Barrande M., Beurroles I. (2005) 7th International Conference on the Characterization of Porous Materials.

CHAPTER 2

Measuring the physisorption isotherm

Introduction—Equipment requirements

There are principally two methods widely used to determine the physical quantities by physisorption. These are the volumetric method and the gravimetric method. The general objective is the same for both. One measures the amount of a gas that adsorbs on the surface as a function of the pressure of this gas. One ends up with a series of paired data, the amount adsorbed versus pressure, from which the physical parameters are extracted. These parameters almost always include a number believed to be the surface area and a quantity related in some way to the strength of the forces holding the adsorbate to the adsorbent. Other parameters sometimes identified are the porosity in terms of pore size and volume. The volumetric technique uses one type of measurement to obtain both data sets. This measurement is of the pressure. The gravimetric technique measures the pressure and the mass gain of the adsorbate with separate instrumentation, using some minor pressure corrections for the weight. Both techniques have their advantages and disadvantages, so it is important to be knowledgeable about both, especially if a decision is to be made as to which one to use or if a purchase is imminent.

There are also other methods to measure, for example, the energy of adsorption, or possibly macroporosity. Firstly, the two principle methods will be looked at in some detail.

The primary differences between the two are:

1) Cost: Usually the gravimetric technique is costlier than the volumetric technique. The volumetric technique requires only high precision pressure transducers and high precision volume measurements. The gravimetric, however, requires a high precision vacuum balance and, perhaps, considerable set-up effort.
2) Capability: Usually the gravimetric technique is more precise and accurate. It is a better research method than the volumetric technique.

The volumetric technique is incapable of some measurements needed in research, but for most routine work, given some important caveats, it is sufficient.

Ideally, one would combine these techniques in one instrument or with other techniques such as calorimetry. This is, of course, much more difficult and expensive, but it has the distinct advantage that the sample in both measurements is exactly the same. Very often, when one changes to a different technique and a different instrument, the samples end up being different in spite of rigorous controls.

The details for the two major methods which follow are required reading for a sound purchasing decision.

In both methods, the adsorption is performed in a temperature and pressure range just below the condensation point of the gas to liquid transition. Usually, the temperature is picked, for practical reasons, as the boiling point of the adsorptive gas. For example, in measuring the adsorption of nitrogen, liquid nitrogen is used to control the temperature of the sample. This is a convenient coolant, which assures a known vapor pressure over the sample, with some caveats. Research work often gets away from this restriction in order to study the adsorption at other temperatures. This latter change could also increase the cost and complexity of the instrumentation. Liquids, however, which have a reasonable vapor pressure at room temperature and above are more easily handled. An example of a fairly well characterized inert gas with a higher vapor pressure would be perfluoro-cyclohexane or sulfur hexafluoride. However, these are rarely used. Use of water, alkanes, and alcohols are quite common and temperature control is only a minimal problem. There is always the question of interpretation with these gasses, however.

For convenience, the following discussions will assume a nitrogen adsorbate and liquid nitrogen as the temperature controlling fluid. The sample will be referred to as a powder, which is not a requirement. Contiguous, open porous samples are also characterized by these techniques.

The volumetric method
Equipment description

The basic volumetric method is shown schematically in Fig. 2.1. System parts are not to scale. This is an idealized system and some of the features may not be present on some commercial instruments. Furthermore, there may be features on some commercial instruments that are not shown here. An example is the matching tube system to compensate automatically for the "dead" volume. Sample chambers are usually constructed from Pyrex using metal vacuum flanges. Even though there are many systems that are all Pyrex, an all-metal system is best.

The following is a description of the system:

1) The powdered sample (P) is contained in the sample tube (H) Note of caution: if the powder sample is produced in the tube, such as hydriding a metal, one should be certain enough room is available for expansion. A rule of thumb is at least a fivefold expansion. Multiple chemical treatments are not recommended in this vertical arrangement due to packing and swelling. For that purpose, a horizontal tube with at least a 10-fold increase in effective sample volume is recommended. In either case metal tubes are recommended.
2) This tube is immersed in a liquid nitrogen bath (L) for temperature control.
3) It is recommended that the room temperature be thermostated to 0.1°C, if this is not possible, consider a temperature controlled air box.

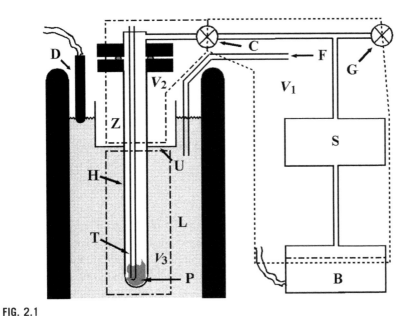

FIG. 2.1

Schematic of the volumetric system.

4) Between the liquid nitrogen cooled area and the rest of the equipment at room temperature is a transition zone (Z). This transition is indicated in the figure as the zone between the dashed lines in the upper (V_2) and lower (V_3) zone.
5) It is recommended that this transition zone be controlled. One way to do this is to have a U-shape cup (U) attached to the sample tube and control the liquid nitrogen such that the level is about halfway up this U-cup.
6) Thus, one has to either watch the liquid nitrogen level carefully or use a level detector (D) to control the flow of liquid nitrogen (F) into the Dewar (large black object).
7) For temperature measurement a thermometer well (T) is provided.
   a) In the case of liquid nitrogen temperatures, a gas ⇌ liquid equilibrium thermometer is recommended. This requires another pressure transducer and additional tubing and valves, but yields the vapor pressure of N_2 directly. It is also very sensitive.

In preparation the system is calibrated:

1) The volume between the two valves (S+tubing+valves+transducer) should be calibrated using PV methods. Any competent standards laboratory can accomplish this. It is quite possible that this calibration has been done by the manufacturer.

To admit gas:

1) The valve to the sample area (C) is closed if it is not already closed.
2) One opens the valve from the nitrogen supply (G) (nitrogen supply should have pressure controller, safety valves, etc.).
3) The pressure is read with a Bourdon or other membrane type pressure transducer (B). With today's technology, there is little reason to use the traditional manometers (and one very good reason not to).
4) The valve C is opened when one wishes to adsorb the gas.

Determination method

Notice in Fig. 2.1 that there are three zones labeled indicating three volumes one must consider: V_1, V_2, and V_3. V_1 and V_2 are the "manifold." The volume of V_1, a calibrated volume, is already known at the beginning of the measurements. This calibration should have been performed either at a standard laboratory or against a secondary standard that is traceable. The first problem is to determine the volumes of V_2 and V_3.

1) Normally V_2 is at a different temperature from V_3 so the total volume cannot be measured correctly. However, an effective volume called the "dead space," V_d, can be measured. This quantity will be then used in subsequent calculations. To determine the dead space, one first does a calibration of the system without a sample
2) with T_2 and T_3 (temperatures of V_2 and V_3) at the operating temperature anticipated.
3) Check to see if the valve to V_2 (C) closed.
4) The adsorbing gas is admitted to the calibrated volume area V_1 (through valve G).
5) The pressure measurement, P_i, is taken.
6) The valve C is then opened.
7) A second pressure measurement, P_f, is taken.
8) The dead space volume is calculated by:

$$V_d = \frac{P_i V_1}{P_f} - V_1 \tag{2.1}$$

9) If the volume of the sample, V_s, is known, then a small correction can be made for its volume. This correction modifies Eq. (2.1) to:

$$V_d = \frac{P_i V_1}{P_f} - V_1 - \frac{V_s T_1}{T_3} \tag{2.2}$$

> **What's the "dead space?"**
> Volumes V_2 and V_3 are at different temperatures, T_2 and T_3. Using the ideal gas equation, the total number of moles in the two volumes, assuming $P_2 = P_3 = P$ is:
>
> $$n = n_2 + n_3 = P/R \, (V_2/T_2 + V_3/T_3)$$
>
> Imagine then that the entire region is a T_2 to yield an imaginary effective volume for the sum.
>
> $$V_d = nRT_2/P$$
> $$V_d = T_2(V_2/T_2 + V_3/T_3)$$
>
> One could therefore think of the dead space (without sample) as a weighted average of V_2 and V_3 according to inverse temperature.

The volume of the sample can be obtained if one knows both the open and closed porosity of the sample and its theoretical density. The closed porosity of the sample should be included in V_s but the open porosity should not be included. These are subtle points, which for many practical applications make little difference and can be ignored. At any rate, it's a good idea to attempt to subtract V_s even with crude data.

It is recommended that if the sample volume is small enough to be ignored, that the dead space be measured with the gas that is being used as an adsorbent. Otherwise, one could use a gas that does not adsorb very much at the temperatures of interest, for example Ne or He. There are at least two problems involved with the inert gas dead space calibration:

1) The laminar-turbulent flow problem is different and attention to this detail is important (see the next section).
2) It is not the case that inert gases at temperature well above their boiling point do not remain on the surface of the sample or in the pores. (For an example of this see the publication by Silvestre-Albero et al. [1].)
3) In the latter case, one needs to figure out what higher temperature out gassing is appropriate, that is sufficient and does not change the sample.

It is recommended that this dead space be measured over a range of pressures. A properly designed instrument will have little variation over a large pressure range. Problems concerning the location of the boundary between V_2 and V_3 can lead to errors. Furthermore, the transport properties of the gas can change with pressure, especially in the low pressure range. In the next section, the errors due to uncertainty in the boundary between V_2 and V_3 and the error due to uncertainty in the gas transport properties are discussed further.

1) Once V_d is determined, one can apply this to the uptake of the adsorbate on the surface. After sealing the sample in the sample tube, with degassing and other preparation steps, the system is evacuated. The details of the degassing depends upon the sample, but it is often a high temperature bake that does not distort the

sample in any way. The degas might be a high temperature treatment in a gas the removes contaminants from the surface, for example a bake with H_2 gas, or maybe the contaminants are part of the problem. For careful scientific investigations degassing procedure, with for instance ceramics, would be to bake the entire system under an ultra-high vacuum (better than 10^{-9} atm.) at temperature of $\geq 200°C$. Routine degassing is often much less vigorous. The interpretation of the data, however, is often dependent upon these steps. For the adsorption process:

2) One starts by first opening valve G.
3) The pressure (P_i) is then measured.
4) Valve G is then closed.
5) Valve C is opened.
6) The pressure drop is followed until a new stable pressure is reached. This may take some time, and it is highly advised to have patience. During this step, it is also advised that the data be recorded and followed as a function of time. Usually an exponential decay is observed, following a curve such as:

$$P_n = P_f + P_i \exp(-0.69t/t_{½}) \quad t \to \infty, P_n \to P_f \tag{2.3}$$

The symbols here mean the following:

a) The subscript on the pressure of "n" is the count number of the gas admission. $n=1$ for the first gas admission just described.
b) P_i, stands for the "initial" pressure just after valve C is opened. This normally need not be recorded.[1]
c) $t_{½}$ is the "half life" constant for pressure decay, so that after a time, P_f is approached.

7) The amount adsorbed for this single data point is given by (subscript $n=0$ means $n_{ads,1}=0$):

$$\Delta n_{ads,n} = n_{ads,n} - n_{ads,n-1} = \frac{P_{i,n}V_1 - P_{f,n}(V_d + V_1)}{RT} \tag{2.4}$$

8) After the first data point is taken, valve C is closed.
9) The procedure is repeated.
10) An additional amount is then adsorbed again and calculated according to Eq. (2.4).

---

[1] This expression could be used for (1) making a decision either manually or in an automatic system, or (2) to expedite to equilibrium for each data point to speed up the analysis.

The values of n_{ads} are obtained by adding the Δn_{ads}:

$$n_{ads,n} = \sum_{1}^{n} \Delta n_{ads,n} \tag{2.5}$$

One should keep a log of the values of $P_{i,n}$ and $P_{f,n}$ as the procedure advances through the n exposures. As one approaches the vapor pressure of the adsorbate in the bulk liquid state, P_{vap}, the amount adsorbed per unit of pressure change becomes larger and larger. At some point, say $P=0.95$ to $0.99 P_{vap}$, the measurements become impractical to perform and the procedure is terminated.

The classical isotherm plot is of $n_{ads,n}$ as a function of $P_{f,n}/P_{vap}$.

For the analysis of this isotherm one should refer to Chapter 1, or, for a more advanced analysis, Chapter 3.

Error analysis for the volumetric method

In this section is a discussion of potential errors for the volumetric technique. Also relevant are the errors analyzed under the General Error Analysis section.

Design errors
Hopefully none of these errors will be encountered. They are listed here so one can be aware of potential problems when constructing or buying equipment.

Boundary errors between V_2 and V_3
Of concern it the potential uncertainty about the boundary between V_2 and V_3 and potential variability. Most problematic with this design error is the sharpness and stability of the temperature transition from liquid N_2 (or other temperature control) to the room temperature region of V_2. The U-cup arrangement for liquid N_2 is recommended. Other temperatures or control methods will require similar thought.

Some instruments come supplied with matching hang-down tubes to compensate automatically for the error associated with the boundary movement. If V_2 and V_3 vary during the measurements, thus varying V_d, then an unknown, unsystematic error is introduced, but with the matching hang-down tubes V_d is automatically compensated. This, however, does not solve the next problem listed, that is the variability of P_{vap}. Even with the matching hang-down tubes, the "U" tube arrangement shown in Fig. 2.1 is still recommended.

Sample temperature control
Poor control of the sample temperature might not seem to be a problem since, after all, the bath is a constant temperature bath. However, the larger the difference between the volume V_3 and the surroundings, such as V_2 and V_3, the greater the potential that the sample temperature will be incorrectly read and controlled. This is mainly due to radiative heating, especially problematic with cryogenic temperature baths. A 1.0 K error in temperature leads to catastrophic errors, especially in the value needed for P_{vap}. For solutions, see the suggestions below and the section on gravimetry for which the solutions there apply equally here.

Poor calibration of V_1

A poor calibration of V_1 is a systematic error and is additive across all n_i. This means that a 10% error in V_1 leads to a 10% error in the amount adsorbed. As errors go, this might not be bad, depending upon application. Furthermore, a post calibration can correct all preceding errors directly. A 10% error in V_1 would also be unlikely, since even a crude measurement of volume should yield a number within 1%.

In determining the pore sizes using the modified Kelvin equation, a poor calibration of V_1 is not critical since the primary size determination is with the pressure measurement. It will directly affect the pore volume measurement. However, the graphical method, which does not depend upon the Kelvin equation, definitely requires a good calibration.

Molecular flow versus viscous flow

The transition between Molecular Flow Versus Viscous Flow can be the source of a large, usually unrecognized error. Proper tube design, that is, proper diameter tubing for the temperature and pressure ranges, is needed to avoid this problem. This is especially true in the low pressure range. This error can be critical for the low pressure work and can lead to incorrect conclusions, including the wrong values of surface area and porosity.

If the tubing cannot be designed to avoid the problem, then one should design within the problem and use the corrections available from the literature

A discussion of the regions of the two realms can be found in most books on vacuum technology, for example, the book by Roth [2]. An example specific to adsorption is by de Dios Lopez-Gonzolez et al. in 1955 [3]. The problem is that in the low pressure range, P_3 is not equal to P_2 but is related to it by:

$$\frac{P_2}{P_3} = \left(\frac{T_2}{T_3}\right)^{1/2} \tag{2.6}$$

At *high* pressures $P_2 = P_3$. The transition between these two regions is governed by the "Knudsen number," which is the ratio of the tube diameter, D, to the mean free path of the gas, λ_f.

- If: $D/\lambda_f > 110$, then $P_2 = P_3$
- If: $D/\lambda_f < 1$, then Eq. (2.6) holds
- If: $1 < D/\lambda_f < 110$, something intermediate and one needs to use calibration curves.

The mean free path, λ_f, can be calculated from some gas equations of state, usually the van der Waal equation. Table 2.1 is a list of some typical values for λ_f which might be of importance. The λ_f is inversely proportional to the pressure, so the particular requirements of a system may be calculated from this table.

Notice from the table that at low pressures (0.001 P_{vap} for the usual N_2 isotherm) the volumetric method breaks down. Although one can attempt corrections, the transition region is very hard to control and should be avoided.

Table 2.1 Values for λ, and minimum D required at 25°C

Gas	λ_f at 10^{-3} atm/m /m	λ_f at 10^{-4} atm/m /m	λ_f at 10^{-5} atm/m /m
H_2	1.2×10^{-4}	1.2×10^{-3}	1.2×10^{-2}
He	1.9×10^{-4}	1.9×10^{-3}	1.9×10^{-2}
N_2	6.6×10^{-5}	6.6×10^{-4}	6.6×10^{-3}
O_2	7.1×10^{-5}	7.1×10^{-4}	7.1×10^{-3}
Ar	7.0×10^{-5}	7.0×10^{-4}	7.0×10^{-3}
H_2O	4.5×10^{-5}	4.5×10^{-4}	4.5×10^{-3}
CO_2	4.3×10^{-5}	4.3×10^{-4}	4.3×10^{-3}
	min. tube dia./cm	min. tube dia./cm	min. tube dia./m[a]
H_2	1.4	13.	1.35
He	2.1	21.	2.13
N_2	0.72	7.2	0.72
O_2	0.78	7.8	0.78
Ar	0.77	7.7	0.77
H_2O	0.49	4.9	0.49
CO_2	0.48	4.8	0.48

[a] Notice, third column is meters.

One could argue that not much is adsorbed below 0.001 atm and therefore the error in the overall amount adsorbed is slight. Unfortunately, this is incorrect and for some high energy materials correction for this error is absolutely critical. For these adsorbents a monolayer equivalent is already adsorbed at this pressure. Not taking this into account can yield a large error.

Notice that with this type of error, a systematic error is introduced in both the dependent and independent variable in the transformed BET equation. As will be seen later, however, this is not as serious an error for the χ theory analysis if one is interested in only the surface area. For the gravimetric method, this correction is only needed for the pressure and not the amount adsorbed if the χ plot is used. A caveat for this last statement is that for porous samples a much more complicated situation exists and thorough degassing and very low pressure measurements may be required. As an aside, note that the Knudsen correction is not necessary when one considers small size pores since immediately outside the pores the temperature is the same as inside the pores.

Equation of state errors

The above discussion on how to calculate the amount of adsorbate was based on the ideal gas law. Whether this holds up or not can be easily checked. For N_2 using a liquid N_2 bath the error is slight. At full P_{vap} the error is only about 0.03 % (in recent developments for measuring into the macropore range, a correction for this deviation may be necessary). This might not be true for other adsorbates and other pressures and temperatures. The usual correction used for this is the van der Waal equation, for which the constants may be found in many handbooks (e.g., the CRC Handbook [4]). A closer approximation would be the Carnahan-Starling [5] equation of state or an empirical virial equation, many of which are available on-line from NIST.

Temperature control of the sample

A temperature error reading at the sample can affect the isotherm considerably in the upper pressure range. Such errors are usually attributed to radiative heating or inhomogeneous temperatures in the sample. Radiative heating is due to the infrared radiation originating in the V_2 section and traveling to the V_3 section. This is difficult to avoid. One possibility is to use baffles in the V_3 section to eliminate this radiation. Baffles, however, can complicate the molecular flow problem previously mentioned and should be carefully designed. Baffles are easier to use with the gravimetric technique. More about this error is presented in the General Error Analysis section.

Limit of detection

Due to limitation of the pressure sensing devices, the very low pressure isotherms are almost never measured by the volumetric technique. The problem can be corrected by employing more than one sensor device to obtain values below 0.001 atm. Two problems are present for this technique. Firstly, the cross-calibration between the two pressure sensors must be very good to avoid large errors. Secondly, the problem of molecular flow becomes important as mentioned in a previous section. This leads to both an accounting problem and a problem in determining an important quantity for isotherm interpretation: the chemical potential of the adsorbate. Theoretically, these problems can be handled. Practically it is easier and more certain not to use the volumetric method in this range.

Advantages and disadvantages of the volumetric technique

One big advantage of the volumetric technique is that it is usually less costly. When one wishes to do more sophisticated work, the cost naturally goes up.

The primary disadvantage of the technique is that it is not as suited for careful research work as the gravimetric technique. Usually, the high vacuum pressure range is not measured with multiple pressure sensors perhaps due to the expense of the system. To use it in this mode, the cost advantage begins to disappear and the amount of effort required to do careful work becomes greater, with many potential pitfalls. This is especially true for the low pressure range of the isotherm, but it can also be true for the upper range where porosity measurements are extracted.

The gravimetric method
Equipment description

The principle of the gravimetric method is simpler than that of the volumetric method. For the gravimetric method, one simply brings in a pressure of the adsorbate and measures the mass gain of the sample. The isotherm is then simply mass gain (or reworked into preferred units such as moles) versus the pressure. In the engineering of the equipment, however, things are not so simple. The vacuum system is usually a conventional metal system but the balance is a very high precision model.[2] The method is usually confined to high quality research work.

The instrumentation which was used to determine the isotherms obtained in researching the χ theory cost around $1 M (1970). Subsequently, five more instruments were built at a cost of less than $500,000 each. Today the cost is considerably less. This is mentioned to indicate that the instrumentation is much more complex and sophisticated than at first appears. It is, true however, that the instrumentation was built for a purpose other than investigating physisorption and was more than was required for physisorption measurements, especially long-term stability. A list of the system requirements for the high quality work is given below. The system requirements are as follows [6]:

1) The balance should have at least a relative sensitivity of 1 µg per gram sample. This would be for argon or nitrogen adsorption on samples with a surface area of $1 m^2 g^{-1}$ or greater. There have been microbalances with a sensitivity or $1 ng g^{-1}$.[3]
2) The system should be all ultrahigh vacuum, including the balance. High outgassing metals, such as platinum, should be avoided. Metals that react with hydrogen should also be avoided. The system should be capable of being baked to 400°C with hydrogen inside at a pressure of 10^{-2} atm.
3) The system should be fastened securely on a heavy table. A machine table is appropriate, but a table made from heavy metal is needed. This table should be fastened securely to a concrete floor, preferably a balance table floor with a foundation on bedrock. No attempt should be made to dampen vibrations to the table as this simply leads to intolerable drifting of the weight. The area where this is installed is hopefully free of excess vibrations through the earth. The area where the above-mentioned instruments were installed was susceptible to an occasional earthquake and to blasting in the distance. A few experiments had to be discarded due to these effects. See disadvantages.

---

[2] ...and microbalance manufacturers seem to go out of business quickly.
[3] One solution that this author is aware of to solve the sensitivity problem is to make a balance that can take a capacity of more than a kilogram sample. Indeed one researcher used a balance in the 100 kg range and read to a mg. By doing this much of the bothersome plumbing problems become simple and the worry about breakage is less, but the sample size and cooling if needed are problems.

4) Some method for long-term data taking is required. The referred to balances were computer-controlled and data taken automatically. Today, this should not be much of a challenge. Long-term data taking is required to determine the thermodynamically valid numbers.
5) Proper temperature baffling is a must.
6) In Fig. 2.2 is a diagram of the baffling used. Indicated in this drawing is a baffle arrangement permanently in place using two copper gaskets on a two-side vacuum flange. To save space, the bottom matching flange had tapped holes for the flange bolts. A lower flange was used to access the hang-down pan. The upper matching flange for this flange also has tapped holes for the flange bolts. The holes in the copper gaskets, except for the holes for the hang-down wire, were staggered both radially (not shown) and transversely, as shown.
7) This arrangement allowed very little radiative heating and an assurance that the temperature read by the gas thermometer was very close to the sample temperature. It cannot be emphasized too strongly that an incorrect temperature reading is a serious, even invalidating, error. The temperature needs to be correct to within 0.01°C, especially if one utilizes any part of the isotherm above $0.3P_{vap}$.
8) Baking in hydrogen and sample admission with a counter flow of inert gas is often required. Therefore, provisions for this are needed. Additionally, it is recommended that a high quality, controlled and a monitored glove box be available for the sample admission side of the balance, since the state of the surface is very sensitive to gas contamination.
9) Pressure gauges should range from 10 to 10^{-12} atm. A combination of Bourdon or diaphragm type gauges and a Bayard-Alpert type gauge would cover this. The diaphragm gauges are used for the pressure measurement for the isotherm. The sensitivity can be as low as 10^{-6} atm. The Bayard-Alpert gauges are needed for vacuum and degassing measurements.

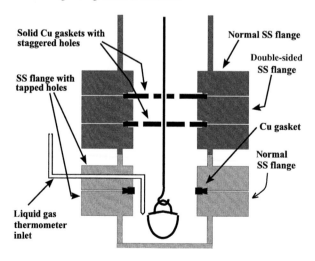

FIG. 2.2

A drawing of the gravimetric method sample area showing the baffle arrangement.

Many of the requirements listed may be loosened, depending on the material being investigated and the quality of the work and pressure range needed.

Fig. 2.3 is an overall view of a typical gravimetric system. Included are provisions for the uniform operation of the cooling bath (L) utilizing a liquid nitrogen coolant. The entire system is also temperature controlled with an air box. Temperature control is not quite as critical with the gravimetric system, excluding of course the sample temperature, since errors in the pressure measurement are not too great. For very precise measurement the use of a "U" cup about the hang-down tube would be advisable in order to minimize pressure variations (this is not as serious error as the dead volume problem without a U tube in the volumetric method). The chamber area should be minimized. If one were to measure two adsorbates simultaneously, minimization would probably be necessary. It may be advantageous to have as the gas inlet valve a controllable valve that can be automatically controlled with a feedback loop from the pressure gauge. This can provide a full automatic, computer-controlled system.

Determination method

For careful work, the gravimetric method also needs to be pre-calibrated. This calibration is for some small corrections. Firstly, there is the buoyancy of the sample and the balance equipment. This correction is usually performed in one of two ways.

Method 1

A fully dense sample of equivalent volume as the anticipated sample is placed in the sample chamber. The system is then sealed and evacuated. All arrangements, such as the liquid nitrogen cooling bath, should be put in place just as if a sample were present. The adsorbate gas is admitted from very low pressures in increments up to nearly P_{vap}. This should yield a very linear plot of "mass gain" or buoyancy, m_b versus pressure. The equation is:

$$m_b = bP \tag{2.7}$$

where b can be either positive or negative.

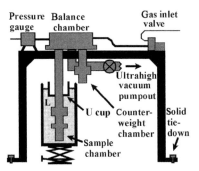

FIG. 2.3

An overview of the system used for the gravimetric method.

A least square routine should be used to determine b, so that the statistical information $(r, \sigma s,$ etc.) is available. For the isotherm, a quantity bP should therefore be subtracted from each isotherm data point.

Method 2

The second method is to determine the buoyancy with a non-adsorbing gas with the actual sample. For example, for a nitrogen adsorption isotherm, use of Ne or He as probe gases would be appropriate. The buoyancy, b, is calculated from the pressure of the probe gas, P_p, by:

$$b = \frac{m_b \overline{M}_{ads}}{P_p \overline{M}_p} \tag{2.8}$$

where $\overline{M}_p$ is the molar mass of the probe gas and $\overline{M}_{ads}$ is the molar mass of the adsorbate. $\overline{M}$ or M_u versus $\overline{M}$ does not matter here as long as it is consistent.

If the baffling or tubing is improperly designed or if one wishes to operate the instrument into and below the crossover from viscous to molecular flow, then a molecular flow correction must also be made. This means that Eq. (2.7) will not be linear in the very low pressure range but will be approached at higher pressures. It is recommended that the first procedure be performed to yield b in the higher pressure range and make this subtraction form from the full range of the calibration. The function left should be sample independent and repeatable for the particular instrument geometry. This should be an even smaller correction than buoyancy. If it is not, the use of different baffles or a longer hang-down tube should be considered. The equation relating the correction for molecular flow, $m_{mf}(P)$, in relation to the mass recording of the trial, m_p, is given by:

$$m_{mf}(P) = m_p - bP \tag{2.9}$$

Therefore, one can determine $m_{mf}(P)$ with a single calibration. The constant b, however, will change with the sample and needs to be determined for each type sample. For routine analysis of similar samples, that is, samples of the same theoretical density and closed porosity, one could initially determine b as a function of sample mass, thus saving some subsequent analysis time. If this is done, one must be sure to use the same counter weights on the other leg of the balance for a particular sample mass. One could also determine b as a function of sample mass and theoretical density, provided the samples contained no closed porosity.

For the measurement of the isotherm, one simply admits the adsorbate to the system at the pressure desired and wait for the mass measurement to settle. This may take some time. For example, for low pressure measurements several hours may be required for thermal equilibrium to be reached. Therefore, it is highly advised to have patience. See the comments before and after Eq. (2.3). For each mass data point, the buoyancy and molecular flow corrections are subtracted. For high-quality work, the $T^{1/2}$ corrections are needed for low pressure. Where this applies is indicated by the ratio of the function $m_{mf}(P)$ to P in Eq. (2.9).

Error analysis for the gravimetric technique

With the buoyancy correction and the molecular flow correction, the data obtained from the gravimetric technique should be very accurate. The limit of detection is the limit imposed by the quality of the balance. Only small pressure corrections are needed in the low pressure range. No pressure correction is required if the hang-down tube has been properly designed for the transition region and the pressures under consideration. However, this may not be possible if very low pressures are to be used, as Table 2.1 would indicate.

The diameter of the hang-down tube need not be restricted and in theory could be several meters wide. In the volumetric method, this would create intolerable dead space problems. The room size could be the limit for the gravimetric method. This may not be as big a problem as at first appears, since the only problem with pressure is the question of what the pressure is in the sample area; that is, what is the true chemical potential? Thus, the pressure in the balance chamber area is irrelevant, albeit related. Molecular versus viscous flow is unimportant so long as $m_{mf}(P)$ is measured. An alternative pressure transducer method for only the sample area is possible by several arrangements.

Advantages and disadvantages of the gravimetric technique

The primary advantage of the gravimetric method is very high precision and accuracy. There is a similar advantage found in normal gravimetric analytical chemistry. High quality research work and pore analysis should be performed with this technique.

There are not many errors associated with the method. The calibration is relatively simple and for routine analysis, trivial. The gravimetric method is usually faster in routine mode than the volumetric method, due to the fact that fewer calibrations are needed.

An exception is if one attempts to dampen the instrument against ground noise. It is best not to do so and to select a location where artificial vibration from traffic, etc. is at a minimum. This may not be possible. In areas of very frequent earth quakes or blasting it may not be possible. For example, in the area around Oak Ridge the frequencies of these incidences (2R or higher including blasting) was about once a week, but the experimental runs lasted less than 24 h, so an occasional experiment had to be discarded. Check with local geologists first to determine what kind of vibrations one might encounter. Check to see if the blasting is announced ahead of time so you can take a justified rest. Earthquakes, which are relatively rare even in California,[4] are not so predictable, so one will have an occasional measurement that needs to be ignored.

Sample preparation, degassing, reacting, and modifying is simpler and can be followed in a straightforward fashion in situ using the mass changes. This is a very important advantage, which is not generally or naturally available with the volumetric method. Switching over to production or preparation conditions and measurements under these conditions is very easy with no removal of the sample. Due to this, other investigations of the sample material can be combined with the surface analysis. Examples of this are the measurements of oxidation kinetics or catalytic activity.

The primary disadvantage is expense. A good balance for use in this system is costly. The sensitivity of the balance, and thus the quality of the work, is related to how much one spends. A second expense is the high quality table and positioning. It was recommended that this table be tied directly to a concrete floor, preferably to a slab meant for a balance. This is an additional expense, but not absolutely necessary. A third expense is the vacuum system and set-up expenses that rarely come with the balance.

General error analysis—Common to both volumetric and gravimetric

In this section errors are presented that one should be aware of regardless of the technique. There will be some duplication from the above discussion and potential errors due to theoretical interpretation are not covered. These will be addressed later. Most of these errors can be avoided with careful instrument design.

Pressure and temperature measurements

It is assumed here that the pressure and temperature measuring devices are properly calibrated. They should be traceable to the National Institute of Standards and Testing (for the USA). The problem is to measure what one thinks is being measured. Here is a list of potential problems and their consequences.

1) Sample temperature problems can arise from inhomogeneous temperature of the sample.

    With respect to this problem, a highly exothermic adsorption can have a significant effect on sample temperature (significant in this case means 0.01 °C or more). The solution to this is to be patient in allowing the adsorption to settle down. Advice about this has already been given in both experimental sections.

---

[4] For California historical records indicate an annual average of 49 earthquakes over 3.5 and 113 for all recorded. This might be somewhat frustrating but can be overcome. Obviously, long term experiments for reasons other than adsorption might not be possible.

2) Sample temperature problems can arise from radiative sample heating.
   With respect to this second problem, design-wise the gravimetric system would seem to suffer from this more than the volumetric potentially. The reason is the volumetric method can have baffling that can completely block "line-of-sight" radiative heating, whereas the gravimetric method must have a hang-down wire, and thus a small path for radiative heating. A baffle on the hang-down wire or on the sample cup could minimize the problem. Furthermore, in the gravimetric technique, the sample takes a while to reach the proper temperature due to the insulating ability of a vacuum. One trick to play to alleviate this slow temperature equilibration problem is to arrange a contact plate slightly below the sample pan. Being sure that the sample side is always a little heavier than the counter weight side, one can simply turn the balance off, allowing the sample pan to make contact with the plate for thermal equilibration. Some manufactures have some ingenious methods of decoupling the sample hang-down from the balance itself and provide such platforms as a part of the system. Alternatively, one can use patience. Volumetric analysis suffers from the problem that baffling is not advised due to pressure problems. However, direct contact with the thermostated walls is normal. It must be remembered, though, that many samples are quite insulating and thermal gradients are inevitable. For such samples, a new arrangement must be made to counter this, such as a horizontal bent tube in the cold zone, or baffling.
   The most likely error is that the sample temperature will be higher than measured or believed. Such an error leads to very large errors in P_{vap} and essentially makes the high end of the isotherm useless. This is the range where analysis of the larger pores is performed. To illustrate this problem, in Fig. 2.4 is a simulation of the effect of incorrect temperature control or measurement. For example, a temperature of only 0.5 K higher than assumed with liquid nitrogen yields an error of 8% in P_{vap}. This translates to an adsorption error at $0.9P_{vap}$ of a factor greater than 2. It could also create problems in analyzing for the surface area. If this error is known to exist, then steps are possible with χ theory to overcome the problem.
3) Insufficient low pressure pump-out and degassing can lead to false conclusions.
   For high energy materials, such as ceramics, a monolayer can already exist on the surface of the sample at 10^{-6} atm (10^{-3} torr) which is low vacuum. Although the surface area can be measured for most samples from this point, false conclusions can be drawn if one takes the data too seriously. A pump-out and degas should be performed to at least 10^{-9} atm and preferably lower. Most gravimetric systems are capable of this and 10^{-12} atm is not unusual.
   Even if the sample has been mainly exposed to an inert gas after preparation and cleaning, there may still be some residual gas adsorbed. This is true, perhaps surprisingly, for He gas. This point is expanded in the next paragraphs.
4) Insufficient low pressure isotherm measurement may lead to the wrong conclusion about the form of the isotherm at these pressures. As will be explained in the theoretical section, the assumption of "Henry's law" due to lack of the low

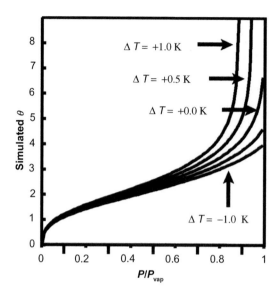

FIG. 2.4

Consequences of errors in temperature measurement/control in the isotherm.

pressure measurements is a bad assumption. In the first edition of this book, several samples were given to illustrate this disproof of Henry's law. This conclusion has not been generally accepted up to now; however, more evidence by other researchers demonstrates the threshold. The very important publication by Silvestre-Albero et al. [1] has also demonstrated this along with the point presented as 5 next.

5) One must be aware that the cover gas used for the determination of the dead space or the buoyancy is removed from the sample before performing the isotherm measurement. Better yet, the dead space of buoyancy should be measured after the isotherm is measured. This is true even if the probe gas used is helium. A strong high temperature, ultrahigh vacuum bake of sample should be performed before the isotherm is measure to eliminate any adsorbed gases.

There are several other points brought out by the publication by Ref. [1] that are listed in the more theoretical section along with some isotherm illustrations.

Kinetic problems

Between each increase or decrease in pressure, one should wait for the adsorption to settle.

There are some instruments based on gravimetric methods that calculate how long this period should be. This computer decision is made on the adsorption behavior and the criteria can be set by the operator. Generally, an equation like Eq. (2.3) is used for the computer decisions. In many cases a decision can be made as to how close to equilibrium to get, and stop the measurement at that point. Alternatively, the process can be speeded up somewhat by assuming that the approach to equilibrium is an exponential decay (Eq. 2.3). Using this assumption, one can extrapolate to the equilibrium value. This has the potential danger of doing the extrapolation too far from equilibrium for this assumption to be a good approximation (there is no guarantee the approach is an exponential decay, so if one is far from equilibrium this can create an error). In either case, this obviously requires some type of pressure or mass recording. Automatic data taking is an ideal solution to this problem, allowing the instrument to run 24 h a day. Many samples have very long settling times and without such a system, much wasted technician time will result.

Sample density problems

The philosophical question sometimes comes up as to what to count as surface. Obviously, closed porosity is not counted in this method. If one has very small pores, they may or may not be counted. If poor degassing or low vacuum is used, then some small pores may already be filled before the measurement is made.

Another problem is what is referred to as "bed porosity." This is the space between the particles and might be a problem when $P/P_{vap} > 0.9$. If bed porosity is present, it would show up as a mesopore or macropore porosity. Bed porosity, however, is not normally a concern for most surface area analyses since it affects the higher portions of the isotherm and the values obtained at low pressure would suffice.[5]

Calorimetric techniques

Calorimetry is conceptually easy, but in practice, deceptively difficult.

Adiabatic calorimetry

In Fig. 2.5 is a schematic of a typical cryostat adiabatic calorimeter. In this case liquid nitrogen is designated as the coolant. The number of walls in the cryostat depends upon the temperature range selected. With helium temperatures, one needs an outer cryostat for liquid nitrogen and an inner cryostat for the liquid helium. The various parts are as follows:

- G—gas inlet and vacuum pump-out port
- HL—heater leads
- I—insulating stand-offs

---

[5] Bed porosity had been blamed in the past for the deviation of the BET from data above $0.35 P/P_{vap}$, but this is now known to clearly not to be the case.

CHAPTER 2 Measuring the physisorption isotherm

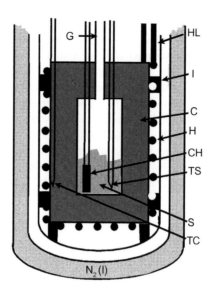

FIG. 2.5

A schematic of a liquid nitrogen cooled adiabatic calorimeter.

- C—copper adiabatic chamber
- H—heater coils for the adiabatic chamber
- CH—calibrating heater
- TS—temperature detector for the sample
- S—powder sample
- TC—adiabatic chamber temperature detector

The equipment shown in Fig. 2.5 is ideally the sample area of a volumetric system. Therefore, in principal, the isotherm and the heats of adsorption can be measure at the same time on the same sample. The reason this is not normally done is the convenience of obtaining the integral heat of adsorption rapidly (see Chapter 4 for details about calorimetry). However, it would be advantageous to let the material cool between aliquots of gas admissions in order to obtain both pieces of information. Letting the sample cool also assures that P_{vap} remains constant. This cooling takes a very long time for a complete isotherm to be obtained, but the data would be much more useful. Such data are very rare in the literature.

The temperature sensors depend upon range and sensitivity requirements. They could include gas-liquid thermometers, platinum resistance thermometers, thermocouples, or thermistors. It is advised to have more than one device in the sample and on the copper adiabatic shield. Very good computer adiabatic controllers are easy to construct. One must take into account in the programming the power required for various temperatures to match the heat capacity of the shield, that is, one needs to adjust the power and "damping" that the power supply puts out according to calorimeter and sample. This may take some preliminary runs to adjust correctly.

Cooling in a vacuum could be a problem at low temperatures. It is traditional to do the preliminary cooling by backfilling both the sample chamber and the insulating spaces with helium (again, see Ref. [1]). During the measurements, however, the cooling will need to be natural. The measurement of the isotherm is a necessary step in analyzing the data obtained from the calorimeter. Most investigators measure the isotherm in a separate apparatus, which brings up the question: is this the same material with the same history? Unfortunately, most publications are not very reassuring on this latter point. Preliminary measurement of the calorimeter without a sample in order to obtain the heat capacity of the calorimeter and of the powder is highly recommended. By doing so, one can obtain the additional information of the heat capacity of the adsorbate.

The calculations required to obtain meaningful information are somewhat complex and tedious. These are described in the analysis section.

Measuring the isosteric heat of adsorption

The isosteric heat, q_{st}, is the heat of adsorption at a constant adsorbate amount. In terms of thermodynamics it is related to the pressure and temperature by:

$$q_{st} = \frac{RT \partial \ln P}{\partial^1 / T \big|_{n_{ads}}} \tag{2.10}$$

Attempts have been made to determine this from the isotherms (for example, see Joyner and Emmett [7]). To do this, one measures two or more isotherms at differing, but fairly close, temperatures. One then fits the isotherms either manually, for example, with a spline fit, or mathematically. Unfortunately, errors accumulate very heavily in this case and the choice of fit can greatly distort that answer. Using the analytical form of the standard curves [8] may aid in this attempt and appears to be successful in some cases but porosity and multiple heats of adsorption makes this unreliable as well.

The thermal "absolute" method

Harkins and Jura described [9] a method of obtaining the surface area in an "absolute way" from a calorimetric measurement. They addressed many of the concerns regarding the method [10] but one must still qualify the method as very limited. Porosity of any type would significantly alter the answer. The explanation by Harkins and Jura for the information obtained—that is, there was initially a liquid film distributed evenly on the powder—was not correct. However, the method is theoretically sound for the χ theory interpretation.

The method and equipment is simpler than the adiabatic calorimeter method but there are some modern concerns with respect to sample reproducibility and initial exposure to adsorptive. The apparatus is schematically represented in Fig. 2.6. The powdered sample (S), which is known to be non-porous, is allowed to equilibrate over liquid water (in principle, this should work for any liquid). It is assumed that a film of water is adsorbed about the particles, as envisioned in Fig. 2.7. The powder is

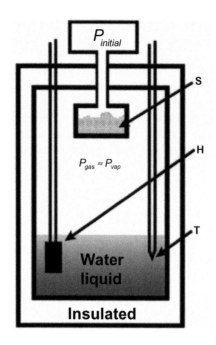

FIG. 2.6
A schematic of the Harkin and Jura calorimeter to measure the surface area of a powder.

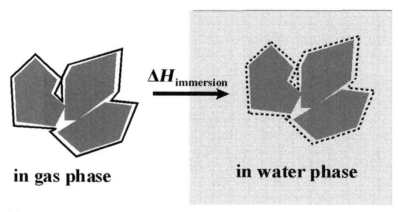

FIG. 2.7
A schematic of how the adsorbed film is destroyed when the powder is immersed in the liquid phase thus releasing its surface energy.

then lowered into liquid water. In the process of doing this, the outer film of the adsorbed water is destroyed, thus releasing the surface energy of this film. The energy released is calibrated against the electrical heater (H) with a temperature measuring device (T-36 junction), in their case a thermocouple. Since the surface tension of water is known, then the surface area may be calculated from the heat evolved, ΔH, or by the simple equation:

$$\Delta H = \gamma_{gl} A_s \qquad (2.11)$$

where γ_{gl} is the surface tension between the gas and liquid phase.

Since the water vapor is nearly the saturation pressure, they assumed that there were at least eight monolayers (probably only 4) of water on the powder initially. This is believed sufficiently thick so that the component of the film energy due to the solid-liquid tension, γ_{sl}, remains constant. In modern terms, it was enough coverage to have the adsorbate collapse and form an outer gas-liquid interface. Thus, this it is the gas-liquid interface surface area that they were measuring and not the adsorbate-adsorbent surface area. This is, however, a very important measurement and with the nonporous materials used, these quantities should be nearly equal. In order to eliminate the possibility that there is additional heat of adsorption they performed a series of experiments to measure the heat of immersion as a function of the exposure pressure [11]. In the process, they produce the exponential decay curve for the E as a function of P, which can be used to back extrapolate to yield E_a. Indeed, at the exposure pressure the value of the heat of immersion was leveling off.

There is not much work performed using this method. Possibly the reason for this is the uncertainty in the interpretation and the difficulty of controlling the experiment. Bed porosity should be a large problem, although one could find samples for experimentation that would minimize this problem. An example of these latter adsorbents would be the rare earth plasters. The measurement of the surface area in this case is at the very high pressure region versus the BET, which is at the low pressure region. Thus, a comparison between the BET and this "absolute" method is somewhat questionable, especially considering that the BET itself is very questionable.

What is more important, for most researchers and engineers, this technique is very limited to special types of powders. With an unknown sample it does not seem to have much utility, as ingenious as it is.

It should be noticed that ΔH is a function of initial coverage or:

$$\Delta H = f\left(n_{ads, intial}, \gamma_{gl}\right) \qquad (2.12)$$

and as:

$$\lim_{n_{ads} \to \infty} (\Delta H) \doteq A_n \gamma_{gl} \qquad (2.13)$$

where A_n is the area of the outer surface layer of the adsorbed film.

Thus, regardless of the intervening function **f**, the final molar energy is what is gas-liquid surface energy. This is in contract to the calorimetry from the gas phase adsorption, where the final molar energy is the molar enthalpy of vaporization, $\Delta_{vap}\overline{H}$ or ε per molecule.

Differential scanning calorimetry[6]

Differential scanning calorimetry is often combined with thermogravimetric analysis of some type, which is thermal desorption or adsorption. The method yields fine details in the analysis. Adsorption experiments are performed by addition of the adsorbate at a rate that is:

1) slow enough that the system is very close to equilibrium but
2) fast enough to obtain a temperature increase enough to measure in the differential mode.

The first criterion can be checked by doing some kinetic studies, either gravimetric or volumetric. The second criterion would probably be obvious during the calorimetry experiment. The calorimetry system has been described by Rouguerol et al. [12] (Fig. 2.8). It provides details of the thermodynamics of adsorption that gravimetric and volumetric methods may not be able to supply and is an excellent complimentary research tool. This is evident, for example, in the study of N_2 and Ar adsorption carbon (Sterling MT 1100) performed by Rouquerol et al. [13]. In this article, there are clear peaks in the heat of adsorption in a region where the isotherm shows only a vague break. The difference between N_2, Ar, and O_2 adsorptions are quite clear.

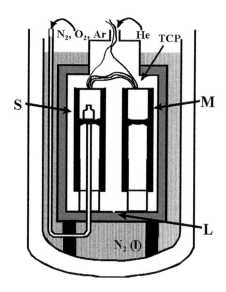

FIG. 2.8

Schematic of the differential calorimeter by Rouquerol et al.

---

[6] I do not have any experience with this particular method nor do I know anyone personally that has. In general, experience with scanning calorimetry has indicated that it is not very precise nor accurate, even with very expensive instruments and technicians that are well trained in the technique. I would suggest that one contact the authors and have discussions about accuracy and precision along with what procedures need be followed.

The differential scanning calorimeter has the advantage that the heat of adsorption or desorption is compared to a standard using a differential temperature measuring method—usually two thermopiles for which the voltage difference between them is measured. Fig. 2.8 is a schematic of the system that Rouquerol, et al. employed ("TCP" indicates the thermopile, "S" the sample chamber, "M" a matching reference chamber, and "L" is a slow He leak).

The analysis requires some calibration and the final equation to obtain fairly rapidly. One needs to do a bit of adjusting of flow rates versus signal strength. However, the great advantage is once setup, one obtains the heat of adsorption as a function of adsorbent pressure quickly. Referring then back to the isotherm, one obtains the heat of adsorption as a function of coverage. The heat evolution is provided by the equation:

$$\frac{\partial \Delta H(T, n_{ads})}{\partial t}\bigg|_{T, n_{ads}} = \frac{1}{k_{ads}}\left(\frac{\partial q}{\partial t} + V_d \frac{\partial P}{dt}\right) \quad (2.14)$$

where ΔH is the enthalpy of adsorption, k_{ads} is the kinetic rate of adsorption, q is the heat transfer, and V_d is the dead volume correction.

More details of this technique are available in the book by Rouquerol et al. [14].

Arnold's adsorptive flow system

A system which apparently is not available commercially is the flow system used by Arnold [15]. This system has some unique advantages, especially for binary adsorptive systems. Although the instruments are somewhat outdated, newer instruments can easily replace these. The original diagram of the system is shown in Fig. 2.9. The following is a description of the parts shown:

A: A capillary flow meter
B: Needle valve (See H)
C: A purifying trap
D: Pressure measuring devices

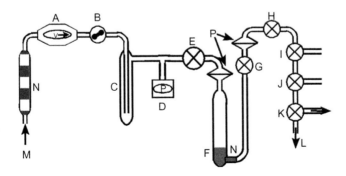

FIG. 2.9

The vacuum flow system developed by Arnold.

E: Isolating stopcock (along with G)
F: Sample tube with capillary hang-down connections
G: Isolating stopcock (along with E)
H: B and H are closed at the same time
I: To collection chambers
J: To mass spectrometer analysis
K: Rough pumps
L: To UHV system
M: Moderate pressure inlet
N: Frit or other containing method
P: Disconnecting flanges
Not shown: a temperature sensor—usually a liquid-gas thermometer

With modern equipment such as ultrahigh valve, pressure measuring devices, etc. this system could be made suitable for ultrahigh vacuum work needed for the low pressure measurements. The operation is really quite simple although the instructions may be a little long.

1) The sample is prepared externally as much as possible. If this is made into an ultrahigh metal vacuum system and the sample cannot tolerate a 300°C bake out, then the sample area would need to be in an appropriate inert box after the entire system, including the sample tube (F), had been baked out.
2) For ultra-high vacuum, bake the system a 300°C and, if need be, use a low pressure of H_2 during the bake to clean contaminates from the steel walls.
3) Rather than using the "U" tube arrangement, shown in F, that Arnold used, there might be other sample chambers that would be more appropriate for a particular application. For example, the sample tube could be arranged so that a flat flange on one end of the chamber would be approximately even with the top of the sample as indicated in Fig. 2.10.
4) Immerse the sample bulb in the cooling area.
5) Adjust the inlet gas pressure a bit above the highest pressure to be measured (usually 1 atm.)
6) Adjust the gas flow during operation to about $20\,mL\,min^{-1}$. To do this, first close the valves to the storage bulbs (I) and analysis train (J), open the valve to the roughing pump (K) and the valves up to the flow valve (H, G and E), and open and adjust the needle valve (B).
7) Allow the flow to continue for 10 min.
8) After the time out, simultaneously close valves H and B.[7]
9) Take the reading on the manometer (D) between these two valves and record the sample temperature (see also the precautions about temperature uniformity and

---

[7] Depending upon the purifier, C, one might want to have a valve between the purifier and the pressure measuring devices, D, and close this instead of B.

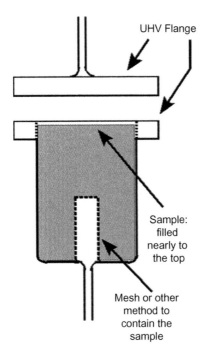

FIG. 2.10

Suggested arrangement for a sample chamber.

laminar versus turbulent flow in the previous sections; these precautions apply here as well).
10) Close valves that isolate the sample chamber (E and G).
11) Allow the sample to warm up—it may be necessary to heat it higher than room temperature depending on the adsorbate-adsorbent system.
12) Open valve H to evacuate the line leading up to valve G. At this point, the instruction may vary depending upon the analysis of the desorbed gases that is used. A good UHV pump-down, L, is recommended before admitting the gas to be analyzed to the analysis train.
13) After pump-down, be sure to close that valve going to the UHV and roughing system (K).
14) Close valves to the analysis section J and open valve I to allow the gas to enter the storage bulb (J).
15) The storage area (I) valve can then be closed to store the sample for analysis. If this bulb were equipped with a flange, one could send this for analysis if one does not have the right equipment to do the analysis close on-site. This option would seem to be tedious and would require many storage bulbs, but for some this might be necessary. Arnold used a chemical reaction to separate the

analysis of the evolved oxygen and nitrogen. For details of this clever technique, see his article.

16) If one has a mass spectrometer, then open valve *J* for on-site analysis.

This description might seem a bit long, but when one sees to logic of it, it begins to seem simple. These steps could easily be automated as well after step 3.

The sample chamber shown in Fig. 2.10 would be an appropriate arrangement for many samples. However, modifications may have to be made for some cases. An example would be the horizontal sample chamber, or possibly better filters to contain very fine powders. Use a horizontal chamber for samples prepared in place with some space for expansion as recommended previously in the section "Equipment description" under "The volumetric method."

Carrier gas flow method

The flow technique, or carrier gas technique, is very similar to gas chromatography. A carrier gas, typically helium, is used to carry an adsorbate gas such a N_2. The sample is cooled down to the adsorption temperature (usually liquid N_2 temperature). During this cool-down, the adsorbate is adsorbed. A downstream detector, usually a heat conductivity detector, picks up the signal indicating that there is a decrease in the adsorbate. The sample is allowed to cool long enough for the signal to return to the baseline. The coolant is then removed, or the sample heated up by some method. A reverse signal is then detected indicating the desorption of the adsorbate. A schematic of the type of signal one observes is presented in Fig. 2.11. The detector is calibrated by the insertion of a shot of adsorbate gas without the coolant. The primary advantages of this technique are:

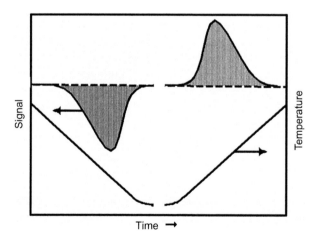

FIG. 2.11

Schematic of the signal observed for the flow system.

1) The equipment is very inexpensive.
2) The throughput is very high.

The disadvantages of this technique are:

1) The precision and accuracy are poor.
2) Normally the isotherm is not accurately obtained.

The technique is probably most useful for rapid throughput for quality assurance purposes, although this should not be the exclusive criterion since identical results can be obtained for very different samples. The engineering aspects, however, are quite important.

Engineering aspects: Carrier gas flow method

This discussion is a very short introduction to the subject of adsorption in the carrier gas system. It may be enough to get one started on the subject, but is definitely not to become competent in the method. Books have been devoted to the subject, so it is much too long to go into detail. There is one item for those interested to consider, that is the application of χ-theory and the differential equation given in Appendix II.D.

For practical engineer aspects, for example, packed column separations, the χ theory, or for that matter any other adsorption theory, is not necessary. Any good empirical fit[8] to the isotherm will work quite well. Indeed, there is good reason to find a function that simulates the isotherm and is easily differentiated. In other words the function D, which is:

$$D(n_{ads}) = \frac{dn_{abs}}{d(P/P_{vap})} \qquad (2.15)$$

is easily determined. The first derivatives for the χ plot are:

$$\frac{\partial \theta}{\partial (P/P_{vap})} = \frac{1}{(P/P_{vap}) \ln(P/P_{vap})} \quad \text{if } P > P_c \qquad (2.16)$$

The inverse of this is:

$$\frac{\partial (P/P_{vap})}{\partial \theta} = +\exp(-\theta + \chi_c)\exp[\exp(-\theta + \chi_c)] \quad \text{if } \theta > 0 \qquad (2.17)$$

However, differentiations of the χ plot, given in Eqs. (2.16) and (2.17), do not serve the flow equations that follow any better than the differentials of an empirical equation. This due to the fact that most practical applications do not deal with very low concentrations.

One caveat to the equations given is that the adsorptive concentration is above the threshold pressure. If the adsorptive in the carrier gas becomes very low this could

---

[8] See Appendices II.A and II.B for some possible programs to use for empirical fits.

break down, with leftover adsorbent in the carrier. For example, it is possible that attempts to remove contaminations from a carrier may fail due to the threshold pressure phenomenon. It is, therefore, important to know the threshold pressure given the identification of the adsorbent, absorptive, the temperature range of operation, and what specification for the contaminants is required. Thus, it is recommended that if levels of less than 1 part per thousand are important, then an isotherm be performed and analyzed using the χ-plot.

In Appendix II.D the required differential equation is present $\partial P_x/\partial P_2$ for the solution to the binary adsorptives. The equation is messy but there are no differentials for the θs that are common for this problem. Again, it is assumed that both adsorptive pressures are above the threshold pressure.

The following discussion will assume a gas flow system. However, many books on the subject do not make a distinction between a gas flow and a liquid flow system since the equations are mostly the same. Therefore, the terms "adsorbent" and "carrier gas" will be used instead of "solute" and "solvent."

For an attempt to determine the behavior of the packed bed, it is convenient to first assume that there are no kinetic effects, that is, the adsorption and desorption happen very rapidly. The question then is; why does anyone need to know, even empirically, what the equation is for the isotherm? If one does not run out of adsorbate, then the front of the adsorption is whatever it is at the beginning of the column.

What is wrong with these questions? Consider the following discrete model: The front of the adsorptive is traveling from left to right. When the front contacts the area labeled "1," as long as there is a supply of adsorptive, the area will fill until it reaches the adsorbate concentration n_{ads} as shown on the isotherm diagram. When area "1" is filled, the front can move on to start filling section "2." Area "2" then needs to be filled before the front can proceed on to section "3." This continues on to section "4," etc. This is the case with the adsorption curve shown in Fig. 2.12.

One can envision that the advancing front is like a flow of water, the capacity of the slab is the size of a bucket that needs to be filled before the next bucket is filled, and the pressure is like the amount of water per second that flows out of a spigot. The filling of the next slab will be called "sinking" versus "sourcing" for the desorption process.

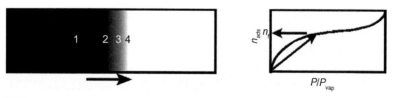

FIG. 2.12

The diffusion-sink situation that one expects from the isotherm type shown on the right. Slabs 1, 2, 3, etc. are filled completely up to n_f before the front moves on.

Examining the situation in this figure one would write that the velocity of the advancing front, v, (not the carrier gas) would be:

$$v = \frac{P}{n_{ads}} \text{ or } v = \frac{\Delta P}{\Delta n_{ads}} \quad (2.18)$$

The "change of" part of this equation on the right expresses the possibility that one does not start with of clean bed but that there is some pre-adsorbed adsorbate present.

Examining the clean bed, according to this analysis Δn_{ads} is either 0 before the front arrives or fills up to n_f the final value before the front can move on. This is due to the negative curvature of the isotherm, and thus the path for the sink is a straight line directly from 0,0 to $P/P_{vap}, n_f$. To do otherwise would violate the basic equations of diffusion.

In Fig. 2.13 is the situation for a different type of isotherm, a Type III form. In this case the curvature of the isotherm is positive. Thus, both diffusion and sinking can operate simultaneously.[9] The implication of this is that, since both the diffusion and the sink are driving the front forward, then the front gets spread out.

For desorption, the process is reversed and under these circumstances the positive curvatures become negative and vice versa. Thus, for adsorption, for Type II isotherms the desorption is spread out and the adsorption is a relatively sharp front, but for Type III isotherms the adsorption is spread out and the desorption is sharp.

Since the first part of Eq. (2.18) cannot be used for the positive curvature, small increments instead are used, as illustrated in Fig. 2.13 on the right (or one could jump straight to the diffusion equation with a sink as mentioned in the footnote).

Some additional equations can be written to describe the behavior. By selecting a particular distance in the bed, at a certain length, d, of x, and time, t, then the amount, $n(x=d,t)$, that is introduced to this position is given by the equation:

$$n(x=d,t) = Y_a v_{(c+a)} A t \quad (2.19)$$

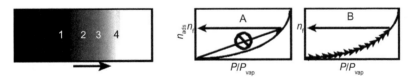

FIG. 2.13

The contrast of a Type III isotherm with the Type II. The unit step function is not allowed but rather small increments of increase happens at the same time that diffusion is present.

---

[9] The second derivative in the diffusion equation must be positive for the "sink" to be positive. This is according to the well known equation: $\partial C/\partial t = \nabla^2 C - S(C)$ where the sink S is often kC. or something similar. If the curvature is negative, then the sink either has to be step function or diffusion is running "up stream." Another way of looking at this is that with a negative curvature the effective diffusion mechanism stops until the bucket is filled.

where X_a is the mole fraction of the adsorbent, $v_{(c+a)}$, is the velocity of the carrier gas including the adsorbent, an A is the cross sectional area of the packed bed.

On the other hand, for any values of x at any particular time, $t=\tau$

$$n(x, t=\tau) = n_{ads} + Ax \qquad (2.20)$$

Thus, at $x=d$ and $t=\tau$:

$$X_a v_{(c+a)} \tau = n_{ads} d \qquad (2.21)$$

Since the average velocity of the front $v = d/\tau$ then:

$$\frac{d}{\tau} := v = \frac{X_a v_{(c+a)}}{n_{ads}} \text{ or } v = \frac{\Delta X_a v_{(c+a)}}{\Delta n_{ads}} \qquad (2.22)$$

where v is referred to as the propagation velocity.

This may not be a familiar equation for some engineers due to the symbolism and the use of moles. According to the symbolism used by Basmadjian [16] this can be written as:

$$V = \frac{G_b}{\rho_b(\Delta q/\Delta Y)} \text{ or } V = \frac{\rho_1 v}{\rho_b(dq/dY)} \qquad (2.23)$$

where:

- V = propagation velocity
- G_b = mass velocity in $kg\,m^{-2}\,s^{-1}$
- ρ_b = density of adsorptive in $kg\,m^{-2}$
- q = mass adsorbate in kg
- Y = mole fraction of the adsorptive
- ρ_1 = density of the carrier gas in $kg\,m^{-2}$
- v = velocity of the carrier gas

The time needed for the front to exit the packed column of length, L, can be calculated from Eq. (2.22). This is referred to as the Breakthrough time, given the symbol t_{bt} here:

$$t_{bt} = \frac{L\Delta n_{ads}}{\Delta X_a v_{(c+a)}} \qquad (2.24)$$

This equation would refer to the case for the sharp front (adsorption Type II, desorption Type III.) The length of time for the desorption "breakthrough" for a Type II isotherm is dependent upon the shape of the isotherm.

Very often Henry's law is used, but now it is known that this might not be reliable, especially if the E_a is high. Under those circumstances, the Henry's Law appears extremely high, whereas, in reality the amount of adsorbate has become zero (in the high vacuum range) even if there is a small residual adsorptive pressure.[10]

---

[10] For example, at liquid nitrogen temperature and a typical 10 kJ mol^{-1} E_a, the slope of the isotherm at the start of the adsorption is 3×10^5 mol atm^{-1} if the P_{vap} is 1 atm. In contrast, the slope at 0.01 atm, which could be the first data point taken, under these conditions is about 4 mol atm^{-1}.

However, this should not be a concern, since that isotherm always starts out with a negative curvature, even for the ultramicropore case, and probably in the high vacuum range. Thus, for a very short distance the line from the origin to the pressure of interest will be above the isotherm. Therefore, one could use a "psudo-Henry's law" arrangement of a line from 0,0 to n_{ads} at some arbitrary low value of p/p_{vap}. Unless the E_a is very low, this will lead to a good approximation of a sharp front. If the E_a is very low, a situation that probably has little application, there will be some noticeable leaking of the adsorptive before the breakthrough occurs.

Unfortunately, the literature lists Henry constants, or something similar, which makes the data suspect, so one needs to investigate more carefully what the various conventions are. The Henry constant has been described in the literature as being "notoriously fickle." This is not at all surprising given the inappropriateness of this concept and the way it is reported. When it comes to practical application, short cuts are almost always used and the Henry constant is an example. It often works well, but there are frequent failures. A further caveat: for reading any literature, one needs to determine what the mathematical derivation is that the author uses and the units used. Therefore, first and foremost, determine what the author means by Henry constant or Henry's law. Here are some of the problems:

- Sometimes the mathematical expression is the n_{ads}/p or maybe $n_{ads}/(p/p_{vap})$.
- Sometimes it is the inverse of these expressions.
- It might be reported in terms of mass instead of moles of adsorbate.
- It might be reported in terms of mass of adsorbent instead of pressure.
- It might be reported in terms of mole or mass fraction of the adsorbent.
- The units vary greatly for adsorbent following: kg, g, mol, mmol mole fraction, mass fraction, etc., even American units.
- Given that this is for a flow system, the adsorbate amount can be transformed to carrier gas values, again with a variety of units.

It should be remembered that Henry's Law is a very old concept and the units and exact mathematical form has gone through many transformations. Therefore, it is wise to investigate carefully how it is used and if the data is appropriate for use. Answer the question: are the units compatible with the mathematical equation you are using? If not, then make adjustments.

For further reading, Basmadjian recommends the book by Ruthven [17] and Yang [18]. He also provides some references from the open literature that could be helpful.

Summary of methods

There are two major method of measuring the isotherm, the gravimetric and volumetric. There are other techniques that measure other properties either instead of these two or injunction with these two. Calorimetric measurements can be done in conjunction, but are more difficult than when performed separately. There are several types of calorimetry and there are several different definitions of heat of adsorption, which will be discussed in a later chapter. Some of the details are discussed.

Another volumetric technique is described for mixed gas adsorption. There is no commercial instrument for the technique. The advantages of this method is that it is relatively fast and could be easily automated.

Details of this chapter are too extensive to summarize. Be sure to read the advantages and disadvantages of the techniques first before deciding on what you will investigate.

References

[1] J. Silvestre-Albero, A.M. Silvestre-Albero, P.L. Llewellyn, F.J. Rodrigues-Reinoso, J. Phys. Chem. 117 (2013) 16885–16889, 1976.
[2] A. Roth, Vacuum Technology, North Holland Publishing Co., Amsterdam, 1976. ISBN: 0-444-10801-7.
[3] J. de Dios Lopez-Gonzolez, F.G. Carpenter, V.R. Deitz, J. Res. NBS 55 (1955) 11.
[4] D.R. Lide (Ed.), Handbook of Chemistry and Physics, 76th ed., CRC Press, Boca Raton, FL, 1995, , pp. p6–48 (or many other editions).
[5] N.F. Carnahan, K.E. Starling, J. Chem. Phys. 51 (1969) 635.
[6] K.A. Thompson, Microbeam Anal. 22 (1987) 115.
[7] L.G. Joyner, P.H. Emmett, J. Am. Chem. Soc. 70 (1948) 2353.
[8] J.B. Condon, Microporous Mesoporous Mater. 53 (2002) 21.
[9] W.D. Harkins, G. Jura, J. Chem. Phys. 11 (1943) 430.
[10] W.D. Harkins, G. Jura, J. Chem. Phys. 13 (1945) 449.
[11] W.D. Harkins, G. Jura, J. Am. Chem. Soc. 66 (1944) 919.
[12] J. Rouquerol, R. Rouquerol, Y. Grillet, R.J. Ward, IUPAC Symposium on the Characterization of Porous Solids, Elsevier Press, Amsterdam, 1988, p. 67.
[13] J. Rouquerol, S. Partyka, F. Rouquerol, J. Chem. Soc., Faraday Trans. I 73 (1977) 306.
[14] F. Rouquerol, J. Rouquerol, K.S.W. Sing, P. Llewellyn, G. Maurin, Adsorption by Powders and Porous Solids, Principles, Methodology and Applications, second ed., Elsevier/Academic Press, Oxford, UK, 2014 ISBN: 9788-0-08-097035-6, pages 49 and 95.
[15] J.R. Arnold, J. Am. Chem. Soc. 71 (1943) 104.
[16] D. Basmadjian, The Little Adsorption Book, A practical Guide for Engineers and Scientists, CRC Press Inc., Boca Raton, USA, 1997. ISBN: 0-8493-2692-3, 1997.
[17] D.M. Ruthven, Principles of Adsorption and Adsorption Processes, John Wiley and Sons, New York, 1984.
[18] R.T. Yang, Gas Separation by Adsorption Processes, Butterworths, London, 1987.

CHAPTER 3

Interpreting the physisorption isotherm

Interpreting the isotherm

The main objective for interpreting of isotherm is to determine the physical properties listed in Table 3.1. The first two physical quantities are minimum parameters to be obtained.

Table 3.1 Specific or molar parameters obtainable from the isotherm

Quantity	Meaning	Interpreted as	Units
n_m	The monolayer equivalence	Surface area, A_s	/mol g^{-1} [a]
$E(\theta)$	Energy of adsorption as a function of θ	Isosteric heat[b], q_{st}	/kJ mol^{-1} [c]
$n_{m,p}$	The monolayer equivalence	Pore area, A_p	/mol g^{-1} [a]
n_{ex}	Non-pore surface area (outer surface)	External area, A_{ex}	/mol g^{-1} [a]
n_p	Amount to completely fill pores	Pore volume, V_p	/mol g^{-1} [a]
$\langle E_a \rangle$	Average energy of first molecule	Energy adsorption	/kJ mol^{-1} [c]
$D(\chi_x, \sigma_x)$	Distribution for adsorbate molecules $x=c$ for initial energy $x=p$ for pores		/mol g^{-1} [a]

[a] of adsorbent.
[b] or some other type of energy.
[c] of adsorbate.

Whatever theory is used to obtain these physical quantities, it should be capable of determining all of these if the adsorbent requires it.

To accomplish the above objectives, one needs to interpret the isotherms using some theoretical formulations which have some solid foundation. Two of the theoretical postulates include two that are classical mechanical, which are very prominent. These are the Bruauer, Emmett, Teller theory (BET), which is the oldest and most often used but least reliable, and various forms of Density Functional Theory (DFT), of which the latest are Non-Local DFT (NLDFT) and Quenched Solid DFT (QSDFT). There are also two nearly identical theories. One is based on quantum mechanics and another, which in reality is the same theory, is based on thermodynamics. These two theories were derived independently [1–3]. The quantum mechanical theory is referred to as χ theory and the thermodynamic theory is referred to as the Disjoining Pressure Theory or Excess Surface Work (ESW) theory. The χ theory has a classical mechanical counterpart, referred to as Auto Shielding Physisorption (ASP) theory, which Dr. E. Loren Fuller used for many years. The χ theory and the ESW theory yield the same equations for the isotherm, and are derivable from each other, lending credence to both.

The NLDFT should be considered a modeling technique and a short description is provided, however, this theory requires an extensive explanation along with other DFT theories. An introductory explanation is provided in Chapter 7 to provide the logic behind the method. A thorough explanation is much too long to be covered in one chapter and the literature is extensive with many forms, modifications, and tweaks. Thus, if one has an interest in this, one needs to refer to the cited literature. A good review has been provided for NLDFT by Ravikovitch et al. [4].

There are other isotherm descriptions of which knowledge is useful. These are various isotherms by Dubinin's group; the Tóth T-equation isotherm, the Freundlich isotherm, and some others are also described in this chapter. These other isotherms do not yield the parameters as described in the list above. The accuracy of the theories are also questionable. This is not an exhaustive list, which would include well over a hundred attempts, including many that are empirical fits in order to express a particular isotherm. Often a good empirical fit is very useful, although not too enlightening. There is some use for purely empirical fits to the isotherm, unfortunately most force the fit to be compatible with Henry's law.

Another technique that fulfills some of the criteria above is the Harkins-Jura Absolute Method. The method yields the surface area and the energy of adsorption with the same measurement. It is very well based upon scientific principles and theory, but it is somewhat limited in choice of adsorbent/adsorbate pairs and it is very difficult (very tricky) to perform the experiments. This might be the reason that it has not been used much.

An alternative to the theoretical approach is the use of standard isotherms. There are several available and some are presented. The principle problem with standard isotherms is that each adsorbent-adsorbate pair has its own standard isotherm. One is not likely to find a standard isotherm of interest in the literature. Even if one does find a standard isotherm that appears to be a match to the adsorbent, it probably isn't. Thus, it is likely that one needs to create a standard isotherm for each particular adsorbate-adsorbent pair that is being addressed. This is only practical if one has to analyze many samples, for example for quality control.

Many of the commercially available instruments have built in calculations based upon NLDFT. However, NLDFT is dependent upon a standard, which is why porous silica is often seen in papers about NLDFT, since there are several standards for silica. Since the silica standard surface area has used the BET for surface area measurements in the past, relying upon these is not valid unless one recalculates the n_m. Such recalculation for some samples has been done in this book. However, some of the data does not look very reliable for such use.

Objectives in interpreting isotherms

It is normally conceded that an interpretation of the isotherm obtained is desirable. Intuitively, one would think that the interpretation of the isotherm would yield some measurement or estimate of the value for those physical properties listed in Table 3.1. Some analyses yield only a value for the surface area and sometimes only for certain isotherm types. In other cases, there are some analysis methods that are most useful for finding the pore volume, but little else.

Unfortunately, none of the conventional methods are reliably based on sound theoretical grounds. Furthermore, to be of practical use, the isotherm should be able to yield the parameters listed in Table 3.1 for an adsorbent of unknown composition. A 2001 quote from Kruk and Jaroniac [5] is appropriate at this point:

However, the practical use of adsorption methods for materials characterization still involves many problems. In particular, a major challenge lies in the identification of suitable procedures of adsorption data analysis among the various methods, (referencing two books on the subject) which in many cases lack proper theoretical background, are inaccurate, inconsistent or even misleading.

Neither author at the time of this writing indicated awareness of the potential of the χ theory or the ESW theory.[1]

To make matters worse, for porous samples one might think that the composition of the pore walls can be mimicked by a non-porous sample of the same chemical composition, or possibly better yet by the external surface of the porous sample. This is not likely to be the case for a variety of reasons. This point is often obscured in the literature with the development of various theories of adsorption. Theories that cannot yield the surface area and adsorption energy independently from some other method are unfortunately of questionable value. Likewise, a theory of adsorption should also not be dependent upon or restricted to the type of adsorbate or adsorbent. For example, a theory that requires a knowledge of the exact nature of the surface atoms and the interactions between these atoms and the adsorbate might yield some insight into the adsorption process but it has little practical predictive power or use. Again the reason for this is usually that the exact nature of the adsorbent surface is unknown.

There are several isotherm interpretations available. The most widely used is the BET [6] (Brunauer, Emmett, and Teller) and its various modifications including the BDDT [7] (Brunauer, Deming, Deming, and Teller). The historical advantage of the BET is that the theory itself can claim to predict a value for surface area. Another widely used isotherm transform, especially for porous material, is the DR [8, 9] (Dubinin-Radushkevich) transform. A modified theory of the latter, the DK

---

[1] I have had discussions with many investigators about this, including one of these authors, who had no knowledge of χ, Disjoining Pressure or ESW theories. Of the perhaps hundreds of investigators I have only found two who had heard of one of these theories. There seemed to have been no interest even in investigating them.

[10] (Dubinin-Kaganer) applies to non-porous surfaces. There seems to be good theoretical reasons that the prediction of the pore volume is theoretically sound in the DR isotherm. The assumption of the DK isotherm is that it yields the surface area; however, this seems to be speculation. There are several other theories of adsorption from the Dubinin group. Standard curves are more useful, especially if one is interested in porosity, although most depend upon a calibration by some other interpretation, again usually the BET. If one has confidence in the standard curve used, then the pore volume can be measured by a subtraction mechanism.

However, the pore radius requires a good answer for the surface area inside, and outside, the pores. Nevertheless, having the pore volume is a useful piece of information by itself. These standard curves include the α_s-curve (see Sing [11]), the t-curve (see deBoer et al. [12]), the Cranston and Inkley standard [13], the KFG [14](Karnaukhov, Fenelonov, Gavrilov) standard fit, and others. The theories based upon density functional theory (DFT) and Monti Carlo simulations had appeared to have had promise, but at the moment must be classified, at best, as a method of generating standard curves. Several have been developed, but so far all require calibrations and are dependent upon the specifics of the adsorbate and adsorbent and knowledge of what the surface area is.

Another isotherm interpretation, the χ curve along with the disjoining pressure theory (or ESW), will be reintroduced here both theoretically and practically, as an analytical standard curve. These theories are self-contained and need no other information such as a calibration with the BET theory.

It is first instructive to look at the general form of a typical isotherm. The general shape of the overall adsorption isotherm curve for the simplest (Type I) case of physisorption may be seen in the upper left graph of Fig. 3.1. The curves simulate three different Type II isotherms. These simulations fit some standard isotherms. Historically, the monolayer was selected as approximately at the position of the "knee" of the isotherm. This position is indicated roughly where the arrow is pointing. In Chapter 1 there was some discussion as to why there is uncertainty in this knee

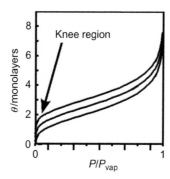

FIG. 3.1

Some typical adsorption isotherm for non-porous materials illustrating the problem of identifying the "knee" due to scaling.

position. This selection was in analogy to the Langmuir isotherm. However, the Langmuir isotherm flattens off to a constant, but since the physical adsorption isotherm does not flatten off, a different criterion had to be selected. Judgement had to be made as to where this knee was, but it was roughly at about 0.03 to 0.1 of the vapor pressure. It turns out that for many materials studied at the time this value gives the equivalent of a monolayer within about a factor of 4. The problem, other than this latter inaccuracy, is that with this approach the shape of the curve in the low pressure range is nearly invariant with scale. Thus if one uses a different scale, say the isotherm from 0 to $0.1 P_{vap}$ instead of 0 to $1 P_{vap}$, one gets a different position for the knee. This was illustrated in Chapter 1 with Fig. 1.9, a Type II isotherm, with the three different magnifications of these curves. To fix this problem, a nonbias analytical method was pursued. Several equations were constructed to describe these isotherms, some of which will be reviewed here.

The natural tendency was to seek an equation that could fit the obtained isotherms fairly well and yield an answer for the surface area.[2] Several equations were available that fit many isotherms but did not yield the surface area or the energies involved. Until recently, the only known equation that could provide an answer was the BET equation. The development of the BET equation was a step forward from the estimation technique, but in the long term it might have been a mistake. Since it was the first theory to claim to arrive at an answer for the surface area, it quickly became set as the correct and only theory. It also didn't hurt to have three famous scientists promoting it. The following discussion is obviously not all-inclusive and the reader is referred other texts, such as that by Rudzinski and Emmett [15], Adamson [16], or Hiemenz [17] for more information.

It would be hoped that within some of the isotherm equations there exist parameters that are identified with some physical quantity such as surface area or pore volume. To extract these parameters, a least square routine of some sort may be used to determine the values. Some isotherm equations such as the BET, for surface area, and DR, for pore volume, restrict the range over which the fit is valid. This range is unfortunately a matter of judgement and such words as "over the linear range" are often used in the literature. In recent years, there has been a general codification for the BET equation to use the range of pressure of 0.05 to 0.35 of the saturation pressure, which is the pressure one would observe over the bulk liquid, P_{vap}. This works well for some ceramic materials but unfortunately poorly for most organic materials. In order to make a judgement as to what the linear range is, one must plot a transformed set of equations. In Figs. 3.2–3.4 are some examples of some transformed plots; the BET and the DR are fitted to data used for a standard silica plot. To obtain the surface area using the BET method, the slope of the transform ordinate versus P/P_{vap} is required.

---

[2] There are many problems converting the monolayer equivalents to surface area. However, I'm using the words "surface area" here rather than the more precise "monolayer equivalents" since that is what is usually referred to.

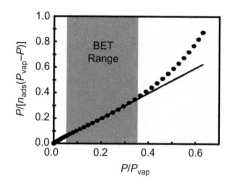

FIG. 3.2

The transformed BET plot to determine surface area typical of silica material. Linear range is assumed to be 0.05 to 0.35 of P_{vap}.

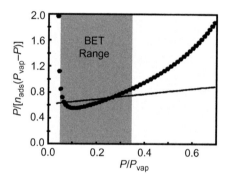

FIG. 3.3

The transformed BET plot for an organic material. The 0.05–0.35 range yields a very poor linear fit, thus a high range should be selected.

For the DR, the transformed plot should yield the amount to fill the pores, n_p. This plot has a very long linear region. The intercept of the ordinate (the $\ln(n_{ads})$ axis) should be an indication of the number of moles needed to fill pore volume, n_p. For the DRK case, it was postulated that it is an indication of the number of moles in a monolayer, n_m. This, however, was speculation, which turns out not to hold up.

In the case of the BET, a linear portion of the curve for the high energy surfaces such as silica and alumina is at about 0.05 to 0.35 of P_{vap}. There are extensive qualifications for this range as noted by Sing [18]. For example, for lower energy surfaces this does not work well for any range of P/P_{vap} as illustrated in Fig. 3.3. In contrast, the DRK-transformed plot usually does have a range that can be easily fitted.

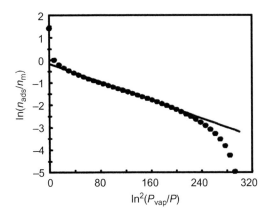

FIG. 3.4

A linear fit to the DRK representation of the adsorption isotherm for a non-porous surface. The fit covers about 2/3 of the $\ln^2$ range.

Almost all of the relative measurements, such as the "standard curves," refer back to the BET surface area measurement. One might say offhand, "What is the point of using the standard plots then, when one could simply use the BET to begin with?" There are two principal reasons to use standard curves. Firstly, one can use them when only a relative answer is needed, for example comparing two samples or for quality control. Secondly, it is generally agreed that the full isotherm contains other valuable information—particularly the mesoporosity and the microporosity. By a comparison with standard curves, which are (hopefully) characteristic of non-porous materials, one can deduce some measurements of porosity and possibly other properties. As you will become familiar with the χ plot, it will become obvious that using the standard curve with the χ transformed abscissa with a standard plot is much more powerful, even if one does not use the χ theory to do the analysis.

From the above discussion, it should be obvious that it is chancy to go on automatic when analyzing adsorption isotherms for the relevant physical quantities. Blindly using the answers that come out of the automatic analyzing instruments is extremely ill-advised. The programs in these instruments today are a reflection of the instrument manufactures bias and the field is still too unsettled to be able to include an unbiased program. In the following section, some more details are presented to enable one to extract some meaningful quantities from the isotherm. Remember, if the material is for publication, leave the raw data available for others to use. It is noteworthy that in the years before the BET theory was accepted, this was the tradition, and the early 20th-century data is still useful and available today.

The interpretation for the adsorption of more than one adsorbate has not been settled upon but the calculation made possible by χ theory is presented in the next chapter as an advanced subject. There are several equations and interpretations in the literature, all of which have either a weak foundation or are simply empirical for the materials at hand. This is fine and may be appropriate for organizing information for the moment, but should not be relied upon for predictions.

First, some analysis methods which cover most of the practical applications for physisorption are presented. The following, then, is a quick description of how to analyze the isotherm of the adsorption of one adsorbate.

Surface area determination from isotherms

There are two methods of obtaining the surface area from the isotherm for an adsorbent with unknown surface character. These two are the BET method and the χ/ESW/ASP theories method. Other theories either need the surface composition specified or use the BET as the basic equation to analyze the surface area. The BET is widely used and has been available since around 1938. Since this analysis is so widely used, much of the information available for materials refers only to the BET surface area. Sometimes the original data for the isotherms have been lost. Therefore, it is important to know what the BET is and interpret the isotherm information given in terms of the BET. Perhaps in the future, this knowledge will no longer be necessary and all the data that has been collected and not recorded so far will have been repeated. For now, a familiarity with this method is advised. The methodologies for both the BET and χ/ESW/ASP methods are presented here and the theories behind them are presented in a later chapter.

The BET analysis

The original form for the BET equation is:

$$\frac{V}{V_m} = \frac{C_{BET} x}{(1-x)[1+(C_{BET}-1)x]} \quad \text{where}: x = \frac{P}{P_{vap}} \quad (3.1)$$

Here V indicates the volume of gas adsorbed at STP, V_m the volume of gas[3] (STP) that is required for a monolayer, P_{vap} is the vapor pressure of the bulk liquid at the same temperature, P is the adsorptive pressure and C_{BET}, the BET constant, is an equilibrium constant. For analysis, the equation rearranges into the transformed form:

$$\frac{x}{V(1-x)} = \frac{1}{C_{BET} V_m} + \frac{C_{BET}-1}{C_{BET} V_m} x \quad (3.2)$$

---

[3] $\overline{V}$ is the new IUPAC symbol for molar volume so V_m may be used for monolayer equivalent volume at STP. The symbol $\overline{M}$ will be used as the symbol for molar mass in $g\,mol^{-1}$.

The general approach to using transformed equations and the BET in particular is as follows:

1) Rework the data according to the transform required. In the case of the BET analysis, this means that the dependent variable (ordinate or computer y) will be:

$$y = \frac{x}{V_m(1-x)} \tag{3.3}$$

The independent variable (abscissa or computer x) is x $(=P/P_o)$.

2) Plot the x-y data and determine the slope and intercept over the region that appears as a straight line (for repeated experiments, be sure to use the same region for consistency). Many spreadsheets have linear regression analysis built in, but be sure to properly specify the range.

3) Equate the determined values of the slope, S_{BET}, and intercept, I_{BET}, with the expression for the slope and intercept in the transformed equation. Thus for the BET analysis:

$$S_{BET} = \frac{C_{BET} - 1}{C v V_m} \text{ and } I_{BET} = \frac{1}{C_{BET} V_m} \tag{3.4}$$

4) Solve for the parameters of interest from these slopes and intercepts. For the BET V_m

$$V_m = \frac{1}{S_{BET} + I_{BET}} \tag{3.5}$$

and from the from the monolayer equivalent volume at STP, V_m, C_{BET} is obtained so:

$$C_{BET} = \frac{1}{V_m I_{BET}} \tag{3.6}$$

5) Relate the parameters obtained to surface area or other physical quantities. For the BET, the V_m is equated to the number of moles of a monolayer. In the case of N_2 and Ar adsorption, IUPAC has set a conversion factor for this. To convert this number to a surface area number, the IUPAC convention settled on a number of 16.2 Å^2 (0.162 nm^2) per nitrogen molecule as a standard. The origin of this number is from a recommendation by Emmett and Brunauer [19]. This recommendation used an equation relating the effective molecular cross sectional area, a, to the liquid density, ρ, and the molar mass, $\overline{M}$

$$a = 1.091 \left(\frac{\overline{M}}{N_A \rho}\right) \tag{3.7}$$

The constant 1.091 is referred to as the packing factor. Unfortunately, according to Pickering and Eckstrom, a depends upon the adsorbate and adsorbent [20]. Furthermore, according to Emmett [21] it is also a function of C_{BET}. Very often the parameter C_{BET} is not reported, which further destroys the data recovery.

In the above analysis, it may be the number of moles, n_{ads}, adsorbed rather than volume being reported. What is reported out may also be in terms of per gram of sample, which is the normal method of reporting. If is not in per gram, then the sample mass should be specified. All the equations stay the same, with n_m number of moles in a monolayer reported out. V_m is usually reported in standard mL per gram sample (so is n_m, i.e., specific quantities) so to convert to moles:

$$n_m = \frac{V_m}{22400} \tag{3.8}$$

and to specific surface area in $m^2\ g^{-1}$, with the sample mass in grams, is given as:

$$A_s = n_m a N_A \tag{3.9}$$

or using $\overline{A}$, the molar area:

$$A_s = n_m \overline{A} \tag{3.10}$$

χ plot analysis

For review, the following physical quantities are defined:

$$\theta = \Delta\chi \mathbf{U} \Delta\chi,\ \Delta\chi = \chi - \chi_c\ \chi = -\ln\left[-\ln\left(\frac{P}{P_{vap}}\right)\right]\ \text{and}\ \chi_c = \ln\left(-\frac{\overline{E}_a}{RT}\right) \tag{3.11}$$

With the definition of $\theta = n_{ads}/n_m$ the basic equation for the χ theory in terms of specific moles adsorbed, n_{ads}, is:

$$n_{ads} = n_m \left(-\ln\left[-\ln\left(\frac{P}{P_{vap}}\right)\right] - \ln\left(-\frac{\overline{E}_a}{RT}\right)\right) \tag{3.12}$$

for the flat surface. The quantity n_m is the monolayer equivalence. See Fig. 3.5 to clarify what the Ss and Is designate.

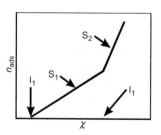

FIG. 3.5

Simulated 2 energy χ-plot and the meaning of the Ss and Is.

If one can assume values for a or $\overline{A}$ to obtain the surface area by Eq. (3.9) or Eq. (3.10), the equivalent steps in the analysis are, with y the ordinate and x the abscissa now labeled "χ":

1) Convert the amount adsorbed into moles, n_{ads}, if not already done.
2) The transformed equation is rather simple. Use as $y = n_{ads}$ and $\chi = -\ln(-\ln(P/P_s))$.
3) Plot the transformed data. This may give more than one straight line segment. If so, refer to later sections as to what the meaning of these segments is. At any rate, if there is more than one straight line segment, analyze each separately.
4) Obtain the slope, S_i, and ordinate (y) intercept, I_i, for the lowest segment. The slope yields the molar monolayer equivalent according to:

$$S_i = n_m \tag{3.13}$$

And the ordinate intercept which is given by the abscissa intercept, χ_c, by:

$$\chi_{c,1} = -\frac{I_1}{S_1} \tag{3.14}$$

This yields $E_{a,1}$ by substitution into:

$$E_{a,1} = -RT \exp(-\chi_c) \tag{3.15}$$

5) These parameters are the important ones, the monolayer equivalents and the energy of adsorption of the first adsorbate molecule. To obtain the surface area, use your favorite conversion factor. For example calculate:

$$\overline{A} = \frac{F_{2D}(\overline{V})^{2/3}(N_A)^{1/3}}{F_{3D}^{2/3}} \tag{3.16}$$

where $\overline{V}$ is the molar volume, N_A is Avogadro's number and Fs are the 2D and 3D packing factors.

The value of $\overline{A}$ for nitrogen is $8.97 \times 10^4 \, m^2 \, mol^{-1}$ and for argon is $7.90 \times 10^4 \, m^2 \, mol^{-1}$ at their normal boiling points. See Table 1.8 for other choices.

6) If there are several segments, the surface areas for each segment, $A_{s,i}$ is given by:

$$n_{m,i} = (S_i - S_{i-1}) \Rightarrow A_{s,i} = \overline{A}(S_i - S_{i-1}) \tag{3.17}$$

where $S_0 = 0$, i.e., there is no segment "0."

At the break between linear segments, read the values of χ which be designate I_i from and calculate the slope by the usual method. The $E_{a,i}$s are given by Eq. (3.15).

Each $E_{a,i}$ is interpreted as the energy of adsorption for the ith type of surface. The total area upon which there is adsorption is the sum of surface areas starting with the lowest in value of χ ($=-\ln(-\ln(P/P_s))$) and summing as the segments, $A_{s,i}$, appear, provided there is no negative curvature in the χ plot. If there is a sudden large increase followed by a sudden decrease to a slope of nearly zero, this is an indication of mesoporosity and needs special treatment.

Strategy of the non-local density functional theory

The calculations involved for the density functional theory (DFT) and the non-local density functional theory (NLDFT) are described in the last chapter. Basically, the molecules and the surface are assigned a size and a potential energy curve (a Leonard-Jones potential) and the configuration assumed by the adsorbate molecules is obtained by minimizing the energy release under whatever conditions are specified. The technique is used primarily for pore size determinations, often in combination with a BET analysis. Another technique which follows the general philosophy is the Monte Carlos technique, that is, a model of the adsorption is proposed and calculations are made to see if a match is obtained with the experimental observations.

The disadvantages of both these techniques are the computational difficulty involved; however, with computers this can be easily done (although there is no guarantee the proper answer will be arrived at for variety of reasons). It would seem reasonable that one could make enough standard NLDFT curves for matching, to cover all isotherm cases and store them, then a comparison can be made to get a match to the isotherm. Indeed, this is roughly the approach: they create Kernels that plug into the integral equations—more about this in Chapter 7. Thus, in Fig. 3.6 is the procedure for modeling.

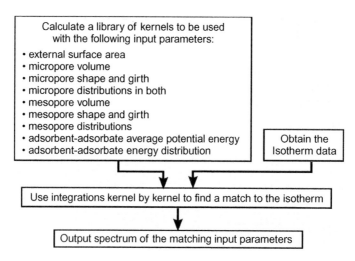

FIG. 3.6

A schematic of the method used for DFT (NLDFT and QSDFT).

The energy of interaction between adsorbate molecules is taken from the literature as a Leonard-Jones potential for the liquid state match. Between the surface and the adsorbate molecule the potential is guessed, or modeled with a non-porous surface, which does not seem to work well. For the external surface a standard curve is used. The modeling needs to use a non-porous material of similar composition and assumes that the porous material has identical underlying isotherms as an accurate description. This assumption, although reasonable, has not been demonstrated to be reliable.

Obviously, with so many parameters to match, this is quite a calculation job. Furthermore, these depend upon "standard kernels"—that is, a kernel is the model of all the parameters—the number of these kernels must run in the thousands and possibly millions. Consider only the adsorbent material. This alone could be in the thousands. Creating standard for everything seems like a very large task indeed. If this could be calculated fast enough, with a minimum search method one should be able to hone in on an answer that would match. There exist fast de-convolution programs [22] that can match the create models, but so far for silica adsorbent only.

Strategy of the χ/ESW theory

Consider, now, the χ theory method for a porous sample. Using only one adsorbate, and not a mix of adsorbates, the χ method follows that given in Fig. 3.7.

Notice that the χ theory method strategy is opposite to the DFT method. It starts with the isotherm or calorimetric data, and analyzes the adsorbent geometry and the energies involved. It needs no models nor energy assumptions for comparison, nor any standard. DFT starts by creating models that might fit and fits the isotherms to the models, making assumptions about the range of the parameters.

Of course, both methods require the interpretation being based on the dimensions assumed for the effective volume and area of an adsorbate molecule. DFT requires

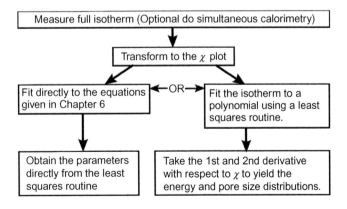

FIG. 3.7

Model of the χ method of analysis for physical adsorption isotherm.

these to model up front and everything is dependent upon these assumptions. χ theory, however, yields useful information without these parameters. The information yielded includes the moles of the monolayer (monolayer equivalents) both outside the pores and in the pores, the amount of material in the pores, the distribution associated with this, and the energy distributions. The effective molecular volumes and areas are needed then to convert these mole amounts to physical dimensions of length, area, and volume.

There have been some attempts to standardize the interconversion from moles to dimensions by comparing the results with other measurements, especially with X-ray pore sizes. However, this conversion question should be considered still to be an open or even an irrelevant question.

Dubinin et al. method of determining surface area

It is questionable as to whether the various isotherms attributed to Dubinin and coworkers yield the surface area. These theories have been described as "semi-empirical" (this is similar to the deBoar-Zwitter theory. An excellent fit was found and then an explanation was devised without a theoretical background). They are definitely useful for finding the mesoporosity volume due to the clear linear extrapolation. According to Kaganer [10], the intercept of the Dubinin- Radushkevich (DR) equation is the monolayer amount. This seems to have been empirically based upon the BET formulation. The modified DR equation, referred to as the DRK equation, for a flat surface is:

$$\ln\left(\frac{V}{V_0}\right) = A \ln^2\left(\frac{P_s}{P}\right) \qquad (3.18)$$

A plot of $\ln(V)$ versus $\ln^2(P/P_s)$ yields a plot which is linear over a fair range of values.

A typical DRK plot has been presented in Fig. 3.4. It has been demonstrated [23] according to χ theory that indeed the value of V_0 should be proportional to the monolayer coverage but not equal. The problem with it being only proportional is that the constant A is given by an equation that has in it unknown quantities, referred to as the "characteristic energy," the "structural constant," and a "scaling factor." One of the problems with this formulation is that both porosity and surface area are dependent upon the intercept value. In other words, there is no way to separate the two physical quantities in this case. Therefore, if one has a sample that is porous and has a significant external area, then the separation of these two physical quantities is not possible.

The methodology for the DRK calculation is as follows:

1) Use $y = \ln V_{ads}$ or $y = \ln n_{ads}$ and use $x = (\ln(P/P_s))^2$.
2) Plot the ys as function of the xs and draw the best estimated fit for the portion that is the most linear. This would be roughly through point at the inflection point of the curve and should cover about 2/3 of the plot.

3) From the intercept ($y=0$) of the plot obtain $\ln(V_m)$ or $\ln(n_m)$.[4]
4) Convert to monolayer coverage or area as was done for the BET.

One of the big advantages of this equation is that the linear portion is very much broader than the BET range. It has the advantage that it does not predict Henry's law, something only a few investigators have noticed and have tried to "fix" in some arbitrary fashion.

Tóth T-equation isotherm

Another treatment that matches the experimental results for many adsorbents is the Tóth isotherms [24], referred to as the T-equation [25]. The basic equation strategy:

$$n_{ads} = n_m \left(1 + \frac{1}{K}\right)^{1/m} \frac{P}{P_s} \left[\frac{1}{K} + \left(\frac{P}{P_s}\right)^m\right]^{-1/m} \left(1 - k\frac{P}{P_s} + kP_{r,e}\right)^{-1} \qquad (3.19)$$

where n_m, K, m, k, and $P_{r,e}$ are fitting parameters, which were designated by the derivation.

The Tóth fitting parameter $P_{r,e}$ is a low relative pressure value and can be ignored with a small amount of distortion. This equation can be rearranged somewhat to yield a simpler looking equation; however, with five fitting parameters probably the best approach is to simply set up a minimum search routine.

It is not clear how useful this equation is, although it is claimed that the parameter n_m yields the monolayer coverage value. The values obtained from this have been compared favorably with the BET values and, with some reservations, the same surface area value is obtained regardless of the adsorbent.

One of the basic assumptions for the theory behind the T-equation is the validity of Henry's law and the notion that the virial equation is a thermodynamic requirement. This latter assumption is approximately correct for many situations but is strictly incorrect.

The Harkins-Jura absolute/relative method

Harkins and Jura [26] describe a method to obtain the absolute surface area of a solid by the following method (see Chapter 2 for more information). The powder is exposed to a high vapor pressure of water. It is best that it be exposed to it in a high sensitivity calorimeter over a reservoir of water. The powder is then allowed to fall into the reservoir and the amount of heat produced is measured. By doing so, one eliminates the outer surface of the adsorbed film, releasing the energy associated with the liquid-gas interface surface tension. Since the liquid-gas surface tension energy is known, one may then calculate from the amount of heat released. This heat

---

[4] It's unclear as to why this should be true.

is an indication of what the area of the powder is (or at least the outer surface area of the adsorbed film before immersion).

The principal problem with this technique is the difficulties involved experimentally. Assuming that these are overcome, there are still the following questions: are the particles well dispersed after immersion? And is there significant porosity or bed porosity in the sample that would lower the observed area due to capillary action? Both of these questions were addressed by Harkins and Jura [27]. One of the unforeseen problems is the variation in heats of adsorption with coverage. Thus, if the adsorbed film thickness is too thin then there will be an additional heat due to the additional adsorption. This can be accounted for by measuring the heat of adsorption as well.

This seems like a simple method and conceptually it is. However, those who have performed calorimetry, especially for physisorption, know full well that such a method is experimentally very difficult and tricky, with many pitfalls and compensating calculations that are needed. This is definitely not recommended for the novice.

Porosity determinations from the isotherm

There are three classifications of porosity. Officially the IUPAC has classified these according to pore diameter as follows:

- below 2 nm: micropores
- between 2 nm and 10 nm: mesopores
- large than 10 nm: macropores

In this section a looser definition will be used.

- Micropores will be the smallest of the pores, which do not cause any positive curvature in χ plot. In other words, in a standard comparison plot, there will be no positive deviation in the plot. Micropores cause, by definition, only negative curvatures in the standard plot.
- For mesopores, there exists a positive deviation due to pore filling, usually referred to as capillary filling, in the intermediate to high end of the standard or χ plot. This increase is then followed by a decrease in the slope to a value less than the slope before the capillary filling. This may or may not be associated with hysteresis.
- Macropores are pores for which the capillary filling is at such a high pressure that it is not practical to observe it on the isotherm. The *official* definition might change as more sensitive instruments become available.

The possibility of a change in the boundary between mesopores and macropores is very likely. Furthermore, the functional definitions presented here and the IUPAC definition may not always coordinate. Another point to remember is that the IUPAC definitions are geared to N_2 adsorption and there is no reason to presume that other adsorbates, for example SF_6, which is much larger than N_2, should behave similarly.

There is another definition for "ultramicropores" that has been used in the literature. There is no official definition. In this book the definition is:

- Ultramicropores are pores present if the isotherm follows the log law, that is, a linear plot of n_a versus $\ln(P/P_{vap})$. This indicates that the adsorption is restricted to a monolayer.

Of course, there are pores that border on the two of these types, especially for the ultramicropores.

Approximate pore analysis method

Exactly how to calculate porosity from the isotherms is a matter of much discussion at this time. The following is one method of interpreting the isotherm with respect to porosity. It is a more detailed and advanced method than presented in Chapter 1, and an even more advanced method and a modeling technique is described in Chapter 6. Therefore, these methods are tagged here as being "approximate." The proposed method in Chapter 6 is based on the 1st and 2nd derivative of the χ plot and a non-linear least squares routine for fit the χ plot. The advanced method is to be used if there is a possibility that the energy distribution and the pore size distribution overlap to some degree. NLDFT is also a possibility for modeling now that it is known how to get a handle on what the monolayer equivalence is from the sample itself.

The analysis is somewhat similar to the Gurvich rule, with the very important difference that the extrapolation is not the intercept of the P_{vap} but rather using a χ-plot, with $\Delta\chi$ as the abscissa, and extrapolating to the ordinated axis. This extrapolation can obviously only be made with a χ plot. This back extrapolation avoids the problem of the contribution of the external surface adsorption creating too high a value for the porosity. It can also apply to more than one distinct pore size, although these may be too close to make a good separation. Very often, however, the adsorption on the external surface area is small compared to the amount adsorbed within pores and the two methods should agree.

Micropore analysis

Fig. 3.8 shows some typical data which might indicate microporosity, although one needs to test for ultramicropores. The data used are from Goldman and Polanyi [28] for CS_2 adsorption on activated charcoal. Not much can be deduced from this isotherm as presented. A transformation of the plot, in Fig. 3.9, using a standard isotherm in the form of the χ plot begins the process.

In the following analysis, r_p is either the radius of a cylindrical pore or, for slit pores, the distance half way from one pore wall to the opposite wall. This is related to the intersection of the two lines in Fig. 3.9, as described in the following procedure.

CHAPTER 3 Interpreting the physisorption isotherm

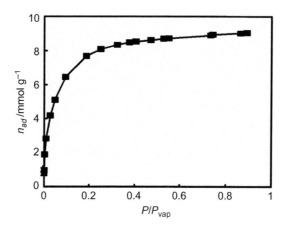

FIG. 3.8

A typical type I adsorption of possibly microporosity isotherm.

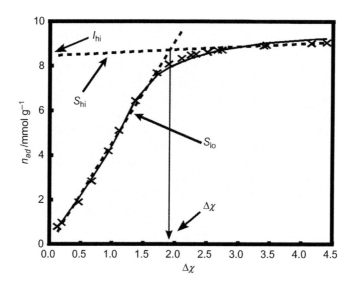

FIG. 3.9

The transformed plot using a standard curve to change the x axis.

1) Transform the abscissa by converting the pressure reading to χ and subtracting χ_c (the abscissa intercept) from χ so the abscissa values become $\Delta\chi$. This is the analytical [29] χ plot used for the standard plot since a reliable carbon standard is not possible.
2) Fit the high values to a straight line and fit the low values to a straight line. These are labeled in the figure as S_{lo} and S_{hi}.

3) Extract the slopes from the high and low lines and the ordinate intercept from the high line.
4) The slope value, S_{lo}, is the monolayer equivalence of the pores plus external area and S_{hi} is the monolayer equivalence of the external area.
5) The value of the ordinate intercept, I_{hi}, is the moles contained in pores.
6) The abscissa value of intersection of the two lines, $\Delta\chi_p$, is the cut-off for the pore filling in terms of "layers."

Example: Fig. 3.9 is an example using data from table #15 by Goldmann and Polayni. The dash lines are the two linear fits. (The solid curved line is the fit to the more advanced Chapter 6 equation.) The parameters extracted from the linear fits are:

$S_{lo} = 4.57 \, \text{mmol g}^{-1}$,
$S_{hi} = 0.046 \, \text{mmol g}^{-1}$ (7.2%)
$\Delta\chi_p = 1.81\text{–}1.90/\text{monolayer}$
$I_{hi} = 8.3\text{–}8.8 \, \text{mmol g}^{-1}$
($\chi_c = -2.217 \Rightarrow E_a = 22.4 \, \text{kJ mol}^{-1}$)

There is some uncertainty in the value of S_{hi}, as is obvious from the graph. In the more advanced treatment of Chapter 6, this calculates to 0.0%. This uncertainty is reflected in the value of $\Delta\chi_p$ as well.

These numbers above should be useful by themselves, but may be philosophically unpleasant. Conversion to the physical quantities or length, volume, and area is more intuitive, although possibly inaccurate. So, using the conversion factor from Eq. (3.16) and the standard curve calibration (or for χ, Eqs. 3.11 and 3.12) with the molar volume $\overline{V} = 6.02 \times 10^{-5} \, m^3 \text{mol}^{-1}$ and molar area $\overline{A} = 1.41 \times 10^5 \, m^2 \text{mol}^{-1}$ the following areas and volumes are obtained:

$A_{total} = S_{lo} \overline{A} = 644 \, m^2 \, g^{-1}$
$A_{ex} = S_{hi} \overline{A} = 6.5 \, m^2 \, g^{-1}$ (A_{ex} includes both the wall edges and the pore openings.)
$A_p = A_{total} - A_{ex} = 638 \, m^2 \, g^{-1}$
$V_p = I_{hi} \overline{V} = 5.0 \times 10^{-7}$ to $5.3 \times 10^{-7} \, m^3 \, g^{-1}$
$r_p = \Delta\chi_p \overline{V}/\overline{A} = 0.77$ to $0.81 \, \text{nm}$

Use the following formulas for cylindrical and slit pores. For slits:

$$r_p = \frac{V_p}{A_p} \qquad (3.20)$$

and for cylindrical pores:

$$r_p = \frac{2V_p}{A_p} \qquad (3.21)$$

In other words, for the same values of r_p and A_p, there is twice the volume for slits. Thus the answer from A_p and V_p the above example is:

$r_p = 0.39$ to $0.42 \, \text{nm}$ for cylindrical pores.
$r_p = 0.78$ to $0.83 \, \text{nm}$ for slit-like pores.

The answer from the intersection of the high line and low line yields:
$r_p = 0.77$–0.81 nm which is in very good agreement with what is expected for slit pores. In this case the more sophisticated analysis of Chapter 6 yielded:

$A_p = 694\,\text{m}^2\,\text{g}^{-1}$,
$V_p = 5.62 \times 10^{-7}\,\text{m}^3\,\text{g}^{-1}$ and
$A_{ex} = 0.0\%$ and a distribution of energy/pores of
$\sigma = 0.186$
(also $\chi_c = -1.977 \Rightarrow E_a = 17.6\,\text{kJ}\,\text{mol}^{-1}$.)

Due to lack of low pressure data the separation of the energy and pore distributions is not possible. There are a couple problems with the method described here:

1) The distributions are not taken into account which also means that
2) the estimate of where the linear portion for the high part of the curve is uncertain.

A more sophisticated analysis has been presented in the literature [30] and in Chapter 6. Using the literature method, which accounts for several other factors, the answer is $r_p = 0.97$ m, which might indicate that the additional effort is not worthwhile. The problem is similar to the questions raised with respect to the surface area, i.e., what is the meaning of pore radius for adsorption? Is it possible that the parameters extracted in unworked form above in terms of (mmol g^{-1}) are the really important parameters?

Mesoporosity analysis

The following is the simple technique to calculate the mesoporosity. Again, a more sophisticated analysis exists and is presented in Chapter 6, but does not seem to be that great an improvement.

Fig. 3.10 presents data by Krug and Jaroniec [31] (KJ) of Ar adsorption on an MCM-41 porous silica material. This figure illustrates the parameters to extract from the χ plot (a plot of n adsorbed versus χ value). The analysis using the χ theory is similar to that used for standard curves such as the α-s [32, 33] except a standard from a similar non-porous material is not necessary. This analysis is referred to as the "simple mesopore analysis." A more detailed analysis presented in Chapter 6 yields the line that nearly follows the data in Fig. 3.10. This detailed analysis made a small correction for the energy of adsorption due to the radius of curvature, which might be the reason for the difference in the answers. The following symbols apply to this simple analysis and to the detailed analysis:

Least squares regression fit to the two linear portions used to yield the following:

S_{hi} = the slope of the higher line,
S_{lo} = the slope of the lower line,
I_{lo} = the abscissa intercept for the lower line of χ plot,

Approximate pore analysis method

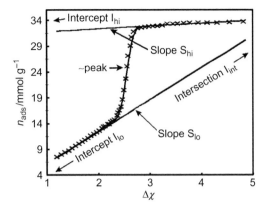

FIG. 3.10
Preliminary parameters obtained from the isotherm to analyze surface area and porosity.
Data by KJ.

I_{hi} = the ordinate intercept for the higher line of the $\Delta\chi$ plot,
I_{int} = the intersection of the higher and lower lines on the $\Delta\chi$ plot.

In Table 3.2 are the parameters extracted from Fig. 3.10. A comparison is made here between an advanced detailed method and the simple method. There are two advanced methods presented in Chapter 6. The one used here inserts a PMF (Probability Mass Function) for the mesopores. The other depends upon obtaining the 1st and 2nd derivative of the isotherm.

To convert these parameters to geometrical meaningful values in Table 3.3, we use the IUPAC convention, which uses the liquid density and both packing factors, F_{3D} and F_{2D}. The quantities calculated from these parameters are:

A_s: the total surface area
A_{ex}: the external surface area
V_p: the pore volume

Table 3.2 Parameters from simple mesopore analysis

Parameter	Determination
S_{hi}	0.19 mmol g^{-1}
S_{lo}	6.41 mmol g^{-1}
$S_{lo} - S_{hi}$	6.22 mmol g^{-1}
I_{lo} ($x=\chi_c$)	−1.998
$I_{hi}(y)$	32.7 mmol g^{-1}
$I(\chi,y)$	(3.26, 33.66 mmol g^{-1})
$\therefore \Delta\chi_p$	3.29 + 1.98 = 5.25

These parameters were extracted from data by KJ of adsorption of Ar on MCM-41.

Table 3.3 Geometric analysis comparing the simple and advanced mesopore analyses

$A_s = 506 \, m^2 \, g^{-1}$
$A_p = 491 \, m^2 \, g^{-1}$
$A_{ex} = 11 \, m^2 \, g^{-1}$
$V_p = 9.35 \times 10^{-7} \, m^3 \, g^{-1}$
#1 $r_p = 3.81$ nm cylinders
#1 $r_p = 1.90$ nm slits
$n_L = 5.25$ monolayers
#2 $r_p = 1.90$ nm
$\sigma_2 = 0.105$

Data by KJ for Ar adsorbed on MCM-41.

r_p: the pore radius
n_L: number of "layers" or partial "layers" possible in the pore.
σ_2: the distribution standard deviation which is a combination of:
  σ_c: the distribution standard deviation of the adsorption energies and
  σ_p: the distribution standard deviation of the pore sizes

according to the non-correlation equation:

$$\sigma_2 = \sqrt{\sigma_c^2 + \sigma_p^2} \qquad (3.22)$$

There are 2 r_ps listed. #1 is calculated from the pore volume and the area and #2 is calculated from n_L, the liquid density and the packing factor F_{1D}. The answer for n_L comes from the $\Delta \chi_p$. Calculation of r_p for #2 assumes all n_L layers are complete. Therefore, the difference between #1 and #2 in the more advanced calculation is related to the pores fractal factor. The match between #1 r_p for slits is an excellent match for #2 r_p indicating a fractal factor of 1.0 that would normally be associated with slit pores. However, MCM-41 has interconnecting pores in 3 directions. This lowers the area with respect to the volume compared to straight cylindrical pores. Even though n_L is 5.25 the pores begin to fill at ~2.5 monolayers due to "pore filling."

The parameter σ_2 includes the distribution for surface heterogeneity and for the pore size distribution. These cannot be separated due to the lack of the low pressure data. This means that the normal distribution for pore size could have a spread of as much as 0.2 of a monolayer. Although this is unlikely, and much of the distribution would be in the adsorption energy, there is no way to determine this with the data presented. This could lead to a substantial uncertainty in the fractal factor.

This uncertainty in the pore distribution, σ_p, is not an unusual problem; indeed, it is normal since the low pressure data is seldom determined. The squares of the PMF distribution value, σ (not an uncertainty), for the heterogeneity and pore size add to yield the square of the observed σ_2 assuming noncorrelation. Thus, one needs the heterogeneity distribution, σ_c, in χ_c in order to calculate σ_p.

Notice that the plot in Fig. 3.10 has as its abscissa $\Delta\chi=0$ and not $\chi=0$. To make this graph, one needs first to do a least squares fit to the lower data of n_{ads} versus χ and find the abscissa intercept which is the χ_c. The subtlety about what part of the external surface area is counted, the openings and the walls or just the walls, is being ignored here since, in any case, it is small compared to the area inside the pores. Then:

$$V_p = \frac{n_p \rho}{M} \text{ (units in mmol, g and mL)} \quad (3.23)$$

This would be calculated assuming liquid density. The following is how to make the calculation. The units will depend on what is being recorded.

$$n_m = S_{lo} \quad (3.24)$$

$$n_{ex} = S_{hi} \quad (3.25)$$

$$n_p = I_{hi} \quad (3.26)$$

$$n_L = \Delta\chi_p \quad (3.27)$$

Thus: n_L is in number of monolayers; in other words, it is the last partial layer possible before either a step function, for slits, or the linear function, for cylinders, drop to zero. It does not necessarily need to be an integer. The n-layer simulations demonstrate this latter conclusion.

For Ar the following are reasonable parameters to convert to distance using the appropriate packing factors: $\overline{V} = 7.90 \times 10^4 m^3 mol^{-1}$, $\overline{A} = 2.86 \times 10^{-5} m^2 mol^{-1}$ so:

$$\therefore f = \frac{\overline{V} A_p \Delta\chi_p}{\overline{A} V_p} \quad (3.28)$$

However, in this case the factor $f=1$ indicating again some intermediate geometry. This is not expected. This looks like a good area of research to resolve the issue.

The Kelvin equation—Correction for curvature

The other estimate, from the modified Kelvin equation, yields the diameter along with the pore size distribution. For the data analyzed here, this distribution is assumed to be a normal distribution in χ. A more detailed analysis does not seem justified by the number of data points in the transition zone.

Obviously, if there is some microporosity present then unless it can be separated in the isotherm then the above answer may be incorrect. Another method of obtaining the mesoporosity is as follows, using the modified Kelvin equation. The χ method is used here, but in principal any well calibrated standard curve should work.

The question is: why check out the porosity with the Kelvin equation or any other equation when the pore volume and radius has been determined? The answer is, one can then get a fractal factor for the pore. This fractal factor might determine whether the pores are cylindrical, slits or something else. In principle, the difference between the two calculations will yield this answer.

The capillary filling equation theory, that is, the Kelvin equation as modified by Cohen [34], can be expressed for cylindrical pores as:[5]

$$-RT \ln\left(\frac{P}{P_{vap}}\right) = \frac{m_c \gamma_{gl} \overline{V}}{r_c} \qquad (3.29)$$

With γ_{gl} = the surface tension of the gas-liquid interface for the adsorptive, r_c = the core radius, which is the pore radius, r_p, minus the "thickness" of the adsorbed layer, t. The value of $m_c = 2$ normally. In light of χ theory this is modified to:

$$e^{-\chi_p} = \frac{m_c \gamma_{gl} \overline{V}}{RT(r_p - t)} \qquad (3.30)$$

Here the χ_p is the value of χ at which the capillary filling takes place. In the case of a distribution of pores it will be the mean value, $\langle \chi_p \rangle$. The value of t is obtainable by using the difference between χ_p and χ_c, or $\Delta\chi_p = \chi_p - \chi_c$, since this would be related to the overall thickness by:

$$t = \frac{\Delta\chi_p \overline{V}}{A_m} \qquad (3.31)$$

Thus, for r_p:

$$r_p = \frac{m_c \gamma_{gl} \overline{V}}{RT e^{-\chi_p}} + \frac{\Delta\chi_p \overline{V}}{A_m} \qquad (3.32)$$

For d_p then $d_p = 2r_p$.

Eq. (3.29) along with the χ equations leads to a pore radius as given in Eq. (3.32). This equation is specifically dependent upon χ and therefore any positive deviation from the straight projected line in the χ plot can be interpreted as capillary filling. Initially, a probability normal mass function (PMF) in χ is assumed. To go beyond this assumption is, in principle, not difficult. However, for the data presented here, this does not seem justified. The PMF, **P**, is:

$$\mathbf{P}(\chi - \langle \chi_p \rangle, \sigma) = \frac{e^{-(\chi - \langle \chi_p \rangle)^2 / 2\sigma^2}}{\sqrt{2\pi\sigma^2}} \qquad (3.33)$$

where σ is the standard deviation in the pore size distribution.

---

[5] Confused about the sign of r_c? OK, here it is picked as positive.

A method of successive approximations is used to obtain $\langle \chi_p \rangle$ and σ. Using an initial estimate for $\langle \chi_p \rangle$ and setting σ to a very low value, a probe value for the fit to the isotherm data, η_i, is created from the equation:

$$\eta_j = \mathbf{P}(\chi_j - \langle \chi_p \rangle, \sigma) + \sum_{i=2}^{j} [(\chi_j - \chi_{i-1})(S_{lo} - S_{hi})(1 - \mathbf{P}(\chi_j - \langle \chi_p \rangle, \sigma))] \quad (3.34)$$

where the subscript I indicates the ith data point between points k and l on either side of the pore filling.

A new value of χ_p is calculated from a weighted average of χ using the square of the difference as the weighting factor. That is:

$$\chi_p = \sum_{i=k}^{l} \chi_i (n_{ads,i} - \eta_i)^2 / \sum_{i=k}^{l} (n_{ads,i} - \eta_i)^2 \quad (3.35)$$

Using this new χ_p, new estimates are made for η. This is repeated until convergence is satisfactory. If the fit at both sides of the transition had similar data scatter, the above method would work very well. However, there are different numbers of data points on two sides of the transition which weigh into the summations. To avoid this problem, it is best to select data points that are judged to be in the transition zone, along with roughly a few additional data points on either side of the transition. In other words, points k and l should be symmetrically located outside the transition zone.

The value of σ is obtained by a similar successive approximation method.

$$\sigma_{new} = \sigma_{previous} + v \sum_{i=k}^{l} \text{sign}(\chi_i - \chi_p)(n_{ads,i} - \eta_i) \quad (3.36)$$

where v is a factor set for the sensitivity of the convergent.

It should be set small enough to avoid oscillations between approximations. In place of the function behind the "Σ" one could use other functions to provide convergence such as "$(n_{ads,i} - \eta_i)^3$." However, this latter function seems to be considerably less stable.

σ for the distribution is in terms of χ and may be converted to distance by simply taking $\langle \chi_p \rangle + \sigma$ and determining its value to give $\langle r_p \rangle + \sigma$.

All of this seems rather involved but it yields the information that one needs; that is, the mean pore radius and the pore radius distribution. This can easily be programmed into a simple spreadsheet to ease the calculations, and more sophisticated programming is not necessary.

The above calculation should yield the correct answer under equilibrium conditions, which is often not obtained. Modifying the Kelvin Eq. (3.29) by setting $m_c = 1$ for the adsorption branch has often been suggested. This assumes that the cylindrical adsorption does not collapse from the ends or from constrictions of the capillaries but rather from the sides. There are reasons to assume either one. Hysteresis is a large problem for mesopore measurements and research by many groups on this subject is ongoing. One possibility is the switch of m_c from 1 to 2.

Isotherm fits which yield relative numbers for the surface area
Langmuir isotherm

The Langmuir isotherm is most appropriately suited for the description of chemisorption. It is especially useful for many kinetic problems involving chemisorption and catalysis. Even though it does not have much utility in physisorption, it is worthwhile to review its basis briefly, since it is very important in a related field. The functional form is also used as a stand-in for the physisorption isotherm for the carrier gas system, since it is simpler than other forms for differentiation. Since most of the carrier gas calculations are concerned with the range 0.1 to 0.5 of P/P_{vap}, this does not lead to serious error. In other words, it serves as a convenient empirical equation.

The underlying assumption for the Langmuir isotherm is that the adsorbate from the gas is in equilibrium with a bonded or tightly held species on the surface. A reaction such as:

$$G + S \rightleftharpoons G\text{-}S$$

for the gas species, G, and the surface sites, S, is assumed. Here the bonded adsorbate molecule is designated by GS. The site assumption is extremely important and restricts the use of this isotherm, as it does for any other isotherm based upon surface sites. The activity of the surface sites is assumed to be important and the activities are proportional to the number of moles, n_{ads}, on the surface. Therefore, by simple equilibrium calculation one gets:

$$K = \frac{n_{ads}}{P_G(n_S - n_{ads})} \tag{3.37}$$

where n_s is the number of surface sites (here expressed in terms of moles) and P_G is the pressure of the gas.[6]

This can be rearranged to:

$$n_{ads} = n_S \left(\frac{P_G}{K' + P_G} \right) \tag{3.38}$$

where $K' = 1/K$.

This isotherm has been widely used for a chemisorption. For dissociative adsorption, for an example diatomic molecular, for example H_2, chemisorption on an active metal where a diatomic molecule will become monatomic on the surface or $H_2 + 2S \rightleftharpoons 2H - S$, and Eq. (3.38) is modified to the Sievert's equilibrium expression:

$$n_{ads} = n_S \left(\frac{P_{H_2}^{1/2}}{K' + P_{H_2}^{1/2}} \right) \tag{3.39}$$

---

[6] For the carrier gas physisorption system, replace P_G with P/P_{vap}.

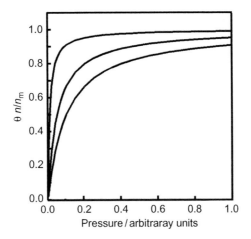

FIG. 3.11

Examples of Langmuir isotherms and the position of the "knee" as it varies with adsorption energy.

The general shape of the Langmuir curve is presented in Fig. 3.11 with different values of K. The value of n_S is set to 1 for this figure. The value of n_{ads} approaches n_S as $P \to \infty$. There are several other potential "Langmuir like" isotherms due to specific stoichiometries, etc. This is a good subject for a book on kinetics.

The Langmuir isotherm cannot yield a surface area number, unless one knows how the surface sites are distributed. If one knows that the approximate area required for one bonding location is $0.2 \, nm^{-2}$, then one can conclude from a calculation of n_S what the area is. An assumption implied in this is that the activity of the surface site is proportional to the number of sites available divided by the original number, i.e., the mole fraction of species on the surface. In bulk calculations, this is referred to as the saturation limit. The assumption that full saturation is the same as the number of original sites may not be valid neither in the bulk nor on surfaces. Some readers may find this statement surprising. The effect where saturation is reached before the number of identical sites are used up has been observed in many solubility measurements with solids. The reasons are multiple, but one must remember that the solute modifies the solvent chemically, i.e., electronically. Use of this method to determine the surface area, or more correctly the specific number of active sites, of a solid is often called a titration, in analogy to solution chemistry.

As useful as the Langmuir isotherm is, due to the site assumption, it should normally not be used for physisorption. There have been some derivations that assume that the sites do not exist; however, these derivations suffer from the unrealistic assumption of localized forces without localization. Implicit in chemical bonding is the assumption of directional, local bonds. Even if for use where the Langmuir isotherms would seem appropriate with specific sites, there is no guarantee that the surface siting is a reflection of the X-ray analysis for the bulk. Some techniques, such are low energy electron diffraction, are more appropriate techniques to work this out (with many caveats dismissed.)

Freundlich isotherm

The Freundlich isotherm was originally an empirical isotherm. There have been numerous theoretical justifications for it for many years up to the present. One is that an exponential energy distribution is attributed to the Langmuir isotherm. It can also be derived by assuming an "ideal gas-like" state for the adsorbate and deriving it from the Gibbs-Duhem equation. The equation for the Freundlich isotherm is:

$$n_{ads} = K_F P^{r_F^{-1}} \tag{3.40}$$

where n_{ads} is the moles of the adsorbate on the surface and r_F is a constant, and will be referred to here as the Freundlich constant.

A special case of $r_F = 1$ is referred to as "Henry's law," which should not be confused with the solution equation of state called "Henry's law." The use of this latter name is certainly confusing to some.[7] The terminology probably should be avoided. It is also often the case that when the Freundlich isotherm is observed with $r_F \neq 1$ that the authors will call it the "Henry's law" limit. This is due to the practice of using a log-log plot in these cases and observing a straight line. Of course this does not mean that the line will pass through (0,0). (As a fact, a log-log plot cannot pass through (0,0).)

Of interest is the derivation of the Freundlich isotherm with $r_F = 1$ from the ideal 2D surface gas. Assuming a surface equation of state similar to the ideal gas law using π in place of P and A_s in place of V one has as an equation of state:

$$\pi A_s = n_{ads} RT \tag{3.41}$$

Often the units for π are dynes/m^2 to yield numbers that are simpler. This is not necessary, however. This author prefers to leave all units in SI for simplicity. Today π is often reported in units of mJ m^{-2}. For most thermodynamic treatments (see Hiemenz [17]) the surface Gibbs-Duhem equation assuming constant surface area would be:[8]

$$-A_s d\gamma = n_{ads} d\mu \tag{3.42}$$

---

[7] ...which is why I will continue to put "Henry's law" in quotes.
[8] Why in the literature is γ assumed to vary and A assumed to remain constant? Everyone seem to miss the point of the basic question, "*What is the A being referred to here?*" Most assume that A is A_s when it is probably A_{ls} the area that is the contact between the liquid and the solid. Thus, not only can A_{ls} vary but A_{gs} would vary in opposite directions. Furthermore, A_{lg} can vary once the outer interface becomes formed. More about this later in reference to the pre-filling collapse.

Or, since $\pi = \gamma_0 - \gamma$:

$$d\gamma = \frac{n_{ads}RT}{A_s} d\ln P \qquad (3.43)$$

Substituting in Eq. (3.41) and integrating and identifying $\gamma_0 - \gamma$ as π, one obtains the $r_F = 1$ Freundlich isotherm with an arbitrary K. Reversing the process with any r_F, will yield in place of Eq. (3.41) one obtains:

$$\pi A_s = \frac{n_{ads}RT}{r_F} \qquad (3.44)$$

with no reason to assume $r_F = 1$. This is difficult to justify (but has been and continues to be worked upon [35]).

The Freundlich isotherm equations do not have the surface area explicitly as a parameter in the equations. There are far too many leaps of faith—not an unusual problem when one discovers a good empirical equation. Therefore, the absolute value of the surface area cannot be determined using these equations.

Polanyi formulations

Polanyi [36–38] basically stated that the free energy of the surface is a function of the coverage of the surface. Thus, the pressure is related to this energy, $E(\theta)$, in the following fashion:

$$\mathbf{f}\left(RT \ln\left(\frac{P}{P_s}\right)\right) = E(\theta) \qquad (3.45)$$

where bold type indicates a function. This is very general and is reasonable and consistent with thermodynamics.

This is often simplified to:

$$RT \ln P = \mathbf{E}(\theta) \qquad (3.46)$$

where θ is the amount on the surface per unit area.

For convenience, θ will be used throughout as the amount on the surface relative to exactly one monolayer equivalent. An especially successful isotherm of this form was found to be:

$$RT \ln P = B\mathbf{e}^{-k_p \theta} \qquad (3.47)$$

where B and k_p are constants.

This form of the equation was known for many years to be an excellent empirical fit to most isotherms, indeed in the judgement of some [16], the best fit by far. This isotherm equation suffers from the same problem as the Freundlich isotherm equation. The surface area is not an explicit parameter in the equation, but is bound up with a multiplicative constant, k_P.

Notice that any formulation based on the Polanyi theory cannot yield the surface area without some additional assumption. This was the primary weakness of the approach.

deBoer-Zwikker formulation

The deBoer-Zwikker [39] polarization theory is a special case of Eq. (3.47). Taking the ln of both sides of this equation and using relative pressures, i.e., compared to the vapor pressure of the liquid state of the adsorbate, yields:

$$\ln\left[\ln\left(\frac{P_{\text{vap}}}{P}\right)\right] = B - k_p\theta \qquad (3.48)$$

DeBoer-Zwikker derived a very similar equation from classical polarization theory, which was:

$$\ln\left[\ln\left(\frac{P_{\text{vap}}}{P}\right)\right] = \ln\left(\frac{\varepsilon_0}{RT}\right) - \left(\frac{aV_1}{v_0 A_s}\right) \qquad (3.49)$$

from which the surface area could be derived. The problem with this formulation is that classical polarization theory yields numbers which are widely incorrect. This theory was generally disregarded and deBoer pursued the standard curve route. Experimentally, however, it was noticed very early by Adamson [40] that this equation fits most adsorption data better than any of the other theories of the day, including the BET. Badmann, Stockhausen, and Setzer used a similar function successfully in a much later publication [32]. This is the isotherm that Fuller noticed fitted almost all the experimental data on simple, nonporous materials and lead to the development of the ASP in the context of the Polanyi formulation and to the quantum mechanical theory, χ.[9]

The Frenkel, Halsey, Hill (FHH) isotherm

The Frenkel-Halsey-Hill isotherm has found much utilization due to the range specified for its application. It seems especially handy for fitting to porosity determinations. It seems to work well between relative pressures of 0.4 to 0.9. The equation is:

$$\ln\left(\frac{P_{\text{vap}}}{P}\right) = k_{\text{FHH}}\theta^{r_{\text{FHH}}} \qquad (3.50)$$

---

[9] What seems to be a missed opportunity is that the various isotherm formulations were never statistically compared. If this had been done, there is no question but that the DeBoer-Zwikker formulation would have been judged to be superior. It would then be a matter of figuring out why this is the case. The reconsideration of the "old discredited" but very superior DeBoer-Zwikker formulation was the inspiration that triggered the discovery of the quantum mechanical explanation.

where k_{FHH} is an empirical constant and r_{FHH} is between 2 and 3. It is easier to make a transform of this equation for plotting purposes so:

$$\ln\left[\ln\left(\frac{P_{vap}}{P}\right)\right] = \ln k' + r_{FHH} \ln n_{ads} \qquad (3.51)$$

where the k' is a constant along with the monolayer coverage value. Notice, then again, this transform is nearly the same form as the DeBoer-Zwikker equation.

Obviously, this cannot be used to determine the monolayer coverage but may be used with caution to interpellate data.

Analysis using standard isotherms

Standard isotherms admittedly do not yield the surface area values by themselves. However, they are probably the most useful of the methods of analysis. The question as to why one would use an analysis that does not yield a value for the surface area may seem puzzling. Firstly, there are times when all one really needs is a relative value. Secondly, the isotherms are useful for extrapolation and as input into various theories, such as porosity calculations. Most absolute numbers for surface area from these isotherms refer back to the BET equation for standardization. With a good, well-characterized standard, which includes a knowledge of the surface area by a method other than BET, which in many cases is very hard to obtain, one can obtain values for surface area and porosity.

There are now several standard isotherms. However, the two used the most are still the α-s standard isotherms and the t-thickness isotherm. The standard t-thickness isotherm on alumina may, however, be inaccurate at higher pressures. There is a tendency today to construct a standard isotherm for the adsorbent-adsorbate pair being used. This is a bit tricky since these standard isotherms are usually used for porosity measurements and to obtain a nearly flat surface that is energetically the same as the porous material seems unlikely.

The standard curve method follows these steps:

1) Measure an isotherm on a known non-porous material with the same chemistry and surface structure as that of the porous material. In the case of silica and alumina and other materials mentioned in a later chapter, this has already been done.
2) Obtain the amount adsorbed as a function, **F**, of relative pressure, $x = P/P_s$ or:

$$n_{ads} = A_s \mathbf{F}(x, T) \qquad (3.52)$$

Normally this curve is measured at only one temperature. If one knows the surface area of this standard, then the value of **F** is scaled so that A_s in the above equation is the surface area value of the standard.

3) Plot the n_{ads} of the "unknown" sample against the function **F**.
4) Calculate the surface area of the unknown as the slope times the known A_s.

The principal use for this is to determine the volume of porosity. This measurement was described previously in the sections "Micropore Analysis" and "Mesopore Analysis" in this chapter.

Principal problem with standard curves

The main problem involved with the standard curves is that the surface area, or monolayer equivalent, is not specified by the technique itself. If one had a standard curve that had a calibration for the surface area, then the implications of these curves, calculations of the porosity and energies, would be valid. This is, however, not the case. The traditional solution to this problem is to use the BET surface area, which introduces a large uncertainty in precision and a very large, at least a factor of 3, error (see Appendix II.H for an explanation and some calculations).

It is a bit ironic that the standard curves normally use the value of $P/P_{vap} = 0.4$ to make comparison, which is slightly outside the BET range of 0.05 to 0.35. However, this introduces a probably small degree of uncertainty, perhaps less than 4%. This is quite small compared to the minimal relative error of about 200%.

Because of these considerations, in the tables that follow both the BET surface area number and the quantum mechanical number for specific surface area are given if possible. In addition QM value for E_a is also provided. There is the proviso for this that these are provided if the physical quantities for the original data are available.

Standard isotherms

Isotherms measured on well characterized material and used for comparison with isotherms of unknowns are referred to as standard isotherms. Tables of a variety of standard isotherms that are described here are presented in the next section. The values of n_m and χ_c are presented with the data tables.

The α_s-curve standard (see Sing, Everett, and Ottewill [11])

The α_s-curve has an advantage that the original data has not been severely reworked. Originally these plots were simple n-plots (i.e., number of moles adsorbed as a function of pressure.) The procedure for obtaining these curves was to obtain a multiplicity of adsorption isotherms on many powders of the same type of material. The value for n_m and the surface area, however, is based on the BET surface area. Therefore, these values carry all the uncertainties of the BET method. An alternative method of determining n_m that is more reliable would be better.

The intention, however, was to avoid the BET theory, so the a_s-curve is not used for those purposes, but rather as a comparison curve to determine how much is adsorbed in pores. These standard curves are very useful for porosity determinations due to the high degree of confidence in the basic standard curve. For the α_s-curve is an averaged and smoothed curve for several similar silica samples is used. Generally, in the literature it works quite well, even in the high pressure range. Curves for both nitrogen and argon are available.

The data in Table 3.4 are some data by Bhambhani, Cutting, Sing, and Turk [41] (BCST) and quoted in the book by Everett, Parfitt, Sing, and Wilson [42] (EPSW). The smoothed α_s curve is given in Table 3.5. From the BET value, one may back calculate the original data. There is some uncertainty in this conversion so the original numbers are present in case someone can discern what the conversion actually is.

Table 3.6 presents some more data by Payne, Sing, and Turk [43] also for silica. In Table 3.7 are the α-s curves normalized to the value of P/P_s of 0.4 [44].

The t-curve

One of the earliest standard curves was the *t*-curve by Lippens et al. [45], which was the adsorption of N_2 on alumina. The data were reported in terms of film thickness in angstroms (unit designator Å and equal to 10^{-10} m). In Table 3.8 are the data for both the smoothed curve and the original data. The conversion from volume adsorbed in $mL\,g^{-1}$ is given by the equation

$$t = 3.54 \left(V/V_m \right) \text{Å} \tag{3.53}$$

The constant V_m is the STP is the monolayer amount in term of adsorptive pressure calculated using the BET method.

IUPAC standards on silica and carbon

The original purpose of the IUPAC (compiled by Everett et al. [42]) round-robin investigation was to create some confidence in the methodology of adsorption isotherm measurements. Standard samples from the same productions batches were used and various laboratories performed the same experiments. The results were not intended as standard curves but the agreement between the various laboratories was generally very good, within 2%. Therefore, these would be as good standards as one would be able to find. Apparently, the archive for these standards no longer exists. The data presented below was extracted from the literature from laboratory "H." This seemed to be a typical data run. The isotherms determined were for Gisil silica, TK800 silica, (silica in Table 3.9) Vulcan 3G carbon, and Sterling FT carbon (carbons in Table 3.10).

RMBM carbon standard

A standard adsorption isotherm curve for activated carbon has been published by Rodriguez-Reinoso et al. [46] (RMBM). The data and the α_s standard are presented in Table 3.11. The carbon studied was an activated carbon form and contained macropores and micropores [47]. The micropores were closed by heating to 2073 K [48]. The value for A_s was obtained from the BET surface area and was reported to be 4.3–4.4 m^2 g^{-1}. This curve is a smoothed curve and at the low pressure range is very different from other standards. In the literature, there are several standards for carbon. There is probably an appropriate standard available for the carbon material of particular interest.

KFG segmented standard carbon curve (by Karnaukhov, Fenelonov, and Gavrilov)

Karnaukhov, Fenelonov, and Gavrilov [14] have presented a standard curve with a segmented least squares fit to the data of α versus P/P_s. The fit is for the equation:

$$n_{\text{ads}} = \sum_{i=0}^{5} C_i \left[\ln\left(\frac{P_s}{P}\right) \right]^i \quad (3.54)$$

Table 3.12 lists the coefficients. C_i. n_{ad} is given here in units of µmol m^{-2} but the surface area per gram of sample is not listed. In order to use this in the usual fashion a table of α-s for this is constructed in Table 3.13. This curve may be useful for determining mesoporosity. It does not extrapolate below $0.10 P/P_s$.

Cranston and Inkley standard for pore analysis

Cranston and Inkley [13] developed a general standard isotherm that did a fair job for a variety of adsorbents including silica and alumina. Basically, the data was averaged and smoothed to yield the standard curve. The data for this curve is not presented in their article but a graph of the averaged isotherm is given. It would be best for those who wish to use this curve to consult the original.

Thoria standard curves

Thoria has the interesting property that it can be fired to a high temperature without changing morphology. Thus, a degassing temperature to clean the surface at 1000°C does not change the surface area. It is, therefore, an interesting research tool and is used for a variety of commercial applications. In Table 3.14 is the standard nitrogen curves for thoria obtained by Gammage et al. [49] for thoria outgassed at 25°C. At this point, the exact conversion from amount adsorbed to t is not certain, but probably the same that deBoer used. For higher outgassing temperatures, the standard curve is the same at high values of χ (high relative pressure) but deviates with a chi-plot break, at a low value of χ. This is due to the degassing of a higher energy plane. The original smoothed curve has been made into an α-s curve. In Table 3.15 is the standard curve for water on thoria. A similar treatment has been used for the smoothed curve. The standard curve for argon adsorption is listed in Table 3.16.

Standard curves for lunar soil

In Tables 3.17–3.20 are the standard isotherms from lunar soil as supplied to NASA [50]. For these samples, the standard curves have been converted here to α-s curves. The first three points were ignored for the α-s curve fit for oxygen adsorption. The reason for the zero values is discussed in the section on the threshold phenomenon. Details about the lunar soils can be obtained in a US government report [51] and addition information is available from an article by Fuller [52].

Summary for standard curves

Standard curves are a useful method of analyzing. The analysis is usually performed graphically. The abscissa values of the standard curve are obtained from a table, and the ordinate values are from the experimental results at the same pressure. The initial slope yields the ratio of the surface area of the sample to the standard. This has several advantages. These include:

1) There is little likelihood that data will be lost by over analysis.
2) There is not a commitment to a value to the surface area in the analysis up to the point described above. This comes when a surface area value is assigned to the standard.
3) The porosity can be determined by the deviation from the standard curve in an unbiased manner.

The third point listed above has been explained in earlier sections. The monolayer equivalence can be determined by using the χ plot which is, in most cases, identical in form to the standard.

The main disadvantage for the use of standard curves is that each adsorbent must have a measured standard curve. Furthermore, not only must one have a standard for each adsorptive but, at least theoretically, a different standard for each adsorbate-adsorbent pair. Since this is often used for porous materials, there is no guarantee that the porous material will mimic the nonporous material. This is especially unlikely in porous materials that have been doped with inhibitors or promoters. Thus, at best assuming that this is not a problem, there need to be thousands of standards or the investigator must create standards and hope they work. There is, at the moment a limited number, concentrating on carbon, SiO_2, MgO and Al_2O_3.

Tables of standard isotherms

Table 3.4 Data analysis of N_2 α-s curves on SiO_2, Fransil-I, by BCST

P/P_{vap}	χ	n_{ads} (μmol m^{-2})	n_{ads} (mmol g^{-1})	P/P_{vap}	χ	n_{ads} (μmol m^{-2})	n_{ads} (mmol g^{-1})
0.001	−1.933	4.0	0.155	0.28	−0.241	13.6	0.526 0.532
0.005	−1.667	5.4	0.209	0.30	−0.186	13.9	0.538 0.544
0.01	−1.527	6.2	0.240	0.32	−0.131	14.2	0.550 0.556
0.02	−1.364	7.7	0.298	0.34	−0.076	14.5	0.561 0.568
0.03	−1.255	8.5	0.329	0.36	−0.021	14.8	0.573 0.580
0.04	−1.169	9.0	0.348	0.38	0.033	15.1	0.584 0.591
0.05	−1.097	9.3	0.360	0.40	0.087	15.5	0.600 0.603
0.06	−1.034	9.4	0.364	0.42	0.142	15.6	0.604 0.615
0.07	−0.978	9.7	0.375	0.44	0.197	16.1	0.623 0.627
0.08	−0.927	10.0	0.387	0.46	0.253	16.4	0.635 0.639
0.09	−0.879	10.2	0.395	0.50	0.367	17.0	0.658 0.664
0.10	−0.834	10.5	0.406	0.55	0.514	17.8	0.689 0.697
0.12	−0.752	10.8	0.418	0.60	0.672	18.9	0.731 0.731
0.14	−0.676	11.3	0.437	0.65	0.842	19.9	0.770 0.768
0.16	−0.606	11.6	0.449	0.70	1.031	21.3	0.824 0.809
0.18	−0.539	11.9	0.461	0.75	1.246	22.7	0.878 0.856
0.20	−0.476	12.4	0.480	0.80	1.500	25.0	0.968 0.912
0.22	−0.415	12.7	0.491	0.85	1.817	28.0	1.084 0.981
0.24	−0.356	13.0	0.503	0.90	2.250	37.0	1.432 1.075
0.26	−0.298	13.3	0.515				

$n_m(\chi) = 0.21$ mmol g^{-1}; n_m(BET) $= 0.40$ mmol g^{-1}; $\chi_c = -2.676$

Table 3.5 The α-s curve for the data by BCST referenced to $P/P_{vap} = 0.4$

P/P_{vap}	α-s	α-s fitted	P/P_{vap}	α-s	α-s fitted
0.001	0.258	0.269	0.26	0.858	0.861
0.005	0.348	0.365	0.28	0.877	0.881
0.01	0.400	0.416	0.30	0.897	0.901
0.02	0.497	0.475	0.32	0.916	0.921
0.03	0.548	0.514	0.34	0.935	0.941
0.04	0.581	0.545	0.36	0.955	0.961
0.05	0.600	0.571	0.38	0.974	0.980
0.06	0.606	0.594	0.40	1.000	1.000
0.07	0.626	0.614	0.42	1.006	1.020
0.08	0.645	0.633	0.44	1.039	1.040
0.09	0.658	0.650	0.46	1.058	1.060
0.10	0.677	0.667	0.50	1.097	1.101
0.12	0.697	0.696	0.55	1.148	1.155
0.14	0.729	0.724	0.60	1.219	1.211
0.16	0.748	0.749	0.65	1.284	1.273
0.18	0.768	0.773	0.70	1.374	1.341
0.20	0.800	0.796	0.75	1.465	1.419
0.22	0.819	0.818	0.80	1.613	1.511
0.24	0.839	0.840	0.85	1.806	1.626
0.26	0.858	0.861	0.90	2.387	1.783

CHAPTER 3 Interpreting the physisorption isotherm

Table 3.6 Data for α-s curves by Payne et al.

	N₂ adsortion data				Ar adsorption data			
P/P_{vap}	χ	n_{ads} (mol m^{-2})	n_{ads} (mmol g^{-1})		P/P_{vap}	χ	n_{ads} (mol m^{-2})	n_{ads} (mmol g^{-1})
0.05	−1.097	34.0	1.518	1.454	0.05	−1.097	23.0	1.027	1.071
0.10	−0.834	38.0	1.696	1.697	0.10	−0.834	29.0	1.295	1.319
0.15	−0.640	43.0	1.920	1.877	0.15	−0.640	32.0	1.429	1.501
0.20	−0.476	46.0	2.054	2.029	0.20	−0.476	38.0	1.696	1.656
0.25	−0.327	48.0	2.143	2.168	0.25	−0.327	41.0	1.830	1.796
0.30	−0.186	51.0	2.277	2.298	0.30	−0.186	43.0	1.920	1.929
0.35	−0.049	54.0	2.411	2.425	0.35	−0.049	45.0	2.009	2.058
0.40	0.087	58.0	2.589	2.551	0.40	0.087	50.0	2.232	2.186
0.45	0.225	58.0	2.589	2.679	0.45	0.225	54.0	2.411	2.316
0.50	0.367	61.0	2.723	2.810	0.50	0.367	55.0	2.455	2.449
0.60	0.672	68.0	3.036	3.093	0.60	0.672	62.0	2.768	2.736
0.70	1.031	77.0	3.437	3.425	0.70	1.031	69.0	3.080	3.075
0.80	1.500	89.0	3.973	3.860	0.80	1.500	79.0	3.527	3.516
0.90	2.250	118.0	5.268	4.555	0.90	2.250	93.0	4.152	4.223

$n_m = 0.927$ mmol g^{-1}; $\chi_c = -2.666$; $R = 0.9986$

$n_m = 0.942$ mmol g^{-1}; $\chi_c = -2.235$; $R = 0.9993$

Table 3.7 Smoothed α-s curve on silica normalized to $V_{0.4}$ as listed by Gregg and Sing

\multicolumn{4}{c	}{N_2}	\multicolumn{4}{c}{Ar}
P/P_{vap}	$V/V_{0.4}$	P/P_{vap}	$V/V_{0.4}$	P/P_{vap}	$V/V_{0.4}$	P/P_{vap}	$V/V_{0.4}$
0.001	0.26	0.280	0.88	0.01	0.243	0.32	0.900
0.005	0.35	0.300	0.90	0.02	0.324	0.34	0.923
0.010	0.40	0.320	0.92	0.03	0.373	0.36	0.948
0.020	0.50	0.340	0.94	0.04	0.413	0.38	0.973
0.030	0.55	0.360	0.96	0.05	0.450	0.40	1.000
0.040	0.58	0.380	0.98	0.06	0.483	0.42	1.022
0.050	0.60	0.400	1.00	0.07	0.514	0.44	1.048
0.060	0.61	0.420	1.01	0.08	0.541	0.46	1.064
0.070	0.63	0.440	0.10	0.09	0.563	0.48	1.098
0.080	0.65	0.460	1.06	0.10	0.583	0.50	1.123
0.090	0.66	0.500	1.10	0.11	0.602	0.50	1.123
0.100	0.68	0.550	1.14	0.12	0.620	0.52	1.148
0.120	0.70	0.600	1.22	0.13	0.638	0.54	1.172
0.140	0.73	0.650	1.29	0.14	0.657	0.56	1.198
0.160	0.75	0.700	1.38	0.15	0.674	0.58	1.225
0.180	0.77	0.750	1.47	0.16	0.689	0.60	1.250
0.200	0.80	0.800	1.62	0.17	0.705	0.62	1.275
0.220	0.82	0.850	1.81	0.18	0.719	0.64	1.300
0.240	0.84	0.900	2.40	0.19	0.733	0.66	1.327
0.260	0.86			0.20	0.748	0.68	1.354
				0.22	0.773	0.70	1.387
				0.24	0.801	0.72	1.418
				0.26	0.826	0.74	1.451
				0.28	0.851	0.76	1.486
				0.30	0.876	0.78	1.527

Table 3.8 Data and smooth t-curve-N_2 on alumina by Lippens et al.

\multicolumn{4}{c	}{t-curve (smoothed data)}	\multicolumn{2}{c}{Original data}
P/P_{vap}	t (Å)	P/P_{vap}	t (Å)	P/P_{vap}	t (Å)
0.08	3.51	0.80	10.57	0.083	3.54
0.10	3.68	0.82	11.17	0.101	3.72
0.12	3.83	0.84	11.89	0.119	3.82
0.14	3.97	0.86	12.75	0.137	3.97
0.16	4.10	0.88	13.82	0.159	4.10
0.18	4.23	0.90	14.94	0.181	4.22
0.20	4.36			0.200	4.38
0.22	4.49			0.227	4.45
0.24	4.62			0.242	4.61
0.26	4.75			0.260	4.72
0.28	4.88			0.285	4.86
0.30	5.01			0.300	5.01
0.32	5.14			0.321	5.14
0.34	5.27			0.339	5.24
0.36	5.41			0.365	5.42
0.38	5.56			0.386	5.55
0.40	5.71			0.408	5.67
0.42	5.86			0.422	5.85
0.44	6.02			0.440	5.98
0.46	6.18			0.458	6.13
0.48	6.34			0.480	6.31
0.50	6.50			0.499	6.44
0.52	6.66			0.520	6.62
0.54	6.82			0.542	6.79
0.56	6.99			0.560	6.97
0.58	7.17			0.579	7.15
0.60	7.36			0.599	7.30
0.62	7.56			0.617	7.51
0.64	7.77			0.635	7.71
0.66	8.02			0.661	7.92
0.68	8.26			0.679	8.22
0.70	8.57			0.700	8.52
0.72	8.91			0.718	8.88
0.74	9.27			0.744	9.24
0.76	9.65			0.758	9.59
0.78	10.07			0.780	10.03

Table 3.9 IUPAC silica isotherms

| \multicolumn{2}{c|}{Gisil silica} | \multicolumn{4}{c}{TK800 silica} |

Gisil silica		TK800 silica			
P/P_s	V /std mL g^{-1}	P/P_s	V/std mL g^{-1}	P/P_s	V/std mL g^{-1}
0.0076	44.1	0.0144	25.3	0.9151	123.5
0.0177	53.6	0.0217	28.2	0.9317	135.3
0.0412	55.9	0.0325	30.1	0.9476	147.4
0.0646	61.0	0.0433	32.5	0.9591	157.9
0.0773	63.4	0.0542	34.0	0.9678	165.7
0.0875	64.4	0.0664	35.4		
0.1394	71.2	0.0953	37.3		
0.1737	74.6	0.1358	40.4		
0.2028	78.0	0.1733	43.2		
0.2586	83.1	0.2167	46.3		
0.3144	88.1	0.3091	51.9		
0.3581	92.2	0.3553	54.8		
0.4202	97.6	0.3958	56.4		
0.4912	106.4	0.4694	61.2		
0.5400	114.6	0.5561	67.1		
0.5711	118.3	0.6406	73.5		
0.6116	127.5	0.7092	79.7		
0.6889	145.8	0.7042	82.6		
0.7276	162.7	0.7352	85.6		
0.7669	189.8	0.7887	94.3		
0.7840	199.3	0.8176	99.2		
0.8227	240.7	0.8486	105.6		
0.8461	288.1	0.8826	114.1		

Table 3.10 IUPAC carbon samples

Vulcan 3G				Sterling FT			
P/P_s	V^a	P/P_s	V^a	P/P_s	V^a	P/P_s	V^a
0.0006	2.13	0.2575	3.63	0.0006	11.7	0.2690	24.3
0.0123	2.50	0.3065	4.16	0.0077	15.1	0.3122	26.7
0.0300	2.56	0.3556	4.61	0.0242	16.1	0.3577	29.3
0.0460	2.62	0.4150	5.07	0.0432	16.5	0.4287	32.6
0.6190	2.66	0.4647	5.37	0.0585	17.1	0.4908	35.3
0.0766	2.72	0.5321	5.78	0.0857	17.9	0.5611	38.5
0.1318	2.89	0.6100	6.29	0.1390	19.5	0.6291	42.2
0.1747	3.09	0.7080	7.30	0.1821	20.7	0.7072	47.4
0.2084	3.24	0.7957	8.47	0.2129	22.0	0.7852	55.9
0.2354	3.50			0.2395	22.8		

[a] Units for V: std mL g^{-1}.

Table 3.11 Standard isotherm for activated charcoal by RMBM

P/P_{vap}	n/n_m	α-s	P/P_{vap}	n/n_m	α-s	P/P_{vap}	n/n_m	α-s
0.005	0.82	0.51	0.18	1.21	0.76	0.44	1.68	1.05
0.01	0.87	0.54	0.20	1.24	0.78	0.46	1.71	1.07
0.02	0.92	0.58	0.22	1.27	0.79	0.50	1.79	1.12
0.03	0.95	0.59	0.24	1.30	0.81	0.54	1.88	1.18
0.04	0.98	0.61	0.26	1.33	0.83	0.60	2.02	1.26
0.05	1.00	0.63	0.28	1.37	0.86	0.64	2.13	1.33
0.06	1.02	0.64	0.30	1.41	0.88	0.70	2.32	1.45
0.07	1.03	0.64	0.32	1.44	0.90	0.74	2.46	1.54
0.08	1.05	0.66	0.34	1.48	0.93	0.80	2.71	1.69
0.10	1.09	0.68	0.36	1.52	0.95	0.84	2.87	1.79
0.12	1.12	0.70	0.38	1.56	0.98	0.90	3.29	2.06
0.14	1.14	0.71	0.40	1.60	1.00	0.94	3.91	2.44
0.16	1.17	0.73	0.42	1.64	1.03			

Table 3.12 KFG coefficients for a standard curve extracted from carbons

			Coefficients (C_i):			
Range	0	1	2	3	4	5
0.1–0.6	27.1667	23.449	16.75	6.5135	0.9971	0
0.55–0.92	46.5644	242.443	1120.65	2884.45	3729.22	1890.9
0.90–0.99	119.463	4983.14	130098	1.792×10^3	1.2438×10^7	3.4279×10^7

Table 3.13 α-s curve using coefficients from Table 3.12

P/P_{vap}	$n/n_{0.4}$	P/P_{vap}	$n/n_{0.4}$	P/P_{vap}	$n/n_{0.4}$
0.1	0.680	0.6	1.219	0.9	1.969
0.2	0.800	0.65	1.287	0.92	2.117
0.3	0.903	0.7	1.374	0.94	2.328
0.4	1.000	0.75	1.471	0.96	2.694
0.5	1.103	0.8	1.582	0.98	3.827
0.6	1.215	0.85	1.734	0.99	5.236
0.55	1.153	0.9	1.977		

Table 3.14 Standard Nitrogen isotherms of low temperature out-gassed thoria

Original data					Smoothed α-s curve			
P/P_{vap}	t (Å)	P/P_{vap}	t (Å)	P/P_{vap}	$n/n_{0.4}$	P/P_{vap}	$n/n_{0.4}$
0.016	1.43	0.602	6.93	0.010	0.221	0.300	0.865
0.027	1.72	0.660	7.38	0.020	0.303	0.350	0.933
0.036	2.30	0.701	7.86	0.030	0.351	0.400	1.000
0.078	2.84	0.758	8.38	0.040	0.394	0.450	1.063
0.104	3.17	0.802	9.06	0.050	0.428	0.500	1.135
0.138	3.42	0.848	9.93	0.060	0.457	0.550	1.202
0.205	3.92	0.898	11.22	0.070	0.486	0.600	1.279
0.248	4.39			0.080	0.510	0.650	1.361
0.358	5.07			0.090	0.534	0.700	1.452
0.402	5.36			0.100	0.558	0.750	1.558
0.462	5.72			0.150	0.649	0.800	1.678
0.501	6.13			0.200	0.726	0.850	1.832
0.558	6.42			0.250	0.798	0.900	2.038
$\chi_c = -1.992$		$n_m = 2.60$ Å		$R = 0.9995$			

Table 3.15 Standard curve to water adsorption of thoria

Original data					Smoothed α-s curve			
P/P_{vap}	t (Å)	P/P_{vap}	t (Å)	P/P_{vap}	$n/n_{0.4}$	P/P_{vap}	$n/n_{0.4}$
0.010	0.92	0.535	5.32	0.010	0.169	0.100	0.526
0.048	1.65	0.555	5.62	0.015	0.216	0.150	0.625
0.068	2.48	0.595	6.18	0.020	0.253	0.200	0.710
0.115	2.82	0.655	6.42	0.025	0.283	0.250	0.787
0.152	3.15	0.711	6.85	0.030	0.309	0.300	0.859
0.205	3.34	0.758	7.35	0.035	0.332	0.350	0.930
0.260	3.68	0.795	8.46	0.040	0.353	0.400	1.000
0.321	4.11	0.850	9.32	0.045	0.372	0.450	1.071
0.355	4.85	0.900	10.42	0.050	0.390	0.500	1.144
0.465	5.08			0.055	0.407	0.550	1.220
				0.060	0.422	0.600	1.301
				0.065	0.437	0.650	1.389
				0.070	0.451	0.700	1.486
				0.075	0.465	0.750	1.596
				0.080	0.478	0.800	1.727
				0.085	0.490	0.850	1.891
				0.090	0.503	0.900	2.114
$\chi_c = -1.855$		$n_m = 2.45$ Å		$R = 0.9996$			

Table 3.16 Argon adsorption on 25°C out-gassed thoria

Original data				Smoothed α-s			
P/P_{vap}	t (Å)	P/P_{vap}	t (Å)	P/P_{vap}	t (Å)	P/P_{vap}	t (Å)
0.011	0.78	0.354	4.94	0.005	0.078	0.650	1.396
0.018	1.13	0.368	5.06	0.010	0.152	0.700	1.496
0.028	1.48	0.378	5.30	0.020	0.238	0.750	1.609
0.038	1.68	0.403	5.32	0.030	0.295	0.800	1.742
0.045	1.86	0.419	5.45	0.040	0.340	0.850	1.909
0.056	2.18	0.444	5.70	0.050	0.378	0.900	2.136
0.064	2.28	0.454	5.75	0.060	0.411		
0.082	2.54	0.468	5.89	0.070	0.440		
0.103	2.81	0.484	5.96	0.080	0.467		
0.118	3.02	0.501	6.18	0.090	0.492		
0.135	3.22	0.520	6.32	0.100	0.516		
0.148	3.30	0.536	6.40	0.150	0.618		
0.158	3.45	0.555	6.52	0.200	0.704		
0.201	3.78	0.561	6.58	0.250	0.782		
0.228	3.94	0.577	6.82	0.300	0.857		
0.235	4.17	0.600	6.96	0.350	0.929		
0.258	4.30	0.652	7.47	0.400	1.000		
0.278	4.46	0.698	7.93	0.450	1.072		
0.302	4.66	0.748	8.55	0.500	1.147		
0.326	4.74	0.802	9.33	0.550	1.224		
0.347	4.88	0.818	9.58	0.600	1.307		
$\chi_c = -1.816$				$n_m = 2.81$ Å		$R = 0.9994$	

Table 3.17 N_2 adsorption of non-porous lunar soil

Original data			Smoothed α-s curve			
P/P_{vap}	n_{ads} (µmol g^{-1})	P/P_{vap}	$n/n_{0.4}$	P/P_{vap}	$n/n_{0.4}$	
0.00051	1.517	0.0005	0.238	0.070	0.616	
0.0036	2.357	0.001	0.272	0.080	0.635	
0.0069	2.815	0.002	0.310	0.090	0.652	
0.013	3.318	0.003	0.335	0.100	0.668	
0.027	3.941	0.004	0.353	0.150	0.738	
0.054	4.505	0.005	0.368	0.200	0.797	
0.106	5.390	0.010	0.418	0.250	0.851	
0.159	5.968	0.015	0.451	0.300	0.902	
0.211	6.374	0.020	0.477	0.350	0.951	
0.267	6.734	0.025	0.498	0.400	1.000	
0.319	7.387	0.030	0.516	0.450	1.050	
0.382	7.470	0.035	0.533	0.500	1.101	
0.419	7.395	0.040	0.547	0.550	1.154	
0.464	7.770	0.050	0.573			

Table 3.18 Argon adsorption on non-porous lunar soil

| \multicolumn{4}{c|}{Original data} | \multicolumn{4}{c}{Smoothed α-s curve} |

P/P_{vap}	n_{ads} (μmol g^{-1})	P/P_{vap}	n_{ads} (μmol g^{-1})	P/P_{vap}	$n/n_{0.4}$	P/P_{vap}	$n/n_{0.4}$
0.029	2.327	0.411	6.869	0.020	0.361	0.400	1.000
0.059	3.416	0.500	7.583	0.040	0.447	0.450	1.061
0.099	3.949	0.600	8.483	0.060	0.507	0.500	1.123
0.144	4.557	0.691	9.234	0.080	0.554	0.550	1.188
0.198	5.210	0.766	10.248	0.100	0.595	0.600	1.257
0.253	5.676			0.150	0.680	0.650	1.332
0.306	6.096			0.200	0.752	0.700	1.415
0.355	6.517			0.250	0.818	0.750	1.510
				0.300	0.880	0.800	1.621
				0.350	0.940		

Table 3.19 Adsorption of O_2 on non-porous lunar soil

| \multicolumn{4}{c|}{Original data} | \multicolumn{4}{c}{Smoothed α-s curve} |

P/P_{vap}	n_{ads} (μmol g^{-1})	P/P_{vap}	n_{ads} (μmol g^{-1})	P/P_{vap}	$n/n_{0.4}$	P/P_{vap}	$n/n_{0.4}$
0.0003	0.000	0.245	4.880	0.00380	0.000	0.100	0.490
0.0006	0.000	0.280	5.631	0.004	0.0051	0.150	0.597
0.0014	0.000	0.352	6.246	0.005	0.028	0.200	0.688
0.0033	0.038	0.397	6.682	0.010	0.106	0.250	0.771
0.0117	0.788	0.452	7.222	0.015	0.157	0.300	0.849
0.0335	1.567	0.523	7.770	0.020	0.196	0.350	0.925
0.065	2.477	0.575	8.281	0.025	0.229	0.400	1.000
0.099	3.078	0.644	8.926	0.030	0.257	0.450	1.076
0.132	3.491	0.713	9.857	0.035	0.282	0.500	1.155
0.161	3.911			0.040	0.304	0.550	1.237
				0.050	0.344	0.600	1.324
				0.060	0.379	0.650	1.418
				0.070	0.410	0.700	1.523
				0.080	0.438	0.750	1.642
				0.090	0.465		

Table 3.20 CO adsorption on non-porous lunar soil

\multicolumn{4}{c	}{Original data}			\multicolumn{2}{c}{Smoothed α-s curve}
P/P_{vap}	n_{ads} (µmol g^{-1})	P/P_{vap}	n_{ads} (µmol g^{-1})	P/P_{vap}	$n/n_{0.4}$	P/P_{vap}	$n/n_{0.4}$
0.0006	2.793	0.219	7.583	0.0005	0.304	0.050	0.610
0.0031	3.378	0.274	8.071	0.001	0.335	0.100	0.697
0.0114	4.204	0.324	8.521	0.002	0.370	0.150	0.760
0.0215	4.557	0.389	8.694	0.003	0.392	0.200	0.815
0.0460	5.541	0.425	9.047	0.004	0.409	0.250	0.864
0.0854	6.119	0.484	9.497	0.005	0.422	0.300	0.910
0.133	6.607	0.538	9.970	0.010	0.469	0.350	0.955
0.177	7.132			0.015	0.499	0.400	1.000
				0.020	0.522	0.450	1.045
				0.025	0.542	0.500	1.092
				0.030	0.558	0.550	1.141
				0.035	0.573	0.600	1.192
				0.040	0.586	0.650	1.248

What about those Type VI isotherms?

There has been some discussion in the literature about the type VI isotherms as being a special case. Usually the steps of the Type II isotherms are interpreted as being due to phase changes, from a liquid-like film to a solid-type film. Indeed, this would be expected below the freezing point of the adsorbent. With the high energy for the first few "layers" one might obtain a liquid-like physisorption followed by a phase transition to the solid.

The characteristic "step-like" isotherm of the Type VI isotherm could also be due to multiple porosity sizes. If one were to have a material that is both microporous and mesoporous there would be at least two steps.

In Fig. 3.12 are some results by Ben Yahia et al. [53] for methane adsorbed on MgO at 87.4 K.

It is interesting to decipher what this isotherm indicates. Is it an isotherm with the following sequence?

1st—liquid-like adsorption with
2nd—micropore filling
3rd—some external area
4th—mesopores filled
5th—remainder of external area
6th—maybe some more mesopores

or is it some other combination of events?

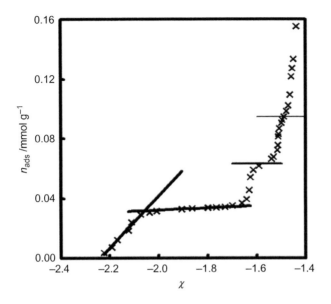

FIG. 3.12

Data by Yahia et al. of CH_4 on MgO at 87.4 K as a χ-plot.

First, the test for this assumption is performed. The initial adsorption is a straight line typical of the liquid phase adsorption. The calculation of the surface area from this slope yielded $n_m = 0.174$ mmol g^{-1} for the total surface. The bend downward is typical of microporosity, since this was followed by an continued upward linear portion. The slope of this second linear portion yielded a monolayer equivalent of $n_{m2} = 0.0076$ mmol g^{-1} for the external area. The intersection of the two lines indicate a shutoff of the pores at $n_p = 0.0316$ for the volume. These values yield by the IUPAC convention of 18.0 and 0.79 m^2 g^{-1} for the surface areas areas respectively and a pore volume of 0.119 mL g^{-1} (or again by IUPAC 1.19×10^{-7} m^3 g^{-1}.) This is OK, there is no apparent contradiction with these numbers.

There is however a problem with the next step. This cannot be the filling of mesopores since the slope at the higher χ values (the slope of the step landing) is obviously greater than before. This is not possible with mesopores according to χ theory. So this step must be something else. Furtnermore, the third landing has an even higher slope.

Since this isotherm was performed at 87.4 K and the freezing temperature is 90.65 K, it is reasonable to assume that second step is of a phase transition to the solid phase. If this were true then the adsorption from this point on would be "in register" and would simulate as a layer by layer adsorption. Backing up a bit then, if this is the case then the value calculated for the second slope does not have the meaning attached to it as stated in the paragraph above but could be, in principle, simply a horizontal line.

Thus, although the first slope. It is for a liquid-like state and could yield the surface area, the rest of the isotherm is either converting to the in-register condition or is already a solid. So checking this possibility, the first step starts at the intersection of the two lines as $0.316 \, \text{mmol} \, \text{g}^{-1}$. The second step should start at $0.632 \, \text{mmol} \, \text{g}^{-1}$ and the third at $0.948 \, \text{mmol} \, \text{g}^{-1}$. These latter two are shown as horizontal lines and certainly indicate that this hypothesis is probably correct.

From this particular example, the analysis agrees with the classical interpretation of the multistep Type VI isotherm. This is especially expected when the temperature of the measurements is below the normal freezing point. Nevertheless, the χ theory probably applies to the first part of the isotherm. An isotherm of this type needs more analysis to determine exactly what is the model that is appropriate. One isotherm at a temperature slightly below the freezing point is not definitive.

Summary of interpreting the physisorption isotherm

DFT and its variants create many models of adsorption in a form that can be stored and recalled in a computer program. The models depend upon the type of material the adsorbent is made from and its starting surface conditions: what type and how many pores the material has. It usually determines the overall surface area using the antiquated and unreliable BET theory, which ends up scaling the entire answer. It then matches the isotherm to one of the "kernals" using an inversion program that is ill-posed.

The χ theory has historical empirical evidence from the equations by deBoer and Zwikker and by Polanyi. It is, however, theoretically based in quantum mechanics; whereas these historical interpretations, although yielding excellent fits, indeed the best of any isotherm, could not yield the correct monolayer equivalent, nor the adsorption energy, as defined by χ theory, E_a. If they had been pursued, the might have yielded scaled pore sizes and other features.

χ theory starts with the isotherm or the heat of adsorption. It then transforms the pressure axis to a form fit for analysis, using the function $-\ln(-\ln(P/P_{\text{vap}})) := \chi$. The amount adsorbed, preferably as moles or millimoles, is plotted against this χ value. For simple, non-porous adsorbents, the slope of this plot is the monolayer equivalency (interpreted to yield surface area). For more complicated adsorbents, the plot is fitted to a function with a least squares routine. Any function will do provided it does a good fitting job. A high order polynomial, a segmented polynomial could be used, but the function as given in this book is most advised. A spline fit is not recommended since data scatter might create a big problem. This function is used simply to obtain the first and second derivative of the transformed isotherm. The data from these three curves yield all the parameters that can be obtained from the isotherm.

An approximation to this with two linear fits is presented in this chapter. This is similar to the method used for standard curves. It has some problems, especially for some mesopore samples, due to the distributions of adsorption energy and pore sizes. Nevertheless, it yields a good approximation. A more sophisticated, although more difficult, method is provided in Chapter 6.

The Disjoining Pressure (Excess Surface Work or ESW) analysis yields the same results for a simple isotherm as χ theory by assuming that the decay constant for the energy is a monolayer equivalent. This theory utilizes a decaying potential from the surface as proposed by Polanyi, and expected from the disjoining pressure theory. The decay constant was fortuitously chosen as the distance of the diameter of an adsorbate molecule to yield the value for the monolayer equivalent of $\theta = 1$. Even though this decay distance is an assumption reinforced by a thermodynamic argument, it is still an assumption. It turns out, however, to be a good assumption as it is confirmed by a derivation of the equations from χ theory, which, indicates the decay constant is a monolayer equivalent.

Both the χ theory and and ESW theory make the assumption that the adsorbed molecules have a tendency to clump in patches, otherwise there would be a factor greater than 1 in free adsorbent area for the first layer coverage. If the molecules were totally separated on the surface, then a factor of 2 would be introduced in the calculation. However, clumping of the adsorbate is most reasonable for two reasons: the interaction between the molecules is certainly strong enough and the factor of 2 in the χ plot slope would change to 1 with increasing coverage, which is not yet observed.

The Langmuir isotherm is one of the "Henry's law" isotherms and is, therefore, not suitable for physical adsorption. It is undoubtedly appropriate for chemisorption.

The Dubinin, et al. isotherms are mostly quite good for fitting to the low pressures. The theoretical basis is mostly a modification of the Polanyi formulation. It would be quite useful for the low pressure range. It has not, however, been tested in the HV range.

The Tóth formulation is an extension of the Langmuir-BET arrangement by assuming more limitations, with similar functions to these two. It yields the Henry's law in the low pressure range so it is unlikely to be a realistic representation.

The deBoer-Zwikker representation was getting very close to the theoretical answer. Unfortunately, the explanation was based on polarization of a uniform liquid film formation, which is not correct. Nevertheless, the criticism by Brunauer [54] was certainly unjustified and premature. The exponential decay used by deBoer and Zwikker, but not the explanation, is now known to be correct and, as an assumption, it is used by the ESW and by the various forms of density functional theory as the assumed potential from the surface. The assumption was also used for the Debye-Hückel theory.[10] The problem was a question of the size of the potential and the decay constant.

---

[10] It has also been demonstrated to be correct by the combination of Poisson's equation and Darken's diffusion equation and determining the equilibrium conditions.

Polanyi's formulation, especially his energy relationship (Eq. 3.46) was right on target. Indeed, this was a amplification of a concept first proposed by Eucken [55] in 1914 and modified by Polanyi [56] in 1916. The main problem was that the functionality testing of the concept was left unfinished. It was not followed through to the implications until many years later by Fuller.

Fuller was familiar with Polanyi's work and realized that, since the deBoer-Zwikker representation was extremely reliable, he was able to put these two works together to yield the proper formulation. He performed hundreds of experiments and amassed a large portfolio of information—much of which is still not published and much of which is hidden by classification—that confirmed this. However, due to resistance, Fuller always had to mask the true interpretation of the isotherm. Due to the overwhelming number of rejections, the obfuscation was used in order to publish in the open literature.[11] He usually used the energy argument instead of the χ theory (or the auto-shielded physisorption theory, ASP) most of the time. By doing so, he was able to publish hundreds of isotherms plotted in such a way that would indicate to the knowledgeable that these were indeed what are now called χ plots.

The Frenkel, Halsey, Hill (FHH) isotherm is an empirical approach that is useful in analyzing porosity. It can yield curves that mimic many isotherms for porous materials.

There are a variety of standard curves for a variety of adsorbents and adsorbates. For each material that is analyzed, and each adsorbate, a standard curve needs to be measured on well characterized material. Such an approach is very reliable in determining relative surface area. Unfortunately, the value of the monolayer equivalent is based on the BET equation, which makes it untrustworthy. Thus, it is useful only for a relative surface area with identical adsorbent material and an identical adsorbate. There is an extensive literature of standard curves; however, there is a serious caveat—much of the data for standardization was perform on equipment that does not meet modern specifications and controls. The answer one gets depends upon someone else having performed the experiments very well.

Finally, for Type VI isotherms when observed below the adsorptive freezing point, the χ theory may apply only in part, or not at all. It would still be worthwhile to perform a χ plot to check this. There is some guidance in the χ plot, for example, the slopes of step runs must decrease with increasing χ to be due to porosity. If it increases, it is probably due to change to the solid phase and subsequent layering. There are other theories that might also be useful for this case.

---

[11] This author has also experienced a great deal of resistance. In one year I had 12 articles rejected on very flimsy and unscientific basis. Examples include: "everyone knows this is wrong" and the "The author is incorrect that the function $\lim (1 - 1/n)^{-n}$ [as n becomes large] goes to $\exp(-n)$ because as n goes to infinity $1/n$ goes to 0 leaving 1." Upon appeal, the editor (W.S.) agreed with the reviewer and advised that I read his book, which I already had.

References

[1] J. B. Condon, The Equation for the Complete Analysis of the Surface Sorption Isotherm-U, D.O.E. Report Y/DV-323, June 16, 1987.
[2] E.L. Fuller, J.B. Condon, Colloids Surf. 37 (1989) 171.
[3] J. Adolphs, M.J. Setzer, J. Colloid Interface Sci. 180 (1996) 70.
[4] P.I. Ravikovitch, G.L. Haller, A.V. Neimark, Adv. Colloid Interf. Sci. 76–77 (1998) 203.
[5] M. Kruk, M. Jaroniec, Microporous Mesoporous Mater. 44–45 (2001) 725–732.
[6] S. Brunauer, P.H. Emmett, E.J. Teller, Am. Chem. Soc. 60 (1938) 309.
[7] S. Brunauer, L.S. Deming, W.E. Deming, E. Teller, J. Am. Chem. Soc. 60 (1938) 309.
[8] M.M. Dubinin, E.D. Zaverina, L.V. Radushkevich, Zh. Fiz. Khim. 21 (1947) 1351.
[9] M.M. Dubinin, Chemistry and Physics of Carbon, vol. 2, Dekker, New York, 1966, p. 51.
[10] M.G. Kaganer, Zhur. Fiz. Khim. 33 (1959) 2202.
[11] K.S.W. Sing, in: D.H. Everett, R.H. Ottewill (Eds.), Surface Area Determination, Butterworths, London, 1970, p. 25.
[12] J.H. deBoer, B.G. Linsen, T.J. Osinga, J. Catal. 4 (1965) 643.
[13] R.W. Cranston, F.A. Inkley, Adv. Catal. 9 (1957) 143.
[14] A.P. Karnaukhov, V.B. Fenelonov, V.Y. Gavrilov, Pure Appl. Chem. 61 (1989) 1913.
[15] W. Rudzinski, D.H. Everett, Adsorption of Gases on Heterogeneous Surfaces, Academic Press, New York, 1992.
[16] A.W. Adamson, Physical Chemistry of Surfaces, second ed., Wiley, New York, 1967.
[17] P.C. Hiemenz, Principles of Colloid and Surface Chemistry, second ed., Marcel Dekker, Inc, New York, ISBN: 0-8247-7476-0, 1986.
[18] K. W. S. Sing, in Adsorption by Powders and Porous Solids—Principles, Methodology and Applications, Rouquerol, F., Rouquerol, J., K.S.W. Sing, eds., Academic Press (ElSevier) Oxford ISBN 978-0-08-097035-6, p. 263.
[19] P.H. Emmett, S. Brunauer, J. Am. Chem. Soc. 59 (1937) 1553.
[20] H.L. Pickering, H.C. Eckstrom, J. Am. Chem. Soc. 71 (1952) 4775.
[21] P.H. Emmett, J. Am. Chem. Soc. 65 (1946) 1784.
[22] M. Jaroniec, M. Kruk, J.P. Olivier, S. Koch, Stud. Surf. Sci. Catal. 128 (2000) 71 (K. K. Unger et al. editors).
[23] J.B. Condon, Microporous Mesoporous Mater. 38 (2000) 359.
[24] J. Tóth, Adv. Colloid Interf. Sci. 55 (1955) 1.
[25] J. Tóth, Colloids Surf. 49 (1990) 57.
[26] D. Harkins, G. Jura, J. Chem. Phys. 11 (1943) 430.
[27] D. Harkins, G. Jura, J. Chem. Phys. 13 (1945) 449.
[28] F. Goldman, M. Polanyi, Phys. Chem. 132 (1928) 321.
[29] J.B. Condon, Langmuir 17 (2001) 3423.
[30] J.B. Condon, Microporous Mesoporous Mater. 55 (2002) 15.
[31] M. Krug, M. Jaroniec, Microporous Mesoporous Mater. 44 (45) (2001) 723.
[32] M.R. Bhambhani, R.A. Cutting, K.S.W. Sing, D.H. Turk, J. Colloid Interface Sci. 82 (1981) 534.
[33] S. Gregg, K.S.W. Sing, Adsorption, Surface Area and Porosity, Academia Press, New York, 1982.
[34] L.H. Cohan, J. Am. Chem. Soc. 60 (1938) 433.
[35] M. Giona, M. Giustiniani, Langmuir 13 (1997) 1138.
[36] M. Polanyi, Verk. Deutsch. Physik Gas 16 (1914) 1012.

[37] M. Polanyi, Z. Elecktrochem. 26 (1920) 371.
[38] M. Polanyi, Z. Elecktrochem. 35 (1929) 431.
[39] J.H. deBoer, C. Zwikker, Z. Phys. Chem. B3 (1929) 407.
[40] A.W. Adamson, Physical Chemistry of Surfaces, vol. 2, Interscience (Wiley), New York, 1967.
[41] M.R. Bhambhani, R.A. Cutting, K.S.W. Sing, D.H. Turk, J. Colloid Interface Sci. 38 (1972) 109.
[42] D.H. Everett, G.D. Parfitt, K.S.W. Sing, R. Wilson, J. Appl. Biochem. Technol. 24 (1974) 199.
[43] D.A. Payne, K.S.W. Sing, D.H. Turk, J. Colloid Interface Sci. 43 (1973) 287.
[44] S.J. Gregg, K.S.W. Sing, Surface Area and Porosity, Academic Press, NY, 1982.
[45] B.C. Lippens, G.G. Linsen, J.H. deBoer, J. Catal. 3 (1964) 32.
[46] F. Rodriguez-Reinoso, J.M. Martin-Martinez, C. Prado-Burguete, B. McEnaney, J. Phys. Chem. 91 (1987) 515.
[47] F. Rodriguez-Reinoso, J.M. Martin-Martinez, M. Molina-Sabio, R. Torregrosa, J. Garrido- Segovia, J. Colloid Interface Sci. 106 (1985) 315.
[48] K.J. Masters, B. McEnaney, Carbon 22 (1984) 595.
[49] R.B. Gammage, E.L. Fuller Jr., H.F. Holmes, J. Colloid Interface Sci. 34 (1970) 428.
[50] R.B. Gammage, H.F. Holmes, E.L. Fuller Jr., D.R. Glasson, J. Colloid Interface Sci. 47 (1974) 350.
[51] E. L. Fuller, Jr., P. A. Agron, The Reactions of Atmospheric Vapors With Lunar Soil. U. S. Government Report ORNL-5129 (UC-34b), 1976.
[52] E.L. Fuller Jr., J. Colloid Interface Sci. 55 (1976) 358.
[53] M. Ben Yahia, Y. Ben Torkia, S. Knani, M.A. Hachicha, M. Khalfaoui, A. Ben Lamine, Adsorpt. Sci. Technol. 31 (2013) 341.
[54] S. Brunauer, The Adsorption of Gases and Vapors, vol. 1, Princeton University Press, Princeton, NJ, 1945.
[55] A. Euckon, Verhandl. Deutsch. Phys. Gesellschaft (1914) 16.
[56] M. Polanyi, Verhandl. Deutsch. Phys. Gesellschaft 18 (1916) 55.

CHAPTER 4

Theories behind the χ plot

The χ plot is shown to be correctly based upon quantum mechanics with the WKB approximation and then by perturbation theory. Perturbation theory is much easier for most physical chemists to comprehend, so that is the presentation made here. For the perturbation derivation, the only assumption used is that the wave equations may be separated into a parallel (to the plane of the surface) part and a normal (perpendicular) part. For the parallel part, the perturbation is small enough that only the first order perturbation is required—simple enough that a senior chemistry major would be able to understand. The results of this calculation are then introduced into the grand canonical partition function with sums becoming integrals due to the large number of molecules being addressed. Changes in the rotational and vibrational modes upon adsorption are not entered,[1] but the change in the translational mode from 3D to 2D (a small contribution seen only with very good calorimetric measurements) is retained. This then yields the overall isotherm and the population density in each "layer." of adsorbate. The details of the normal direction are determined by superimposing a Lennard-Jones potential to the normal direction and solving the concatenated quantum mechanical harmonic oscillators approximation for the ground state.

The connection to the thermodynamic theory, which was independently arrived at and referred to here either as Disjoining Pressure Theory or Excess Surface Work (ESW) theory, is demonstrated. The assumption that there is a potential parameter that follows an exponential decay with adsorption is proven correct by a mathematical proof, and the assumption that the decay constant has the value of one monolayer equivalent amount is determined by χ theory to be correct. This makes the χ theory and the ESW theory essentially the same theory, even though the interpretation of the potential seems to differ.

The rest of this chapter concentrates on how to modify the χ theory to account for other phenomena, such as heterogeneous surfaces—that is, surfaces that have a distribution of adsorption energies, porosity measurements, and binary adsorption.

The heat of adsorption is addressed. There are several definitions for heats of adsorption and these are reviewed. The predictions of the χ theory are then compared to data, for which both the isotherm and the calorimetric data was obtained on the same sample. Even supposed identical samples can have variations in the energy

---

[1] Entering these back in would not be difficult if one knew what modes are being modified. In principle, one could use spectroscopy to determine this.

of adsorption. Thus, one cannot use separate publications to make the calculations. This is true even with different sample that have identical names. The isotherm derived numbers and the calorimetric numbers are in excellent agreement, providing more supporting evidence for the quantum mechanical and thermodynamic theories.

The quantum mechanics of the normal direction is present in order to obtain the density profile in the normal direction. This profiling can be used in the analysis of porosity. For this purpose, the quantum mechanical harmonic oscillator is used in conjunction with the n-layer normal distribution. This method is much simpler than, for example, density function theories, and yields nearly the same results if the same parameters are used. However, χ theory uses the energy parameters derived from the isotherm itself. This will be used in subsequent chapters to analyze porosity.

The first of these topics, heterogeneous surfaces, is relatively easy to address, since χ theory generates no anomalies in the process, unlike other isotherm theories, for example BET. The topic of porosity measurements may be handled in an approximate way by using the χ plot as a standard curve. Since χ theory and the χ plot do not require a separate standard curve, this yields internally consistent results, even if the energy of adsorption varies greatly. Chapter 6 is a presentation of a more sophisticate method.

Use of the χ equations for binary adsorption is demonstrated, with some interesting and important practical results demonstrated with actual data. A method which uses the isotherms of the pure adsorbates to predict the mixed adsorbate isotherms is presented. The method, used as a screening method, would be a great method to save effort when making decisions about the correct adsorbent and pressures to use in, for example, separations and catalytic processes.

Since much data has been described by equations other than χ, a comparison and interpretation for these isotherms is preformed. These isotherms include the Freundlich and the Dubinin-Polanyi class of isotherms, thus relating their parameters with those of the χ theory.

The final section discusses some thermodynamic details related to physisorption, specifically the Gibbs' phase rule when applied to surfaces, the derivation of the spreading pressure and another thermodynamic observation referred to here as the "Olivier monolayer criteria."

Introduction—A short historical background

In this section, the history behind the use of the χ plots is presented. As early as 1929, deBoer [1] recognized that what is being referred to here as the χ plot was an excellent fit to adsorption data. The accuracy of the χ plot has been known for many years, starting with the deBoer-Zwikker [2] equation. Adamson [3], in his book, described it as the best description for the entire isotherm ever devised. The deBoer-Zwikker theory, which was created to explain this plot, depended upon polarizability in its derivation. In spite of its obvious advantages, the theory behind it seemed to be very weak, according to Brunauer [4]. It was claimed that polarizability could not account for the high energies observed. This claim was justified on the basis of assuming that one layer had to be complete before the next layer would start to form. It did not take into account the some of the adsorbent would not be covered as the second layer formed. What was also ignored by critics is that London forces are not the only forces operating for strongly adsorbed molecules. For example, metal oxides have very strong polar bonds, metals have image potentials, and the list go on. Therefore, the forces could be, and are now known to be, much greater than initially assumed, especially by Brunauer. Furthermore, the mathematics of the deBoer-Zwikker formulation did predict very high energies, especially for the first layer.[2]

Two derivations to explain the χ plot will be presented. These include the disjoining potential theory (or excess surface work theory (ESW)), and the quantum mechanical derivation (or χ theory.) The classical derivation [5], or auto shielding physisorption (ASP) theory [6], is very similar to the quantum mechanical derivation, and was actually formulated to make it easier for reviewers to understand and accept. Although the ASP derivation is simpler to understand, it seems less plausible and there is a temptation, for chemists at least, to attempt to revert to classical chemical reaction equilibrium mathematics, upon which the BET theory is based. This process, which is quite understandable, is incorrect, and simply creates confusion.

Theory behind χ plots
The quantum mechanical derivation of the "simple" χ equation

The quantum mechanical derivation of the χ theory is quite simple. It is based on a few principles that are well understood and widely used in both adsorption and other areas. For this discussion, the normal direction to the surface is designated as the "z" direction and the directions parallel to the surface, including the surface, are designated the "x and y" directions.

- Firstly, the wave functions of three-dimensions are easily separated with the wave component for the z being independent of the wave equation x and y directions.
- The wave equation in the z direction is determined by intermolecular forces. This is simulated by a Lennard-Jones 6–12 potential (although any reasonable potential that is solvable will do in this direction.)
- The x and y portion of the wave function is a simple "particle in the box" with the perturbation involving other adsorbed particles.

In the derivation it will be demonstrated that:

- Each adsorbed molecule will contribute a very small perturbation and thus only a first order perturbation calculation is required.

---

[2] At the time most investigators assumed that the adsorbate was layered, that is the first layer is compete before the second and so forth. At the time the BET broke this idea but no-one tried to break this "rule" for the deBoar-Zwikker situation.

- The shape of the perturbation does not matter. So long as it is small, the magnitude of the perturbation is equal to the energy-x-y volume only.
- Given that only the first order perturbation is required for each atom, a simple principle may be stated regarding perturbation theory in this case. Given the solution to the wave equation, that is, the energy as a function of quantum number, if one introduces a perturbation, then for most wave numbers, the energy is a volume average or, in this 2D case, of a surface an area average, of the original energy and the perturbation. This was pointed out in the first edition of this book that this is the known result of a first order perturbation. This was used as the in the first edition as a starting point to derive the χ theory, since it seemed to be obvious. However, this was apparently not obvious to the reader so in this edition the narration backs up a bit to demonstrate that this particular perturbation is indeed derivable.
- In addition, this to a first approximation, this perturbation technique can also yield a *maximum* value for the *uncertainty* of the energy shift for an individual adsorbate molecule as a function of wave number.

Some of the conventions used are:

- The depth of the potential box is designated as E_q from the gas phase. However, the standard state is, for practical purposes, the liquid vapor pressure at the temperature of interest, and is designated as ε.
- It is assumed that the temperature of adsorption is such that the adsorbed molecule will behave much as a liquid molecule would behave. Otherwise there would be no formation of the liquid state at the saturation pressure. This means that the specific potential wells corresponding to surface atoms on the surface (that is, at the bottom of the empty well) are small compared to translational energy of the molecule. Thus, the details of the surface potential on an atomic scale may be ignored. Recent modifications of the DFT theories have needed to succumb to this reality. There might be, however, larger areas where there are a variety of energies, but here for simplicity sake, a homogeneous surface is used. Therefore:
- For the first adsorbate molecule to arrive at the surface one can treat it as simply a particle-in-a-potential-box. The energy of the first molecule will be in a state consistent with its deBroglie wave length, designated E_0. That is, E_0 is the energy released for the first molecule ($|E_0|$ is greater than $|E_q|$ by a certain amount depending upon the wave number; however, at the temperatures of interest it is usually the ground state and $|E_0|$ is only slightly above $|E_q|$.) The energy E_a is the energy release referenced to the liquid source or $E_0 - \varepsilon$.
- For the second particle, it will arrive at the surface and it will experience one of two potentials. One of these potentials is E_q. The classical[3] interpretation of this is

---
[3] This is the classical analog to the addition of the wave functions and their exchange to yield symmetric and antisymmetric wave functions.

that if it were to encounter the first molecule, with an effective area of a, it might "roll under" it. On the other hand if it were to "roll over" the other molecule then the energy would be an area average of E_q and the energy of interaction between the adsorbate molecules, ε. The first molecule's energy is also modified in the same manner due to the presence of the second molecule. In addition to this, there is now the interaction energy between the molecules regardless of which one "rolls over" or "under." Quantum mechanically, the energy for this second addition would be $E_0 - \varepsilon$ for area not covered by the "tooth" for the first molecule or $(A_s - a)$, where a is the area occupied by the first molecule and A_s is the surface area. The energy for the area covered by the first molecule would simply be ε. Thus, for the two molecules:

$$\sum_{m=1}^{m=2} E_m = (E_0 - \varepsilon) + (E_0 - \varepsilon)\left(\frac{A_s - a}{A_s}\right) + 1\varepsilon \qquad (4.1)$$

- This logic is repeated for the third molecule:

$$\sum_{m=1}^{3} E_m = E_a + E_a\left(\frac{A_s - a}{A_s}\right) + E_a\left(\frac{A_s - a}{A_s}\right)^2 + 2\varepsilon \qquad (4.2)$$

- and for the Nth molecule:

$$\sum_{m=1}^{N} E_m = \sum_{m=1}^{N} E_a\left(\frac{A_s - a}{A_s}\right)^{m-1} + (N-1)\varepsilon \qquad (4.3)$$

This takes into account all the possible interactions that could be present including adsorbate-adsorbate interactions designated with the final εs. Notice that the sequence of the addtion does not matter, since these are all identical particles. Typically, the thermal wave length of the adsorbate molecules is about 1/20 of the size of the molecule itself, but the model takes into account even the long range interactions. Another way of looking at this is the Copenhagen convention, in which the spread-out molecular wave "sees" a large portion of the surface before its collapse to a molecular particle.

These, then, are the primary assumptions and the implications. The primary assumption of a mobile adsorbed molecule and the classical interaction of the "roll over" effect are the designation of a liquid-like phase. As stated in Chapter 1, without the "roll over" the adsorbate would act like a gas phase and without the mobility assumption, the adsorbate would behave as a solid. Thus, the mobility is concluded by a process of elimination. This mobility is often referred to as the "non-localized," or "NL," assumption in other formulations, so it is not now unique or unusual and should no longer be a stumbling block. This was a critical break from the Henry's law types of theory, including the BET. The "localized" or solid state has been widely observed under different conditions and confirmed with various diffraction techniques.

For the gaseous state there seems to be some evidence for it, especially at pressures approaching the critical pressure, but for the conditions of adsorption that this book addresses, it is undoubtedly beyond the limit of detection by either gravimetric or volumetric techniques.

The following analysis is the quantum mechanical derivation of Eq. (4.1), which makes the "roll over" and "under" assumption irrelevant. The "roll over" explanation was originally used by Fuller to make the ASP theory understandable and acceptable. For those who trust the derivation's conclusions, which are after all quite reasonable, skip this next topic and go on to "Continuing the derivation."

A Particle in a box with a "Tooth"—Perturbation theory for adsorption on a surface with a molecule present

The quantum mechanical derivation needed for Eq. (4.1) will, for simplicity here, assume a 1D box of length L. Two-dimensions is an obvious extension. However, the 2D derivation simply doubles the indices and makes for confusion, but does make the approximation even more valid since the consideration of Eq. (4.8) below is squared.

The following is a summary of the treatment that can be followed in almost any textbook on quantum mechanics (see, for example, the book by Sherwin [7].) This treatment assumes non-degenerate perturbation, which is logical. The zero energy level is taken as the adsorbate on the bare surface, thus the energy, E_a, is to the top of the potential well.

The energy, ε, is a positive perturbation which is considerably smaller than E_a. There are two obvious observation made in the following derivation. The first observation is that the potential wall is higher than ε. This makes it obvious that adsorption should take place, but it is not a necessary condition. The second insight is that the width of the well is extremely wide compared to an adsorbate molecule and the number of cycles for molecular wave. These two observations lead to the conclusion that the infinite wall for the potential well is a good approximation for the wave functions, though not the energies, to the wave function for the finite wall. Thus, for ease of calculation, an infinite wall will be used. Both of these assumptions are not necessary,[4] but it simplifies the derivation considerably.

The diagram in Fig. 4.1 illustrates the justification for using an infinite wall in place of the finite wall for the wave functions. The error basically has to do with where the virtual sine wave crosses the zero amplitude. For a finite barrier the wave curves upward before $x=0$ and most of the wave may be simulated as a sine wave that goes to zero with $x<0$. However, the difference is trivial compared to the overall wave that may exist—in this case, more than 1000 cycles.

---

[4] I have solved this both numerically and doing the normal finite wall and not surprisingly come to the same result, although with a lot of mess in between. Hint: The digital is easier.

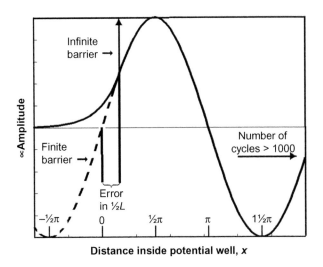

FIG. 4.1

Illustration of the error made by assuming an infinite potential well depth versus a finite potential well depth.

In Fig. 4.2 is a diagram that indicates the arrangement of the potential well with the reference state for the adsorbent, which is the bulk liquid state. A potential perturbation is shown, which is energy of ε below the *top* of the potential well. This value is equal to the heat of vaporization of the pure liquid, the standard state. The overall depth of the well is E_q with the average energy state E_0. The width of the perturbation is l as indicated by the dashed lines. The height from the bottom of the well, labeled α, is the perturbation that the first adsorbed molecule will create. The increase in observed energy of adsorption for the next adsorbed molecule is labeled as al/L, as will be determined by the following derivation.

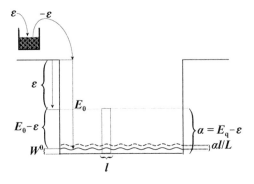

FIG. 4.2

Diagram of the surface potential well and the energy definitions.

CHAPTER 4 Theories behind the χ plot

Starting with the unperturbed equation with various states n:

$$\hat{H}^0 \psi_n^0 = W^0 \psi_n^0 \tag{4.4}$$

where $\hat{H}$ is the Hamiltonian operator, ψ_n^0 are the wave functions and W^0 are the energy eigenvalues.

The unperturbed eigenfunctions, individually labeled with integer n, are:

$$\psi_n^0 = \sqrt{\frac{2}{L}} \sin\left(\frac{n\pi x}{L}\right) \tag{4.5}$$

where L is the length of the potential box and x is the position within the box ($0 \leq x \geq L$.)

The energy eigenvalues for the unperturbed system are:

$$W_n^0 = \frac{n^2 \pi^2 \hbar^2}{2mL^2} \tag{4.6}$$

This is the off-set from the bottom of the well to the first ground state. Energies will be referenced against E_a and not E_q. A perturbation of $\varepsilon > 0$ is introduced with a width of l which changes $\hat{H}^0$ to $\hat{H}^0 + \hat{H}'$ and W^0 to $W^0 + W'$. This leads to the use of the original wave functions in combinations to yield[5] an n by n matrix of corrections, $\mathbf{H}'_{mn}$. This matrix is used to correct the original equation:

$$\mathbf{H}'_{mn} = \int \psi_m^{0*} \hat{H}' \psi_n^0 ds \tag{4.7}$$

Since

$$l < 1 \times 10^{-6} L \quad \therefore \quad \alpha l <<< E_q L \tag{4.8}$$

then the first order correction ($a_n = 1$ and $a_{m \neq n} << a_n$) should be an extremely good estimate of the overall perturbation energy.

The first order approximation is obtained in the following fashion. For each original n there is now a ψ that is modified in the perturbed system; call it ψ'_n. These ψ'_n consist of linear combinations of the original ψ_n^0s. The perturbed system to the first order yields then:

$$\left(\hat{H}^0 + \hat{H}'\right) \psi'_n = \left(W^0 + W'_n\right) \psi'_n \tag{4.9}$$

and rearranged:

$$\hat{H}^0 \psi'_n - W_n^0 \psi'_n = \left(W'_n - \hat{H}'\right) \psi'_n \tag{4.10}$$

Using the linear combinations of ψ_n^0s for each ψ'_n and Eq. (4.4) for each j:

$$\hat{H}^0 \psi'_n = \hat{H}^0 \sum_j a_j \psi_j^0 = \sum_j a_j W_j^0 \psi_j^0 \quad \text{also} \quad W_n^0 \psi'_n = W_n^0 \sum_j a_j \psi_j^0 \tag{4.11}$$

---

[5]The original wave functions were ψ_n^0, where n is enough wave functions either for practical purposes or the energy of the next function would be the free particle. The corrected wave function due to the perturbation for the mth level is then: $\psi_m = a_1 \psi_1^0 + a_2 \psi_2^0 + \ldots (\sim 1) \psi_m^0 + \ldots a_m \psi_n^0$. The index, however, was discontinued at the nth level, so m then runs from 1 to n, thus creating an $n \times n$ matrix of equations.

Theory behind χ plots

$$\sum_j a_j W_j^0 \psi_j^0 - W_n^0 \sum_j a_j \psi_j^0 = \left(W_n' - \hat{H}'\right)\psi_n^0 \tag{4.12}$$

$$\sum_j a_j \left(W_j^0 - W_n^0\right)\psi_j^0 = \left(W_n' - \hat{H}'\right)\psi_n^0 \tag{4.13}$$

Multiplying both sides by the complex conjugates, ψ_n^{0*} s:

$$\psi_n^{0*} \sum_j a_j \left(W_j^0 - W_n^0\right)\psi_j^0 = \psi_n^{0*}\left(W_n' - \hat{H}'\right)\psi_n^0 \tag{4.14}$$

and then integrating over all space, the values of the integrals are either 1 or 0, depending upon the values of j and n except for the integral with $\hat{H}'$.

$$\sum_j a_j \left(W_j^0 - W_n^0\right) \int \psi_n^{0*}\psi_j^0 d\tau = W_n' \int \psi_n^{0*}\psi_n^0 d\tau - \int \psi_n^{0*}\hat{H}'\psi_n^0 d\tau \tag{4.15}$$

$$\int \psi_n^{0*}\psi_j^0 d\tau = 0, n \neq j, \quad \& \quad \int \psi_n^{0*}\psi_n^0 d\tau = 1 \tag{4.16}$$

$$0 = \sum_n a_n \left(W_n^0 - W_n^0\right) = W_n' - \int \psi_n^{0*}\hat{H}'\psi_n^0 d\tau \tag{4.17}$$

or:

$$W_n' = \int \psi_n^{0*}\hat{H}'\psi_n^0 d\tau \tag{4.18}$$

Thus, for the first order approximation, only the matrix elements $\mathbf{H}'_{nn}$ are used to make the calculation. This is almost always justified for even fairly large perturbations. It certainly should be justified in this case, since the perturbation is less than a millionth of the overall well, as indicated by Eq. (4.8). The diagonal matrix $\mathbf{H}'_{nn}$ should yield all of the W_n' values required for this calculation.

Concentrating on the nth ψ and using the real part of the ψ_ns, and noting from the arrangement illustrated in Fig. 4.2 that $\hat{H}' = \alpha$ between distance $x - \frac{1}{2}l$ and, $x + \frac{1}{2}l$, where x is an arbitrary position at $(0 + \frac{1}{2}l) \leq y \leq (L - \frac{1}{2}l)$ (the $\frac{1}{2}l$, $<<<L$ and can be ignored also), then:

$$W_n' = \alpha \int_{x-\frac{1}{2}l}^{x+\frac{1}{2}l} \psi_n^{0*}\psi_n^0 d\tau \tag{4.19}$$

Using Eq. (4.6) This yields the perturbation on the original energies of:

$$W_n = \int_0^L \psi_n^{0*}\hat{H}^0\psi_n^0 d\tau + \alpha \int_{x-\frac{1}{2}l}^{x+\frac{1}{2}l} \psi_n^{0*}\psi_n^0 d\tau \tag{4.20}$$

CHAPTER 4 Theories behind the χ plot

Now for a little transformation. The first integral as given by Eq. (4.20) is the starting W^0s. The second integral yields:

$$\frac{2}{L}\int_{x-\frac{1}{2}l}^{x+\frac{1}{2}l} \sin^2\left(\frac{n\pi\tau}{L}\right) d\tau = \frac{1}{L}\left(\tau - \frac{L}{n\pi}\sin\left(\frac{n\pi\tau}{L}\right)\cos\left(\frac{n\pi\tau}{L}\right)\right)\Bigg|_{\tau=x-\frac{1}{2}l}^{\tau=x+\frac{1}{2}l} \quad (4.21)$$

But what is n? n is the original n for the solution to the entire potential well, but between $\tau = x - \frac{1}{2}l$ and $\tau = x + \frac{1}{2}l$ there will be only ~6 to ~30 wave nodes according to the deBroglie wave equation. Therefore, expressing this within the perturbation the following change in the integer may be made:

$$n = \frac{kL}{l} \quad 6 \tilde{<} k \tilde{<} 30 \quad (4.22)$$

Substituting kL/l for n into Eq. (4.21), evaluating the integral, substituting into Eq. (4.20) and using the trigonometric relationship for sine-cosine, the following is obtained:

$$W_n = W_n^0 + \frac{\alpha}{L}\left(l - \frac{l}{k\pi}\left[\sin\left(\frac{2k\pi x + l}{l}\right) - \sin\left(\frac{2k\pi x - l}{l}\right)\right]\right) \quad (4.23)$$

or using the minima and maxima of the sine terms:

$$W_n = W_n^0 + \frac{\alpha l}{L}\left(1 \pm \frac{1}{k\pi}\right) \cong W_n^0 + \frac{\alpha l}{L} \quad (4.24)$$

Recognizing that the convention for the Ws are from the bottom of the potential well as opposed to the top referenced against the vapor pressure potential. The $\pm 1/k\pi$ is small compared to 1 and the exact value depends upon the position x and the number of nodes, whether an even integer, an odd integer or somewhere in between.

Notice that Eq. (4.24) is the starting point for the χ theory derivation. This can be seen by making the association $\alpha = E_q - \varepsilon$ and recognizing that the third molecule "sees" both of the previous molecules as perturbations:[6]

$$W_{n,1} = \alpha \quad \text{and} \quad \alpha = E_0 - \varepsilon \quad (4.25)$$

$$W_{n',2} = \alpha + \alpha\frac{1}{L} = \alpha\left(1 + \frac{1}{L}\right) \quad (4.26)$$

$$W_{n'',3} = \alpha\left(1 + \frac{1}{L}\right) + \alpha\left(1 + \frac{1}{L}\right)\frac{1}{L} = \alpha\left(1 + \frac{1}{L}\right)^2 \quad (4.27)$$

Notice that there is a very small correction since α is reference against E_q and not against $E_0 := W_1^0$. This however is a constant for α throughout the derivation and does not alter the basic results.

The primes in this equation indicate slightly different quantum numbers. For each of the molecules added, however, there is an additional energy of ε that needs to be added since the accounting for the definition of α begins with release of ε. This adds

---

[6]Since the second (and the first) adsorbed molecules are also probability waves, the base level in the box includes them both as part of the box.

the term $N\varepsilon$ at the end of this series (3ε for the third equation.) By extension, to 2D, this continues with larger numbers of adsorbate molecules to yield Eq. (4.3).

The primary problem is variation in l and how that effects the precision. In Eq. (4.24) are some maximum errors that could be seen from differences in l. In Table 4.1 are listed the maximum errors associated with a single molecule. The adsorbate N_2 has a thermal deBroglie wave length of about 12 nodes[7] per molecule diameter, so the maximum error from this effect for a particular molecule is given as 1.6 %. Since the molecules move around, this value varies between $+1.6\%$ and -1.6% depending upon x and y. With enough particles it would seem reasonable that the error averages out. Thus, by chance, the error should be insignificant.

An important point to keep in mind is that the quantity al/L is an area (in 3D, a volume), and the specifics of the term al, whether a is small or l is small, does not matter for the perturbation except for the maximum error, and a large a and a small l yields the highest error. If the pre-adsorbed molecules are expressed as a wave, the effective l becomes larger and a must become small for the term to remain a constant.

Another important point to notice is that the surface free of any adsorbate molecules is given by:

$$\left(1-\frac{1}{L}\right)^N \text{ or for 3D}: \left(1-\frac{a}{A_s}\right)^N \quad (4.28)$$

Notice that the second term is the same as the function inside the summation of Eq. (4.3).

Summary of the QM perturbation theory

From this derivation the following may be concluded:

- Due to the small size of the perturbation of each molecule adsorbed compared to the size of the adsorbate surface, a first order perturbation is all that is needed for determining the dependence of the energy release.

Table 4.1 Error expected from the QM for the χ equation

Nodes across the adsorbate molecule	Maximum error $\pm l/k\pi$
6	5.3%
8	3.9%
10	3.2%
15	2.1%
20	1.6%

---

[7] The thermal deBroglie wavelength is given by the equation $\lambda = \frac{h}{\sqrt{2\pi mkT}}$ where h is Plank's constant, k = Boltzmann's, m is the mass of the molecule and T is temperature in kelvins. Assumning 77.2 K and Nitrogen radius or either 0.23 nm by IUPAC convention or 0.155 nm as van der Waal, the wave length, λ, calculates to be 0.038 nm. This yields the ratio of diameter to wave length of 12.2 and 8.2 respectively.

- The ground state correction for most systems is very small, according to Eq. (4.5), for most adsorptions. For example, N_2 it is about $70\,J\,mol^{-1}$ for $0.1\,nm$ surface versus the energy of adsorption in the kJ range.
- These conclusion may not be the case for such extremely small areas, such a biologically contained nano-space and very large molecules.
- The perturbation in the 2D space is the 3D energy-area space of the molecule. This means that although the derivation derived in this section used a rectangular solid to simulate the perturbation, this does not really matter. The energy-area space of the molecule may be any shape, including a deBroglie wave.

The 3D function given by Eq. (4.28) incorpoated into Eq. (4.3) will be multiplied by the first molecule energy of adsorption and will be entered into the grand partition function in the next section.

Continuing the derivation

Defining a quantity called the coverage, $\theta = Na/A_s$ and recognizing that for large N the first part of Eq. (4.3) involving the sum may be replaced with an integral thus:

$$E_N = E_a \int_0^N e^{-\theta} dm \qquad (4.29)$$

Here the *definition* of the e function has been taken advantage of in taking the limit of high value of A/a^8. Since the energy and combinational considerations are settled, one can proceed by various paths to arrive at the isotherm. For an open system[9] such as this, the grand canonical ensemble is the best method and most convenient method for this purpose. The reason that it is the best is that the the grand canonical ensemble is designed for an open system, where material can enter and leave the system as required. This is then given by:

$$\Xi = \sum_N (\lambda Z)^N \exp\left[-\left(\frac{E_a}{kT}\right)\int_0^N \exp(-\theta)dm - \frac{N\varepsilon}{kT} + {}^1/_2 N \pm \frac{Nf(T)}{kT}\right] \qquad (4.30)$$

The terms in the third and fourth terms of the exponential function are small terms, which include:

1) $\frac{1}{2}kT$ = the loss of some translational modes for the molecules near the adsorbent leading to a difference in heat capacity of $\frac{1}{2}kT$ for low coverages [8]; and
2) $f(T)$ = possible changes in rotational modes, vibrational modes, etc., for heat capacity effects.

---

[8] OK, for the sceptics and critics: $e := \lim_{x\to\infty}(1+1/n)^n$ thus, substituting $-A/a \equiv n$, raising both sides to the Nth power to match Eq. (4.3) and noticing a/A is extremely small (typically $<10^{-4}$ for an error of 0.02%) and $\theta = NA/a$ then one ends up with this equation.
[9] The definition of the "system" is the combined adsorbent and adsorbate combination.

These latter terms are almost always small, especially for simple adsorbent molecules. However, the first one has been observed with the heats of adsorption [5].

The usual method is to take the natural log of Ξ (which yields the grand canonical potential) and then differentiate with respect to N the maximum term, designated here by Ξ_{max}, obtained from the natural log, the grand potential. This maximum is found by setting the resultant equation to 0. The canonical ensemble term λZ is replaced by the fugacity, $\tilde{p}$,[10] which is equal to simply P at low pressures,[11] which is known to be true.

$$0 = \frac{\partial \ln(\Pi_{max})}{\partial N} = \ln(\tilde{p}) + \frac{\{-E_a e^{-\theta} - \varepsilon + \tfrac{1}{2}kT \pm \mathbf{f(T)}\}}{kT} \quad (4.31)$$

Keep in mind that the system for adsorption experiments is the adsorbate-adsorbent and thus $E_a < 0$. The switch to chemistry type molar quantities, $\overline{E_a}$ and $\overline{\varepsilon}$ are now used. For the moment, without the last two terms, this can be rearranged:

$$\frac{\overline{E_a}}{RT} e^{-\theta} = \frac{\overline{\varepsilon}}{RT} - \ln(\tilde{p}) \quad (4.32)$$

Assuming ideal gas for the adsorptive so that $\tilde{p} \approx P$ and recognizing that as $N \to \infty$ ($\theta \to \infty$) then $P \to P_{vap}$ (and $\varepsilon > 0$.) This implies that:

$$\ln(P_{vap}) = \frac{\overline{\varepsilon}}{RT} \Rightarrow \overline{\varepsilon} \equiv -\overline{E}_{vap} \quad (4.33)$$

Substituting then:

$$0 = \ln\left(-\frac{\overline{E_a}}{RT}\right) - \ln\left[-\ln\left(\frac{P}{P_{vap}}\right)\right] \quad (4.34)$$

Noticing that as $\theta \to 0$, then

$$\frac{P}{P_{vap}} = \exp\left(-\frac{\overline{E_a}}{RT}\right) \quad (4.35)$$

Thus it is required that a finite, although perhaps a small pressure, called the threshold pressure with the symbol P_c:

$$P_c := P_{vap} \exp\left(-\frac{\overline{E_a}}{RT}\right) \quad \therefore \frac{P_c}{P_{vap}} = \mathbf{g}(T, \overline{E}_a) \quad (4.36)$$

The threshold pressure is needed according to thermodynamics to form the adsorbate, which is being addressed here. This is interpreted as being the phase change from the postulated surface gas (or no adsorbate) to the liquid phase starting on the surface.

The surface gas phase, if it exists, would normally be very small and possibly not even detectable with most apparatuses used to measure the isotherm. A quick calculation of the amount of gas adsorbate would indicate the density would be less than a

---

[10] $\tilde{p}$ is the latest IUPAC symbol for fugacity.
[11] This pressure needs to be referenced against a standard state. This will be straightened out below by referencing to P_{vap}.

thousandth of a monolayer equivalent[12] (see Appendix II.F for the evidential calculations for these statements.)

A simple calculation for this pressure at a typical $\overline{E}_a$ of $10\,\text{kJ}\,\text{mol}^{-1}$ (and $P_{\text{vap}}=1$ bar) yields a pressure of 2.0×10^{-7} bar. Although such a pressure is easily measured, it hardly ever is for adsorption experiments. This has been done, rarely, only recently. Looking at it another way, to see a 1 torr offset (1.3×10^3 bar), barely visible in the usual isotherm, the energy of adsorption would have to be $4.3\,\text{kJ}\,\text{mol}^{-1}$ or less; a very low value, indeed. If, on the other hand, high, or even better, ultra high vacuum measurements were performed, the isotherm would appear to have a threshold pressure for adsorption. The threshold pressure concept has been very contentious, but it can be demonstrated experimentally, and has been by several investigators. Needless to say, the Disjoining Pressure Theory also predicts a threshold pressure as well, but this is usually not mentioned.

A compact and simple form of the χ isotherm equation is created by defining the following:

$$\chi_c := -\ln\left(-\frac{\overline{E}_a}{RT}\right) \equiv -\ln\left[-\ln\left(\frac{P_c}{P_{\text{vap}}}\right)\right], \chi = -\ln\left[-\ln\left(\frac{P}{P_{\text{vap}}}\right)\right] \text{ and } \Delta\chi \equiv \chi - \chi_c \quad (4.37)$$

The molar energy of adsorption for the first adsorbed molecule, $\overline{E}_a$ is given by:

$$\overline{E}_a = -RT\exp(-\chi_c) \quad (4.38)$$

and recognizing that negative values of N and therefore χ are not allowed then this simple form of the χ theory equation is obtained:

$$\theta = \Delta\chi \mathbf{U}(\Delta\chi) \quad (4.39)$$

where $\mathbf{U}$ is the unit (Heaviside) step function.

This equation is itself useful. It is capable of yielding an analytical expression for standard isotherms. Since θ is the adsorbate coverage then logic indicates that the surface that remains free of adsorbate is given by the equation:

$$\theta_{\text{free}} = e^{\Delta\chi \mathbf{U}\Delta\chi} \quad (4.40)$$

This equation and similar ones are important for the density profile in the normal direction. This will be expanded upon later in this chapter.

The theory also predicts the heats of adsorption when reported in terms of moles of material adsorbed rather than coverage. Notice that the value $\overline{E}_a$ is not the function $\mathbf{E}_a(\Delta\chi)$, the heat of adsorption as a function of coverage or pressure. This will be addressed later in this chapter with the various definitions of the "heat of adsorption."

The theory establishes very firmly a method to calculate the monolayer equivalent moles, n_{ads} and the "heat of adsorption" as a function of moles of adsorbate with

---

[12]It must be emphasized that the experimental conditions being referred to here are the normal physical adsorption experiments using an adsorptive with the $P \leq P_{\text{vap}}(T)$ where the sample temperature is $T > T_{\text{fusion}}$. It could possibly be extended beyond these conditions; however, at much lower temperature it should not be surprising that a solid phase forms on the adsorbent. There are also conditions where a gas-like phase *might* be detectable.

consistency between the isotherm and the enthalpies of adsorption and presents no possibility for an anomaly. It also fulfills the "thermodynamic criterion" given by Dubinin that has been tested by many measurements of the isotherm. The two parameters, the monolayer equivalency and the heat of adsorption that are yielded directly should be the primary quantities to be sought. However, the parameters that are most often quoted are not these, but rather the surface area and possibly the BET constant or other constants. At any rate, regardless of what theory is used, the isotherm must include two constants as a minimum in order to fit the experimental data.

The connection with surface area, however, is not established at this point since the value for a, the effective area occupied by an adsorbate molecule, is yet to be determined here. However, since it has already been established that the onset at χ_c is the start of the liquid phase, the a must be related to the van der Waal's radius and close packing or liquid density, which is the assumption made for the BET. For the BET, this seems rather contradictory, since the theory is more applicable to epitaxy. In other words, it is a localized theory, more applicable to a solid than to a liquid.

It has been established here that a monolayer equivalent is defined by $\theta = 1$. It should be noted that at $\theta = 1$ there is still unoccupied adsorbent surface. This "free surface" can be calculated from Eq. (4.28) assuming a small a/A_s and is given by:

$$\theta_{\text{"free"}} = \exp(-\Delta\chi) \Rightarrow \therefore \theta_1 = 1 - \exp(-\Delta\chi) \qquad (4.41)$$

where θ_1 stands for the coverage of the 1st "layer."

Thus the coverage of the first layer when $\theta = 1$ is not 100% but rather only about 63%. In the section on porosity, this equation will be used for ultramicroporosity since "layers" greater than the 1st are excluded due to geometry. Additional calculation for $\theta = 1$ for other layers indicates that 31% of the adsorbate molecules are once removed from the surface, about 6% are twice removed, and very little is left for the 4th "layer." This is an interesting point that the distribution of the molecules is over roughly three molecular "layers." This does not take into account the wave function in the normal (z) direction. When this is added into the theory, the physical distinction between layers after ~5 to 10 layers, depending upon the Lennard-Jones potential, is totally missing.

A quick comment: the definition of "layer" is a little different in this theory. Classically speaking, a 1st layer molecule is one that is in direct contact with the adsorbent surface as the minimum path. A 2nd layer has one molecule intervening between itself and the surface in the minimum path. A 3rd layer molecule has two molecules intervening in the minimum path, etc. The QM "exchange" however is very rapid and energy conservative.

The heats of adsorption can also be derived from this theory, placing the translational energy back in. The loss of ½RT is a deviation from the "thermodynamic criterion," but a very small one normally unobservable in the isotherm. The term **f**(T), however is more difficult and varies with each adsorptive and geometry of adsorption. For most simple molecules **f**(T) = 0 but for more complex molecules the surface might restrict some vibrational and rotational modes; for example, fatty acids adsorbed on an "organic" surface. Thus, **f**(T) for simple molecules it is not an observable deviation from the "thermodynamic criterion," but could be quite large for large complex molecules. Derivations for the heat of adsorption using the χ theory are presented the subsection Reformulation for a Distribution of E_a Values in this chapter and presented in more detail in Chapter 5 in the section on Heats of Adsorption.

Summary of χ theory derivation

In this section, the "simple χ theory," that is, the isotherm equation that describes adsorption on a flat surface, was derived from quantum mechanics using the perturbation technique. The derivation yields a simple equation, Eq. (4.39) with definitions Eq. (4.37), that has two parameters as output from a simple linear regression. These two parameters are the monolayer molar equivalence, n_{ads}, and the energy of adsorption for the first adsorbed molecule, $\overline{E_a}$.

The calculation of the coverage of the molecules in direct contact with the surface is presented, and demonstrates that a monolayer equivalent is not that same thing as a completed monolayer in direct contact with the surface. This yields an energy of adsorption that could be misinterpreted as a potential penetrating into the adsorbate, as deBoar and Zwikker postulated. This, however, could lead to other misinterpretations.

The energy of adsorption as a function of coverage is contained within the equation but requires further explanation presented in a later section.

For this derivation, the change in the translation mode from the 3D liquid state and the surface confined phase, which is quite small, was ignored. Since simple adsorbates were considered, the changes in internal vibrations and rotations of the adsorbates were also ignored.

The concept of the "threshold pressure," P_c, was derived, thus explaining its origin.

Converting χ_s to different conditions

Change in vapor pressure basis

There may be a reason to convert the χ and $\Delta\chi$ values to a different value for P_{vap}. This can happen when there is a change in temperature, which changes the value or ε/RT but also slightly for E_a, and also for mixtures of gasses. For a gas mixture, normally a binary mix, there is a change due to ideal mixing and also ε that is observed in the binary phase diagram of the bulk liquid. This requires adjustment in the standard state bases pressure since the vapor pressure of each gas is not the same as the total vapor pressure. For this latter case, some theoretical solution equations, such as the ideal solution, regular solution, or some other function can fulfill the need. Lacking this, a digital description or an imperial equation for binary phase diagram would be needed.

The following is the method to convert a χ that is related to a vapor pressure over the bulk solution designated $p_{vap,x}$ to another base vapor pressure designated $p_{vap,y}$ (here the lower case p is used in recognition that this might be a partial pressure.) The χs for these two environments are designated as χ_x and χ_y. The following derivation is for only one component for the binary adsorbent, so no index is used to designate which component is being addressed. Assuming that there are no other interactions between components on the surface than those present in the bulk fluid, then the equation for all the components are identical.

The relationship between $p_{\text{vap},x}$, χ_x and the partial pressure, p is:

$$\chi_x = -\ln\left[-\ln\left(\frac{p}{p_{\text{vap},x}}\right)\right] \tag{4.42}$$

Using the exponential operation twice and solving for p:

$$p = p_{\text{vap},x} \exp(-\exp[-\chi_x]) \tag{4.43}$$

The threshold pressure, P_c, is a function of E_a/RT only. It does not vary with the solution vapor pressure at constant temperature (recall that P_{vap} and now p_{vap} the standard state basis pressure selected for χ theory.). So that:

$$P_c = p_{\text{vap},x} \exp(-\exp[-\chi_{c,x}]), \quad P_c := \exp\left(-\frac{\overline{E}_a}{RT}\right) \tag{4.44}$$

There is then a similar equation for the condition y. For solutions, p_{vap}, can usually calculated from the solution concentrations either by some known method, such as the ideal solution equation, the normal solution equation, by the phase diagram, or by digital data. At any rate, assume the ideal solution, so the gas concentrations are available from:

$$p_{\text{vap},x} = X_x(g) P_{\text{total}} \tag{4.45}$$

for component x. Since P_c is a function of E_a and T only, then the right side of Eq. (4.44) may be set equal to the right side for the "y" condition and P_{total} cancelled out:

$$X_x \exp(-\exp[-\chi_{c,x}]) = X_y \exp(-\exp[-\chi_{c,y}]) \tag{4.46}$$

It should be obvious that this is not a simple proportionality or other simple conversion. Either the above equation needs to be used or alternatively one of the logarithm forms:

$$\chi_{c,x} = -\ln\left\{+\exp[-\chi_{c,y}] + \ln\left(\frac{X_x}{X_y}\right)\right\} \tag{4.47}$$

Notice that the units for the Xs do not matter, so long as they are consistent. Therefore, the p_{vap}s may be substituted.

Change in temperature

This is the dependence on the temperature under the isosteric assumption. In this case, the bulk liquid bath (or virtual) is at the same temperature as the sample. If the bath temperature is assumed constant, then this leads to the normal isosteric heat dependence. Starting with the simple χ equation and since:

$$\frac{P_{c,x}}{P_{\text{vap},x}} := \exp\left(-\frac{\overline{E}_a}{RT_x}\right) \quad (\text{Recall}: \overline{E}_a < 0) \tag{4.48}$$

If the heat capacity is negligible then:

$$\ln \frac{P_{c,y} P_{vap,x}}{P_{c,x} P_{vap,y}} = \frac{\overline{E_a}}{R}\left(\frac{1}{T_y} - \frac{1}{T_x}\right) \tag{4.49}$$

Since from Eq. (4.37)

$$\exp(-\chi_c) = -\ln\left(\frac{P_c}{P_{vap}}\right) \tag{4.50}$$

then:

$$\chi_{c,y} = -\ln\left(-\ln\left(\frac{P_{c,y} P_{vap,x}}{P_{c,x} P_{vap,y}}\right) + \exp(-\chi_{c,x})\right) \tag{4.51}$$

so:

$$\chi_{c,y} = -\ln\left(\frac{\overline{E_a}}{R}\left(\frac{1}{T_x} - \frac{1}{T_y}\right) + \exp(-\chi_{c,x})\right) \tag{4.52}$$

This expresses the shift in the critical χ and if the ratio of P/P_{vap} is held constant χ would not change. However, if one were to hold the pressure of the adsorptive constant and change temperatures, then a correction for χ is needed since P_{vap} changes. This correction is made using the Clausius-Clapeyron equation:

$$\chi_y = -\ln\left(\frac{\Delta_l^g \overline{H}}{R}\left(\frac{1}{T_x} - \frac{1}{T_y}\right) + \exp(-\chi_x)\right) \tag{4.53}$$

The symbol $\Delta_l^g \overline{H}$[13] is the enthalpy of vaporization.

Summary

The basic functions of χ and the constant χ_c are dependent upon temperature and selection of standard state. The conversions may be required for changes in vapor pressure, which are dependent upon temperature. The value of the energy of adsorption, and thus the value needed to calculate χ_c is only modestly dependent upon temperature, however, if there is a large temperature change, then this needs to be accounted for as well.

What is *a*? ... and is the hard sphere approximation appropriate?

As the reader may have gathered by now, the question of the surface area metric is a psychological-philosophical question for most practical purposes. It would be pleasing to have a calibration between the physical quantity n_m and the metric surface area. The metric, however, still brings up the question of the fractal nature of the

---

[13] 2014 IUPAC convention—GB3 2nd printing.

measurement, a point not overlooked by many investigators. Nevertheless, this section proceeds with all the doubts about the value of this with the caveat of an unresolved metric. A similar question arises for porosity measurements, but again the question might be, how many molecules are in direct contact with the pore surfaces and how many molecules are in subsequent 'layers'? In other words, the actual geometry, although it might be approximately correct, is usually not relevant to the actual problem. These next few sections address the possibility of some geometric answer that might be useful, even though it might only be an approximation with uncertain meaning.

To relate Eq. (4.39), or θ, to the surface area, A_s, a value for a, the effective molecular cross area, needs to be determined. This question vexes all the adsorption theories. The question of the value of n_m, the moles in the monolayer equivalence, is firmly answered by χ theory. However, with regard to surface area, A_s, firstly, the hard sphere approximation to an adsorbed molecule needs to be determined. Assuming that the molecules are hard spheres and that they stack in the liquid in a close packed arrangement (FCC or HCP) then there is a 3D packing factor, F_{3D}, to be accounted for. F_{3D} is the ratio of the volume of the hard spheres to the total volume. For close packing $F_{3D} = 0.7405$. The effective volume for molecule is the molar volume divided by Avogadro's number or:

$$v_+ = \frac{\overline{V}}{N_A} \tag{4.54}$$

where the "+" indicates that the space not occupied by the hard spheres, or "empty" space, is included.

Since the density, ρ, is usually given in g cm^{-3} and the molar mass, $\overline{M}$, is in units of g mol^{-1} a conversion factor of 1×10^6 cm^3 m^{-3} (exactly) is needed to convert to the units of m^3 using the density, ρ. For the hard sphere calculation, the density of the solid, ρ_s, is used. Thus:

$$v_+ = 1 \times 10^{-6} \frac{\overline{M}}{\rho_s N_A} \tag{4.55}$$

Taking into account the 3D packing factor, the volume of one isolated molecule, v_- with the "−" indicating the empty space has been subtracted, becomes:

$$v_- = 7.405 \times 10^{-7} \frac{\overline{M}}{\rho_s N_A} \tag{4.56}$$

from the formulas for the hard sphere, molecular radius, r, may be calculated:

$$r = \left(\frac{3v_-}{4\pi}\right)^{1/3} = 5.612 \times 10^{-3} \left(\frac{\overline{M}}{\rho_s N_A}\right)^{1/3} \tag{4.57}$$

the cross sectional area, a_-, of an isolated molecule can be calculated from this:

$$a_- = \pi r^2 = 9.894 \times 10^{-5} \left(\frac{\overline{M}}{\rho_s N_A}\right)^{2/3} \tag{4.58}$$

However, to convert back to the effective area, a, one needs to divide by the 2D packing factor,[14] $F_{2D}=0.9069$:

$$a_+ = 1.091 \times 10^{-4} \left(\frac{M}{\rho_s N_A}\right)^{2/3} \tag{4.59}$$

Sometimes the quantity "molar area," $\bar{A}$, is used. Multiplying by Avogadro's number yields:

$$\bar{A} = 1.091 \times 10^{-4} N_A^{1/3} \left(\frac{M}{\rho_s}\right)^{2/3} \tag{4.60}$$

There is, however, a problem with the hard sphere assumption. The close packed arrangement of the solid is assumed, which is not observed with liquids, but the liquid densities are used. The close packed arrangement is often observed with the solid phase and a different density for the liquid at the same temperature, which would have to be the melting point[15] or triple point. In Table 4.2 are some densities of some potential adsorbates for both the liquid and the solid phases at the triple point. In Chapter 1, in the section "The advantages of the χ theory formulation," a table was presented to indicate alternative methods of calculating a. In the Chapter 1 table and in Table 4.2 it is interesting that the molecules that are nearly spherical have a liquid to solid ratio of the densities of about 0.85. Nitrogen, which is slightly aspherical, has a lower ratio, probably due to the increased packing in both states. What is needed then is to correct the solid density that one starts with by multiplying initially by the liquid to solid ratio. This will divide the $\bar{A}$ by this ratio to the ⅔ power to obtain the corrected $\bar{A}$.

$$\bar{A} = \frac{9.192 \times 10^3}{R_{l:s}^{2/3}} \left(\frac{M}{\rho_s}\right)^{2/3} \tag{4.61}$$

Table 4.2 Densities of the liquid and solid state for some potential adsorbates

	M_u	ρ_l (g cm^{-3})	ρ_s (g cm^{-3})	Ratio "$R_{l:s}$"
N_2	28.01	0.808	1.03	0.784
Ne	20.2	1.21	1.44	0.835
Ar	39.95	1.39	1.62	0.858
Kr	83.80	2.42	2.83	0.854
Xe	131.3	3.06	3.54	0.863
CH_4	16.04	0.423	0.494	0.856
C_6H_6	78.1	0.877	1.012	0.867

---

[14] There is also a 1D packing factor, $F_{1D}=0.8165$. This may seem puzzling but later you will find out where this is used.
[15] Which is a function of the hydraulic pressure.

It should be noticed that this factor makes a considerable difference in $\overline{A}$.

To make things a little more confusing, DFT calculation depends upon the Lennard-Jones 6–12 potential, or some other potential, to calculate the equilibrium position, r_σ. This may be used to calculate the effective area given by:

$$a = \frac{\pi r_\sigma^2}{4 F_{2D}} \tag{4.62}$$

(The LJ r_σ is actually $2\times$ the hard sphere radius.)

Which yields the values in Table 4.3 for a compared to the IUPAC areas for the noble gasses. This non-agreement is disturbing.

Typically, the correction going from Eqs. (4.60) to (4.61) is about 10%. Using the full ratio of $R_{l:g}$ rather than the 2/3 power would yield a difference of about 15%. Changing Eq. (4.61) to:

$$\overline{A} = \frac{9.192 \times 10^3}{R_{1:s}} \left(\frac{M}{\rho_s}\right)^{2/3} \tag{4.63}$$

which is a possibility. The reason for this possibility is due to the higher energy of the adsorption in the normal direction the "layers" are counted in the following fashion:

- The first layer consists of molecules that are in contact with the adsorbent surface.
- The second layer consists of molecules that are in contact with at least one first layer molecule and not in contact with the adsorbent surface.
- The third layer consists of molecules that are in contact with at least one second layer molecule and not in contact with any first layer molecules nor the surface.
- This then continues likewise to the fourth and higher layers.

Thus the vacancies can only be counted within the layers and not between layers in this 2D accounting scheme.

To make this clearer, consider that extra space required for the liquid is due to vacancies in the perfect lattice of the close packing. A ratio of 0.875 ($\approx$0.856 that seems to be dominate) would indicate that 1 of every 8th adsorbate molecule is absent, i.e., there is a vacancy. Without these vacancies, the material would not behave as a liquid. As defined here, an adsorbate molecule in the 1st layer is in direct contact with the adsorbent surface. A molecule in the 2nd layer is in direct contact with at least one molecule in the 1st layer. A molecule in the 3rd layer is in contact

Table 4.3 IUPAC vs LJ 6–12 calculated molecular area

	IUPAC a (nm^2)	a from $r_\sigma (\equiv d)$ (nm^2)
Ne	0.095	0.065
Ar	0.166	0.100
Kr	0.167	0.115
Xe	0.187	0.137

with at least one molecule in the 2nd layer. Thus a molecule in the *n*th layer is in contact with at least one molecule in the the (*n* − 1)th layer. Therefore, when calculating the distance from the surface, the only thing that is important is how many molecules separates the molecule of interest from the surface. This distance is $2nr + r$ to the center of the molecule. Therefore to have 1 in every 8th molecule be missing, there must be, on an average, 1 in every 8th molecule in each layer that is missing. A schematic would, at this point, be helpful to illustrate this (Fig. 4.3).

Another potential question is: why not use the van der Waal radius obtained from the van der Waal gas law?[16] The problem is that if one uses the van der Waal radius from the gas law to calculate the effective volume, which includes F_{3D}, the wrong answer is obtained.

The van der Waal radius is twice the radius one expects from the liquid and the area that one molecule excludes another molecule is four times what one would expect from the liquid area. This, however, assumes that the adsorbate molecules are isolated from each other. It would be more realistic to assume that the molecules clump into liquid-like patches, as illustrated in Fig. 4.3. If this is the case, the excluded area is only double at the edges of the patches.

Using only the hard sphere approximation, the relationship of a to $\overline{A}$ is possible to provide. The hard sphere approximation since:

$$\theta = \frac{n_{ads}}{n_m} \quad \text{and} \quad n_m = \frac{A_s}{\overline{A}} \tag{4.64}$$

then the χ equation, Eq. (4.39) becomes:

$$\frac{n_{ads}\overline{A}}{A_s} = \Delta\chi U(\Delta\chi) \tag{4.65}$$

From the slope of the χ plot, that is, number of moles adsorbed, n_{ads}, versus χ, one may obtain the surface area for any particular coverage:

$$A_s = \overline{A}\left(\frac{\partial n_{ads}}{\partial \chi}\right) \tag{4.66}$$

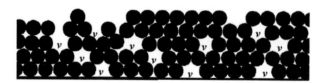

FIG. 4.3

A Schematic of the layering of the adsorbate molecule and the vacancies, *v*, demonstrating how the adsorbed liquid state dimensions are expanded in the plane of the adsorbent but not in the normal direction.

---

[16] Notice that the van der Waal radius from the Lennard-Jones potential is different from the gas law.

The hard sphere geometry and Lennard-Jones correction

This section was presented in the first edition of this book. It is from the classical point of view, so it may not be very useful. In this treatment, the Lennard-Jones expression is used only in a ratio. Thus, the specific values or the Lennard-Jones parameters are irrelevant.

In classical modeling using the hard sphere approximation, the perturbation assumption that the classical molecular radius is the only geometrical interference needs to be re-examined. In the classical model the molecule, which is "rolling over," must leave the surface before its center is over the edge of another molecule.

This situation is illustrated by the diagram in Fig. 4.4. This requires a step function up and down. This is unrealistic and ignores some energy considerations. This is not very important for the n_m calculation according to perturbation theory, but changes the calculation of the effective value of $\overline{A}$. The exact form of this small perturbation is not obvious, but here a Lennard-Jones 6–12 (LJ 6–12) potential will be used. This will be used to correct $\overline{A}$.

This LJ 6–12 potential is assumed for both the adsorbate and the surface atoms. Since the adsorbate molecules are free to travel over the surface, the 6–12 potential is considered a uniform average in the parallel plane of the surface. By referring to Fig. 4.4, the following geometrical arguments may be made. The L-J potential has a distance, r_o, designated in the 6–12 equation by:

$$E_{LJ} = \varepsilon\left[\left(\frac{r_\sigma}{\mathbf{r}}\right)^{12} - 2\left(\frac{r_\sigma}{\mathbf{r}}\right)^{6}\right] \tag{4.67}$$

and is related to the other r values by:

$$r_{vdW} = 2^{-1/6}(r_\sigma + 2r_t) \tag{4.68}$$

where r_{vdW} is the LJ van der Waals radius and r_t is the radius of the immobile surface atom or ion, that is, center-to-edge.

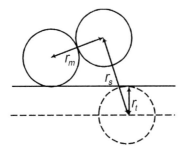

FIG. 4.4

The arrangement of an adsorbate molecule "rolling over" another and the distances defined for the treatment of the energy correction.

By simple geometry, from Figure 4.4 the distance between the average surface molecule or ion and the molecule that is rolling over is:

$$r_s = \frac{r_{vdW}\left(\sqrt{r_{vdW}^2 - r^2} + 1/2 r_{vdW} + r_t\right)}{r} \tag{4.69}$$

where r is the distance between centers in the plane of the surface.

Using this, the effective fraction of the excluded area compared to the hard sphere ratio, s, may be calculated from the expression:

$$s = \int_0^{r_\sigma} \varepsilon\left(\left[\left(\frac{r_\sigma}{r_s}\right)^{12} - 2\left(\frac{r_\sigma}{r_s}\right)^6\right](2\pi r) dr\right) \bigg/ \left(\int_0^{r_\sigma} E(2\pi r) dr\right) \tag{4.70}$$

Making the appropriate substitution for r_s and evaluating numerically s is given as:

$$s = -0.0967 \frac{r_t}{r_\sigma} + 0.9653 \tag{4.71}$$

Using the considerations of Eq. (4.66) and including s for a geometric correction:

$$\frac{n_{ads} s \overline{A}}{A_s} = \Delta \chi U(\Delta \chi) \tag{4.72}$$

This correction of s is then included in the value of θ. It could be included in $\overline{A}$ except it is preferred to keep $\overline{A}$ a function of the adsorbate only, but s is also a function of adsorbent. The surface area is then:

$$A_s = \frac{s\overline{A}}{\partial \chi / \partial n_{ads}} \tag{4.73}$$

Thus the form of the simple χ equation remains unchanged:

$$\theta = \Delta \chi U(\Delta \chi) \tag{4.74}$$

The ratio of r_t to r_σ is always greater than 0 and is unlikely to be greater that 0.5. Therefore, the reasonable range for s is from ~0.92 to 0.965. This value is independent of the value for E_a. It depends upon the ratio of radii but not upon the absolute values of the individual radii. The recommendation, if nothing is known about the adsorbent surface, would be to use the lower number for this factor, which is 0.92. Or:

$$A_s = \frac{0.92 \overline{A}}{\partial \chi / \partial n_{ads}} \tag{4.75}$$

Summary

The conversion of the monolayer equivalence to surface area is discussed. There are many problems with this conversion that has been more extensively discussed in the literature. It is advised to report the monolayer equivalence every time with surface area as an option with the caveat that there are multiple problems with the conversion.

Several conversion factors are given along with the IUPAC convention. These are presented in some tables.

In summary of χ theory so far

For a single type of adsorbing plane with no porosity considerations, the χ plot is a plot of the moles of adsorbate, n_{ads}, as the ordinate (y-axis) and the $-\ln(-\ln(P))$, defined as the χ value, as the abscissa. The slope of this plot yields the monolayer equivalence, n_m, which can be related to the surface area, A_s, with the value of $\overline{A}$ as given in Eq. (4.60). However, there are still questions, not really answered here, nor elsewhere for that matter, about what is the value of $\overline{A}$ and possibly s, an energy correction, which would allow one to calculate A_s, the surface area. The answers, even given all the considerations here, are probably only good to 10%, and part of that is the question of what is the meaning of surface area, as Prof. Kenneth Sing [9] has pointed out.

The energy release of the first adsorbate molecule, is derived from the abscissa intercept to yield the value of χ_c. The molar energy of the first molecule is then $\overline{E_a}$ of Eq. (4.38). $\overline{E_a}$ is not the energy of adsorption as a function of adsorbate amount, $\mathbf{E}(n_{ads})$ or $\mathbf{E}(\Delta\chi)$, but it can be seen from the equation at this point that as $n_{ads} \to \infty, \mathbf{E}(n_{ads}) \to \Delta\overline{H}_l^g$. This will be derived later in this chapter.

This means that the χ theory predicts a consistency that can be checked between the isotherm and the heats of adsorption, provided one is careful in specifying which "heat of adsorption" is being used. There is one extra note for this statement, notice that by Eq. (4.31) there is the additional term of $\frac{1}{2}kT$, which should off-set the heat of adsorption a slight amount. This is a slight deviation from the Dubinin thermodynamic criterion. The observance of this off-set requires very accurate and precise data. The term $\mathbf{f}(T)$ being rotational and internal vibrations observed in the liquid state is not likely to change from the bulk liquid to the adsorbed state, but it is possible for more complex adsorbates. For example, it might be observed with medium to large chain amines and carboxylic acids.

In the next section, another theory based on a continuum model and thermodynamics is presented. This modern model, published in 2000, also makes one reasonable assumption different from the χ theory assumption and ends up with results identical to the simple χ theory.

The disjoining pressure derivation (ESW)

An alternate derivation to the quantum mechanical, or χ theory, is the disjoining pressure theory (recently renamed Extra Surface Work of ESW theory.) This theory, derived by Churaev et al. [10], independently roughly 20 year after the χ theory, is much more understandable and is based only on one assumption, which is the decay of a potential, whose decay constant is one monolayer thick. This is really two assumptions: (1) the potential drop-off is an exponential decay; and (2) the decay constant is one monolayer. It turns out that the first assumption can be derived from the mathematics itself, as will be seen in this section. This leaves only the decay constant assumption. This potential is similar to the arbitrary "external potential" for the several DFT theories, except the decay value is not arbitrarily adjustable. It uses classical thermodynamics of surfaces and the ideal gas law with the additional reasonable assumption about the energetics of a liquid film. Of course, to use the latter assumption, one must also assume that the adsorbate is similar to the liquid phase, in contradiction to the BET theory.

It begins with the definition of the disjoining pressure, Π, a well established although little known thermodynamic concept proposed by Derjaguin [11]. There is a quantity that is a function of the coverage, Γ, or adsorbed film "thickness,"[17] t, defined by the equation (for the theory t and Γ may be used interchangeably):

$$\Pi(t) = \left.\frac{\partial f(t)}{\partial t}\right|_T \qquad (4.76)$$

For those unfamiliar with disjoining pressure, a quick explanation may be in order. In thermodynamics of films, the definition of the Helmholtz free energy includes the area of the film times the film tension, γ. Mechanically, γ consists of two components, the interface tensions, which is in the plane of the surfaces and the disjoining pressure, Π, which is normal to the surface. These add to yield the film tension:

$$\gamma = \sigma_1 + \sigma_2 + \Pi(t)\overline{V} \qquad (4.77)$$

There are two σs listed here, since the film has two different interfaces. When using the Gibbs-Duhem equation one needs to keep in mind that the variable is the area for σs and these are usually constants unless the material changes, or in the case of adsorption, the possibility that one of the surfaces is too ill-formed to have an interface tension. The disjoining pressure, however, is a function of thickness and in general is a tensor, thus it does have a derivative.

This physical quantity is interpreted to be the pressure needed to separate two parallel plates from each other when there is an intervening liquid phase. The liquid phase is, in this case, interpreted to be the adsorbate, which is in equilibrium with the adsorptive. Thus, the chemical potential of this intervening phase may be specified by the adsorptive pressure in the gas phase. Π can be related to the difference in the chemical potential, $\Delta\mu$, between the pure liquid phase at the saturation pressure, μ_{vap}, and the chemical potential of the adsorbate, μ_{ads}, or:

$$\Delta\mu = \mu_{ads} - \mu_{vap} \qquad (4.78)$$

relating the chemical potential to adsorptive pressure and using the vapor pressure as the standard state and assuming low pressures where $\widetilde{p} = P$ then:

$$\Delta\mu = RT \ln(P/P_{vap}) \qquad (4.79)$$

In the film, because of equilibrium conditions, the change in chemical potential must be given by the expression (See Ref. [12]):

$$\overline{V}\Pi(t) = -\Delta\mu \qquad (4.80)$$

---

[17]The meaning of film "thickness" on a nearly atomic scale is somewhat questionable. Nevertheless, it is a convenient concept from our macroscopic, continuum viewpoint.

(Notice that since $\Delta\mu$ is negative then by definition Π is positive.)

The excess surface energy, Φ, is obtained from the product of the surface excess, Γ, and the change in chemical potential, provided the surface is flat. Using the above equation then:

$$\Phi(\Gamma) = -\Gamma\overline{V}\Pi(\Gamma) \tag{4.81}$$

(Φ is negative since it indicates it is exothermic.)

Up to this point, no modeling has been introduced, merely thermodynamics and definitions. The functionality of Π becomes important to proceed. The dependence of Π upon the film thickness has been found [13, 14] to reliably follow an exponential equation or:

$$\Pi(t) = \Pi_0 \exp\left(-\frac{t}{\lambda}\right) \tag{4.82}$$

where λ has been referred to as a "characteristic length."

This then is an assumed equation of state. Other than thermodynamics (and in this derivation for simplicity the ideal gas law,) these are the only assumptions made by the disjoining theory.

- The function is an exponential decay, and
- the decay constant seems to be one monolayer distance for gas adsorption.

The second assumption is not provable from this derivation.[18] It does, however, have considerable experimental support. The first assumption is also the assumption that Debye [15, 16] made for his famous theory of adsorption. In the case of the Debye theory, later it was demonstrated that the exponential assumption can be derived from thermodynamics, electrostatics, and Darkens equations of diffusion [17]. A similar argument could be made here, thus eliminating this as an assumption as well. This, however, is not necessary—see below.

The assumption that λ is one monolayer in distance, or rather one monolayer equivalence, will be proven correct with the χ theory. Substituting into Eq. (4.81) and replacing t/λ with the equivalent type expression in Γ, i.e., Γ/Γ_m one obtains:

$$\Phi(\Gamma) = -\Pi_0 \overline{V}\Gamma \exp\left(-\frac{\Gamma}{\Gamma_m}\right) \tag{4.83}$$

In Fig. 4.5 is a sketch of what this function looks like. The function starts out a $\Phi=0$ and goes to a minimum at $\Gamma=\Gamma_m$. This can be demonstrated by differentiating Eq. (4.86):

$$0 = \frac{d\Phi(\Gamma)}{d\Gamma} = -\Pi_0 V_m \left[\exp\left(-\frac{\Gamma}{\Gamma_m}\right) - \frac{\Gamma}{\Gamma_m}\exp\left(-\frac{\Gamma}{\Gamma_m}\right)\right] \tag{4.84}$$

---

[18]This is not a unique problem for the Disjoining Theory. There is also the same problem when one assumes a surface ideal "gas" equation and tries to obtain "Henry's law." Going the other way works, however.

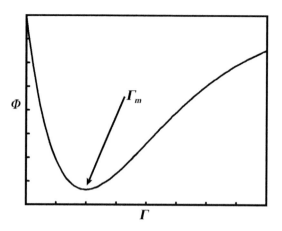

FIG. 4.5

The functionality of surface excess energy, $\Phi(\Gamma)$, with coverage, Γ.

From which one obtains $\Gamma(\min\Phi) = \Gamma_m$. The question is, "What is the molecular meaning of Γ_m? and how is it related to surface area" It has been assumed that Γ_m corresponds to the monolayer equivalent. Some reasonable argument would seem to indicated this, however, these arguments are not really proofs. The proof is demonstrated in the discussion of the χ theory and the derivation of the disjoining pressure theory from it.

Notice that the equilibrium requirement of Eq. (4.84) could use a undetermined function **f** in place of **exp** in Eq. (4.83). If this is then set to 0 to get the minimum as expected by thermodynamics for equilibrium then:

$$\mathbf{f}\left(\frac{\Gamma}{\Gamma_m}\right) - \frac{\Gamma}{\Gamma_m}\frac{\partial \mathbf{f}}{\partial \Gamma}\left(\frac{\Gamma}{\Gamma_m}\right) = 0 \qquad (4.85)$$

The general and particular solution to this equation is:

$$\mathbf{f}\left(\frac{\Gamma}{\Gamma_m}\right) = \exp\left(\frac{\Gamma}{\Gamma_m}\right) + C_{DP} = 0 \qquad (4.86)$$

where C_{DP} is simply an arbitrary constant.

If C_{DP} were to be anything other than 0, then Eq. (4.78) would need to be modified by the addition of a constant. This would simply shift the basis to the energy, and makes no difference in the overall scheme.

This leads to a very important conclusion: the exponential decay although originally thought to be an assumption, is not; the exponential decay is the natural result of a differential equation derived only from thermodynamics. Thus, really the only assumption is that the decay constant is one monolayer equivalent.

This is an amazing result and in itself invalidates all other isotherm theories that do not have this form since thermodynamics is more basic than quantum mechanical modelling. In a later section, the discussion about the Gibbs' phase rule is further evidence that many of these same theories are in conflict with thermodynamics.

There is one other item to notice in this theory. Going from Eqs. (4.82) to (4.86), the variable and parameter have been changed from t/λ to Γ/Γ_m. This would normally be interpreted as if the adsorbate adsorbed one complete layer of molecules at a time. This, however, is not an assumption of Disjoining Pressure theory.[19] Otherwise, this substitution would be an identity substitution, but it is not an identity. This is a very important point in the derivation and indicates that "potential" from the surface is an exponential function, but it is not a function of distance as assumed by DFT, but rather it is a function of the number of adsorbate molecules directly in contact with the surface.

This assumption is the justification for the classical mechanical ASP model by Fuller. It does not matter whether the potential is produced by a distance from the surface or by the masking of the surface, the effect would be the same for the overall adsorption. Thus according to χ theory the integration constant, C_{DP}, must be zero since it is impossible to have a negative adsorption (the basis for **U** in χ equations.) From the point of view of the Disjoining Pressure (ESW) theory one may simply set C_{DP} equal to zero as a new definition for the position of the adsorbate-adsorbent interface. Thus, **$C_{DP}=0$** is a new thermodynamic definition for the position of the interface. This is, of course, consistent with the modeling used for the χ theory.

It follows that, in the case of Disjoining Pressure (ESW) theory, it doesn't matter if the following model is incorrect. It does, however, matter greatly for other theories which rely upon the modeling. In the case of χ theory, the modeling is dictated by the grand potential, which is that no model needs to be entered other than the perturbation theory. For many of the DFT theories and Monte Carlos calculations the locations of the molecules, that is the 3D molecular density must be part of the input into the grand partition function and continuously adjusted for a certain number of molecules. This adjustment is performed to obtain the minimum in the grand canonical partition function for any particular number of molecules. This throws into question the validity of the various DFT formulations that assume a distance potential. This means that not only must the adsorbent surface be de-localized but also the adsorbate molecules.

Back to the final isotherm, given the information about the exponential decay, combining Eqs. (4.79), (4.80) and (4.82) one obtains:

$$RT \ln\left(\frac{P}{P_{vap}}\right) = \overline{V}\Pi_0 \exp\left(-\frac{t}{\lambda}\right) \quad (4.87)$$

Substitution of t/λ to Γ/Γ_m, means that the argument of the exponential function is θ. By dividing by RT and taking the ln one immediately can recognize the χ form of the equation:

$$-\ln\left(-\ln\left(\frac{P}{P_{vap}}\right)\right) = \theta - \ln\left(-\frac{\overline{V}\Pi_0}{RT}\right) \quad (4.88)$$

---

[19] It needs to be emphasized that the Disjoining Pressure theory is model-less. A continuum model would work fine.

Proving the connection between $\theta = t/\lambda$ and associating $\overline{V\Pi_0}$ as being equal to $\overline{E_a}$. This is proves the ESW theory in terms of quantum mechanics.

Why this theory works may not seem obvious or intuitive, and QM often does not provide an intuitive explanation. The next section might provide some insight in terms of classical mechanics. This picture is not a proof nor is it exactly the situation, but it may provide some intuition (the ASP theory by Fuller is also more intuitive, but one must think "statistically.")

The meaning of Γ_m in the hard sphere model

Although one could argue whether the exponential assumption of Eq. (4.86) is or is not part of thermodynamics but rather modeling, it is clear that the meaning of Γ_m requires an explanatory model. This could be a continuum model, but most chemists would balk at this; after all, the distances are on a molecular scale. There is no clear connection up to this point between Γ_m and the actual surface coverage. The following model should be a fairly intuitive picture of an atomic scale model. It is important to realize that the modeling is based upon the hard sphere model for the adsorbate molecules. Again, modeling on the classical basis is not accurate but is for psychological/philosophical purposes only. In the χ theory formulation the specifics of the molecules does not effect the outcome for the value of n_m. This should apply equally to Γ_m.

The maximum incremental energy released by the adsorption process should be at this minimum point. In other words, for two plates held together by the intermolecular forces of the liquid, this Γ_m is the point at which the maximum force operates. Assuming the molecules are hard spheres, this should occur when there is exactly one monolayer between the two plates.

One may be able to visualize this by referring to Fig. 4.6. The maximum incremental energy released will be at the point for which the intermolecular forces exert maximum force. This point is Γ_m, as illustrated in the diagram. For example, below a monolayer, "A" in Fig. 4.6, there are fewer molecules between the plate than in the monolayer case "B." Thus, there are fewer intermolecule forces for "A" than for "B" since the vacant positions do not produce these forces. When a monolayer is exceeded—shown as "C"—then the Π falls off due to the fact that there must be more than one adsorbate molecule between the two plates at some positions. Since forces between the adsorbate molecules are weaker than the forces between the adsorbate molecule and the plate molecules, then a relative easy separation can occur between adsorbate molecules that are stacked between the plates. Thus, the force between the plates for situation "C" are less than for "B." The conclusion is that there is a minimum in Φ, i.e., higher energy release, when there is exactly one monolayer of adsorbate between two plates.

This, however is for the opposing plates. What if there is only one plate and the force to remove the molecules is simply the chemical potential difference between the top layer of molecules and the vacuum? This is illustrated in Fig. 4.7 with the contrasting images "D" through "F" schematics. Of these arrangements, "F" is

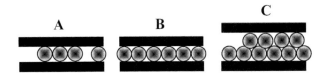

FIG. 4.6

Adsorbate molecules between two plates to account for the size of the force between them.

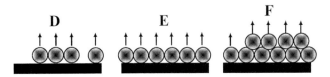

FIG. 4.7

Adsorbate molecules between two plates to account for the size of the force between them.

trending toward the easiest. Again to remove the molecules of a partial monolayer from the surface—case "D"—it is obvious that it takes less energy by shear numbers than from the full monolayer—case "E." To remove the top layer of molecules from the case "F" where there is more than a monolayer, it take less energy due to the fact that the forces between several molecules are between adsorbate molecules and are not overcoming the heterogeneous intermolecular forces. A subtle point here is that what is meant by force is the gradient of the chemical potential.

Relationship to χ-theory

Using the symbol Γ_m for a monolayer surface excess, then within the first approximation of the hard sphere approximation, $\Gamma_m = \Gamma_1$. Using this together with Eqs. (4.79), (4.80) and (4.82) one arrives at:

$$\exp\left(-\frac{\Gamma}{\Gamma_1}\right) = -\frac{RT}{\overline{V}\Pi_0} \ln\left(\frac{P}{P_s}\right) \tag{4.89}$$

Although in this form it looks different from the χ theory equation, it is identical if $\overline{V}\Pi_0 = \overline{E}_a$ of the χ theory and, of course $\Gamma/\Gamma_m = n_{ads}/n_m$ and the ln of both sides is taken:

$$\frac{n_{ads}}{n_m} = -\ln\left[-\ln\left(\frac{P}{P_s}\right)\right] - \ln\left(\frac{\overline{E}_a}{RT}\right) \tag{4.90}$$

or in short form:

$$0 = \Delta\chi \mathbf{U}(\Delta\chi) \quad \Delta\chi = \chi - \chi_c \quad \chi_c = \ln\left(-\frac{\overline{E}_a}{RT}\right) \quad \chi = -\ln\left[-\ln\left(\frac{P}{P_s}\right)\right] \tag{4.91}$$

since $\Delta\chi \geq 0$.

One might expect that this theory should not work for anything less than a monolayer since it depends upon the concept of a film and is therefore incorrect. Firstly, the intermolecular forces between adsorbate molecules should bring the molecule into patches of molecules, rather than individual molecules, thus creating a patchy monolayer (this argument invalidates the previously proposed factor of $f = 1/2$ in the 1st edition.) Secondly, if it did not form patches there seems to be no reason that the forces to the surface would be any different. It will be further demonstrated in the next section that this criticism is unjustified. The quantum mechanical considerations validate the theory down to the very first adsorbed molecule. Furthermore, the reasoning that indicates that the constant $C_{DP} = 0$ or something else doesn't matter, indicates that the assumption that $t = \lambda$ is exactly the point of one monolayer equivalent by Churaev et al. was not necessary. The theory stands alone with no assumption except what is being formed on the surface will end up being the liquid phase at $P = P_{vap}$.

Another conclusion from the above discussion is that a shift in the energies with C_{DP} is somewhat inconvenient and a bit unrealistic since it shifts the standard state to something other than the vapor pressure of the adsorbent. If one keeps the energy adsorption constant and still uses a non-zero C_{DP}, then position of the starting Γ shifts. Therefore, by setting $C_{DP} = 0$, as had been done, the boundary of the liquid-solid interface is given a fixed position.

This then, is a new definition of the interface boundary. This solves a thorny philosophical question of: where is the surface? Both the thermodynamics and the quantum mechanics create this definition.

It should not be surprising, in retrospect, that the above is the logical conclusion about disjoining pressure, for it is entirely consistent with other aspects of thermodynamics, especially the Gibbs' phase rule.

Summary of disjoining pressure theory

The derivation of Disjoining Pressure Theory (Excess Surface Work or ESW) was presented. This theory is based on the thermodynamics of surface chemistry and is well grounded. The entire theory, except for one decay constant, is a consequence of thermodynamic laws, the ideal gas law, or more generally the fugacity, and the definition of the system's boundaries.

The equations have an originally assumed decay constant, which can be proven with differential equations to be correct except for the value of the constant. The value of the constant can be proven using the quantum mechanical derivation of the χ theory. The resultant equations of the Disjoining Pressure Theory are identical to the χ theories, lending credence to both.

In the derivation of the decay equation, a constant, C_{DP}, is introduced. The original assumptions of Disjoining Pressure started the decay at the surface, but in the proof that the decay of the potential was an exponential decay, this constant as integrating constant is introduced. χ theory says that this should be zero, but it could be anything, and the Disjoining Pressure theory is valid, with a shift in the surface boundary. Thus, by setting $C_{DP} = 0$ the equations for Disjoining Pressure theory and χ theory are identical. This then is a new definition for the surface boundary, based on a thermodynamic definition and the result of the quantum mechanical perturbation theory.

An explanation of a possible reason that the decay constant was selected to be one monolayer thick is given, however such reasoning is not needed and presents an unrealistic picture of the density distribution. The potential decay that is specified has been used by DFT in the past, but by using the assumption of one monolayer needing to be complete may explain why the various forms of DFT miscalculate the external layer distribution.

Finally, a mathematical proof is presented that demonstrates that the Disjoining Pressure theory and χ theory are indeed identical.

Given the basic χ equation, the rest of this book will assume that the adsorbate may be treated as a continuum and the measuring device will be the radius of an adsorbate molecule. There will be answers for depth profiles and pore fillings, which may not make sense on an atomic scale. One might look at the results this way: if the molecules can accommodate the results of the calculation, they will, or come as close as possible. Thus, if a pore radius is calculated as 1.5 monolayers thick, the molecules will stager in such a way as to make the nearest approach to this value.

Heterogeneous surfaces
The additive principle of χ plots

One of the nice features of the χ plots is that for several mixed surfaces the χ plots add. This is quite obvious because the dependent variable in the χ equation is the amount adsorbed, which, of course, must add experimentally. An important feature of the χ theory is the unit step function in Eq. (4.72). If there are several surface planes of different energies they would simply add:

$$n_{\text{ads}} = \sum_i n_{\text{m},i} \Delta \chi_i \mathbf{U}(\Delta \chi_i) \tag{4.92}$$

Thus, for the various slopes:

$$\frac{\partial n_{\text{ads}}}{\partial \chi} = \sum_i n_{\text{m},i} \mathbf{U}(\Delta \chi_i) \tag{4.93}$$

One can take this one more step and determine the second differential:

$$\frac{\partial^2 n_{\text{ads}}}{\partial \chi^2} = \sum_i n_{\text{m},i} \delta(\Delta \chi_{c,i}) \tag{4.94}$$

The usefulness of this last equation is that the sum of δ functions as an expression of the distribution of χ_cs and thus the distribution of the various energies of adsorption. This fact will be utilized when a distribution is detected, which is χ plot Feature 1.

According to Eq. (4.93), when several surfaces are present with distinct energies of adsorption, the χ plot will start at low pressures with the highest energy surface.

The slope then yields the surface area. After the appropriate χ_c for the next surface, the slope yields the sum of the two surfaces. This addition is continued until all the χ_c values have been exceeded. Thus, at least in the early portion of the χ plot, an upward bending of the χ plot is an indication of more surfaces becoming active in the adsorption process. An upward bend can also be indicative of capillary filling in mesopores, however, this happens at higher values of $\Delta\chi$. As a rough rule, below $\Delta\chi \approx 2.5$ an upward bend may be due to additional surfaces adsorbing and an upward bend above $\chi \approx 0$, especially a large upswing followed by a leveling off, is due to capillary filling. This leaves, unfortunately, some overlap and judgement may be required to distinguish the two.

It is not common to find pure materials with more than two distinct energies of adsorption. It may be common to find energy distributions, as will be illustrated later. A couple examples of two distinct energies of adsorption are found with carbon and with some ceramics that have distinct crystallographic planes on the surface.

In Figs. 4.8–4.10 are some examples of χ plots where it appears that two or more energy surfaces are involved. These are Vulcan and Sterling FT carbon [18] and high fired thoria [19]. The adsorption on thoria has an addition feature due to mesoporosity, which can be separated out from the simple surface adsorption. This separation will be used as an example in a later section. Fig. 4.10 has an additional feature: a negative curvature section. This is indicative of porosity as mentioned in the section on porosity on standard curve interpretation in Chapter 1 and in more detail in Chapter 6.

Insensitivity for $\chi \geq \max \chi_c$

It should be obvious from Eq. (4.93) that, after the last break in the χ plot, the slope of the line yields the total surface area. Mathematically, this can be written:

$$A_s = \overline{A} \frac{\partial n_{ads}}{\partial \chi}\bigg|_{\chi \geq \chi_{max\,c}} \tag{4.95}$$

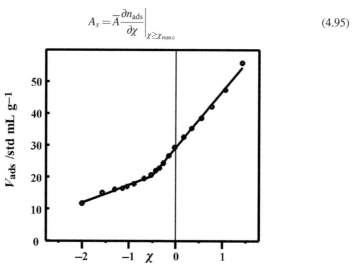

FIG. 4.8

χ plot of N_2 adsorbed on Vulcan C with two $\overline{E}_a$ straight line fits.

Heterogeneous surfaces   179

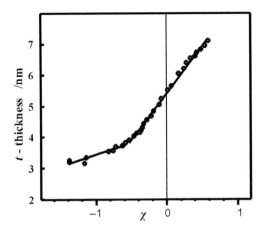

FIG. 4.9

χ plot N$_2$ adsorbed on Sterling FT C with two $\overline{E}_a$ straight line fits.

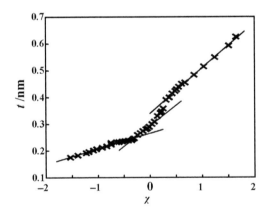

FIG. 4.10

Chi plot of nitrogen adsorbed on high fired thoria indicating two energies of adsorption and another feature by the multiple straight line fits.

Provided other complications are absent, such as capillary filling or bed porosity capillary filling, the final answer is the surface area of the total sample.

Reformulation for a distribution of E_a values

Eq. (4.94) is the starting point for treating surfaces that have a distribution of energies. In place of the sum of δ functions, one may insert a distribution function. Any distribution function could be allowed, both continuous or a series of δ functions, or a combination. One of the more common distributions in energy [20] is the ln-normal distribution which is the same as a normal distribution in χ_c. Therefore, the modified Eq. (4.94) is:

$$\frac{\partial^2 n_{ads}}{\partial \chi^2} = \frac{n_m}{\sigma\sqrt{2\pi}} e^{\frac{(\chi-\langle\chi_c\rangle)^2}{2\sigma^2}} \tag{4.96}$$

where the symbol $\langle\chi_c\rangle$ indicates the mean of the χ_c values.

σ is the standard deviation in the χ_c distribution. When $\sigma = 0.48$ the low pressure Freundlich isotherms are generated, whereas $\sigma = 0.23$ generates the low pressure range for the Dubinin sets of isotherms. The demonstration of this is in a later section. Without any addition complications, such as porosity, Eq. (4.96) may be integrated from $-\infty$ to χ twice to yield the shape of the isotherm as:

$$\theta \equiv \frac{n_{ads}}{n_m} = (\chi - \langle\chi_c\rangle)\left[\frac{\sigma}{2\sqrt{2\pi}} \exp\left(\frac{(\chi-\langle\chi_c\rangle)}{2\sigma^2}\right) + \frac{1}{2}\mathrm{erf}\left(\frac{(\chi-\langle\chi_c\rangle)}{\sigma\sqrt{2}}\right)\right] \tag{4.97}$$

This is basically the same equation as Eq. (4.65) with the quantity in the square brackets replacing the step function and indeed Eq. (4.97) becomes Eq. (4.65) as $\sigma \to 0$. The shape of the isotherm was given in Chapter 1 as the χ representation of type III.

Summary of application to heterogeneous surfaces

The χ-plot is additive with moles of adsorbate for the same value of χ. This is the basis of the non-anomalous nature of the theory. Thus the first derivative, which yields to monolayer equivalence for any particular plane can be determined, provide there are differing energies of adsorption. The second derivative yields the energies of adsorption as either a delta function or, if there is a distribution of energies, yields these distributions.

Heats of adsorption
Isosteric heat of adsorption, q_{st}

Dubinin [21], in order to derive features of the isotherm, postulated what he referred to as the "thermodynamic criterion," which is:

$$\left.\frac{\partial RT \ln\left(P_{vap}/P\right)}{\partial T}\right|_{n_{ad}} = 0 \tag{4.98}$$

There doesn't seem to be any justification for putting this forward, but one can make the following interpretation. This partial derivative is the same as ΔS going from the bulk liquid phase to the adsorbed condition. Thus, the molecular arrangement in the adsorbed phase is identical to the molecular arrangement in the liquid phase (this contradicts the BET formulation, which requires a phase transition at high coverages.) The justification for this became clear with the development of the χ theory [22]. If one performs this operation on the simplified χ equation, Eq. (4.39), an identical result is obtained. If one ignores only the internal modes in Eq. (4.31) represented by $\mathbf{f}(T)$, then one has for the partial of $\ln(P/P_{vap})$ with respect to $1/T$:

$$-RT\frac{\partial \ln\left(^{P}/_{P_{vap}}\right)}{\partial\left(^{1}/_{T}\right)}\bigg|_{n_{ads}} = (E_a - \tfrac{1}{2}RT)e^{-\theta} \qquad (4.99)$$

And the partial, with respect to T, is relatively small but finite and possibly measurable by calorimetry:

$$\frac{\partial RT \ln\left(^{P_{vap}}/_{P}\right)}{\partial T}\bigg|_{n_{ads}} = \tfrac{1}{2}R \qquad (4.100)$$

Dubinin referred to the quantity $RT \ln(P_{vap}/P)$ as the "adsorption potential" and gave it the symbol "A." Since the obvious interpretation of Eq. (4.98) is that the entropy of the adsorbate is the same as that of the bulk liquid, the adsorbate must be a liquid phase that is stabilized by the "potential" of the surface. The deviation, which is barely measurable, from the Dubinin thermodynamic criterion, Eq. (4.98) is thus due to the loss of the one degree of freedom, which the surface, being 2D and not 3D, does not allow, at least for the first layer.

One of the problems when one looks at the literature, or when calorimetric quantities are reported, is the variety of definitions of "heat." Hopefully, the following will aid in clearing up the confusion (and hopefully not create more confusion.) The quantity derived in Eq. (4.99) is what is often referred to as the isosteric heat of adsorption, which causes some confusion with the experimental quantity, which references a 1 atm standard state. Here the heat referenced to 1 atm standard state will be referred to as the heat of the liquid-adsorbate transition or $\overline{q}_{1a}$. Therefore, by χ theory:

$$\overline{q}_{1a} = (\overline{E}_a - \tfrac{1}{2}RT)e^{-\theta} \qquad (4.101)$$

The isosteric heat should include this plus the molar enthalpy of vaporization.

$$\overline{q}_{st} = \overline{q}_{1a} + \Delta_l^g \overline{H}(T) \qquad (4.102)$$

$\Delta_l^g \overline{H}$, or $\overline{\varepsilon}$, might have a significant temperature dependence, so it should be clear that this quantity is at the temperature of the sample (the temperature correction for $\overline{\varepsilon}$ is provide by Eqs. (4.43) and (4.44).)

The integral heats of adsorption

Experimentally $\overline{q}_{st}$ is very difficult to measure directly. Attempts to find the partial of $\ln(P/P_{vap})$ with respect to $1/T$ by measuring the isotherm at two or more temperatures have not been very accurate. Furthermore, as mentioned above, even if χ theory is used, without the base pressure correction the answer is not correct. Usually the problem is due to the uncertainty in the shape of the isotherm compared to the precision that is acceptable. Direct calorimetric measurements have been more successful. Calorimetric measurements are more precise but they measure the integral heat of adsorption, Q', and the molar heat of adsorption, $\mathbf{Q'}$ ($\overline{Q'}$ by present convention[20]), as defined by Morrison et al. [23]. Another quantity, the integral energy of adsorption, Q, was defined by Hill [24, 25] for constant volume conditions. These quantities can be obtained with more accuracy and precision than the isosteric heat. Nevertheless, the isosteric heat is often reported.

From these experimental quantities, the isosteric heat is obtained by the "usual method." This usual method is as follows:

1. Q' is measured up to a certain amount of adsorption. The calorimetric details involve steps to calibrate the calorimeter and determine the heat capacity of the calorimeter, the adsorbent, and the adsorbate and adsorptive up to the pressure corresponding to the n_{ads} that Q' corresponds to. The isotherm must also be measured. Thus one has, after significant mathematical manipulation, a set of Q'_i, n_i, and P_i.
2. It is assumed that the $\overline{q}_{st}$ for an average of two points $\langle n_i$ and $n_{i+1}\rangle$ is given by:

$$\overline{q}_{st \langle n_i \text{ and } n_{i+1} \rangle} = \frac{Q'_{i+1} - Q'_i}{n_{i+1} - n_i} \tag{4.103}$$

Unfortunately, there are a few problems associated with this method. The first problem is critical in terms of archiving.

1. Information is lost and cannot be recovered if the original data are not presented somewhere. This is because the number of points is one less than measured. Although this may seem to be a minor problem, none of the original data can be recovered since this is a threaded string of calculations.
2. Problem 1 would not be so serious if it were not for the fact that this method introduces errors due to the averaging effect. There is no guarantee that it is linear, as implied by Eq. (4.103) and indeed may suddenly change. Thus, the reported Q' will be different from the actual value.
3. An additional problem is the usual introduction of scatter when one tries to digitally differentiate data as implied in Eq. (4.103).

Given the problems associated with this method, it would be highly advised to report Q' and not $\overline{q}_{st}$. After all, Q' is just as useful both theoretically and practically as $\overline{q}_{st}$.

The molar integral heat, $\overline{Q'}$, is defined as the integral heat per mole of the adsorbate or:

$$\overline{Q'}(n_{ads}) = Q'/n_{ads} \tag{4.104}$$

---

[20]The notation $\mathbf{Q'}$ was used by Morrison, Los, and Drain from previous IUPAC convention.

Both of these quantities may be referenced to the liquid state rather than to 1 atm. Using subscripts "la" to indicate this, the following may be derived by substituting into Eq. (4.101):

$$Q'_{la} = n_m \left(\overline{E}_a - \frac{1}{2}RT \right) \left(1 - e^{-\theta} \right) \quad (4.105)$$

and the molar integral heat is:

$$\overline{Q}'_{la} = \theta^{-1} \left(\overline{E}_a - \frac{1}{2}RT \right) \left(1 - e^{-\theta} \right) \quad (4.106)$$

One may also derive the expected heat capacity, $\overline{C}_{p,ads}$, by differentiating $\overline{Q}_{la}'$ with respect to T. Thus:

$$\overline{C}_{p,ads} \big|_{n_{ads}} = \frac{(1 - e^{-\theta})}{2\theta} R + \overline{C}_{p,l} \quad (4.107)$$

where $\overline{C}_{p,l}$ signifies the heat capacity at constant pressure for the liquid phase.

Since the first term is small, ($\leq 1/2 R$) one expects the heat capacity of the adsorbed film to be about the same as the bulk liquid.

The next section contrasts the results of the BET theory calculation versus χ/ESW theory and experimental data.

Heats of adsorption from BET plot

The BET heats of adsorption is examined here because it is widely reported in the literature. It is important that one understands the following. The heats of adsorption using the BET plot are basically useless and lead to incorrect values for the entropy of adsorption. The form of the isosteric heat in no way resembles the observed value. To calculate the isosteric heat one solves the BET equation in terms of θ or n_{abs}. The equation for this is:

$$\frac{P}{P_{vap}} = \frac{\sqrt{C_{BET}(C_{BET}(\theta^2 - 2\theta + 1) + 4\theta))} + C_{BET}(\theta - 1) - 2\theta}{2\theta(C_{BET} - 1)} \quad (4.108)$$

The C_{BET} is related to an energy $\overline{E}$ by:

$$C_{BET} = \exp\left(\frac{\overline{E}_{BET}}{RT} \right) \quad (4.109)$$

The $\overline{E}_{BET}$ value is not the energy of adsorption, nor is it the isosteric heat. Eq. (4.108) is easily used to obtain the isosteric heat by differentiating the ln of it numerically. Regardless of the parameters used, the shape of the curve obtained is a classic downward "S" curve, as illustrated in Fig. 4.11 for three different starting energies labeled $\overline{E}_{BET}$. This calculation never matches the calorimetric data, which is usually an exponential decay curve, or something near an exponential decay. This discrepancy has long been known but ignored by making adjustments to the entropy term. Of course, these adjustments conflict with the Dubinin thermodynamic criterion. These points

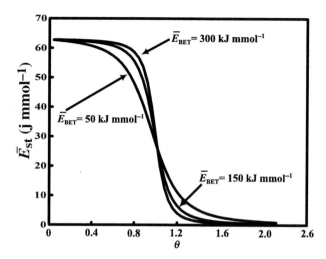

FIG. 4.11

Dependence of calculated isoteric heat from the BET equation for various input energies.

become obvious when one looks at the calorimetric data seen in Chapter 5 examples. The heats of adsorption is another indicator that the BET theory is far from correct even in the "BET range."

Depth profiles and n-layer calculations with χ theory[21]

χ theory, with the original postulate, cannot calculate the depth profile. This should not be surprising since the assumptions made dealt with the 2D energy profile on the surface and ignored the third dimension. The molecular areal density of each "layer" is calculated, but the details of the density profile cannot be calculated. This is in contrast to the calculations by non-local density functional theory (NLDFT) where the profile is an integral part of the calculation and, therefore, is one of the resultant outputs used in the model matching. However, the principal disadvantages of NLDFT are the dependence upon the specifics of the surface, and the use of the BET, either directly or indirectly, to obtain the surface area. For example, in the first case, the "external potential" profile, which is usually unknown, must be adjusted in the model. In the second case, even if a standard curve such as the α-s or t-curve is used, these in turn depend upon the BET. Another previous disadvantage of the DFTs was the difficulty of the calculation, which is now being overcome, with some instruments having the embedded calculations. However, such embedded programs are

---

[21] These next two section, binary adsorption and depth profiles, have not yet been published and no doubt additional research is required, both theoretically and exprementally.

usually not very versatile and the assumptions made may not be appropriate for some of the measurements. For most, this is an embedded unknown in the interpretation.

Assumptions used to calculate the Normal Direction Profile

The situation for the χ theory regarding the normal direction, however, can be rectified with additional assumptions. There are four assumptions for the total normal distribution of each layer about its average center position. These are as follows:

1. The vibrational modes of the adsorbent surface itself are quite small. Justification: this is likely to be quite small since the vibrations in the solid state are restrained and combined as phonons. The vibrations are determined by higher potentials such as the Morse potential that are more restraining. At any rate these amplitudes are probably unknown, so a "hard wall" assumption is used.
2. There is the possible heterogeneity of the E_a. This can be determined if the very low pressure data is recorded in order to get the amount of positive curvature at χ_c.
3. The magnitude of the vibrational distribution is in the ground state. Justification: most adsorption experiments are at low temperatures. However, the small distribution from ground state vibrations becomes magnified from layer to layer. In the discussion that follows it will be demonstrated how this is calculated.
4. The uncertainty about the position of adsorbate molecule from the second layer onward. That is, is the second layer molecule sitting in the tetrahedral position, the position directly above two molecules or above only one. Logically, the last possibility is the least likely, but there must be a combination of the first two, otherwise the adsorbate could not be fluid. This means that the layer distances have an uncertainty of the difference between the tetrahedral position and the two point position.

The original assumption is that the normal direction's wave function may be calculated independently from the plane of the surface. In other words, the solution of the overall wave function is separable into a wave equation for the plane of the surface and equations normal to the surface. This would seem to indicate that only population of the first layer of atoms can be calculated. This is an incorrect conclusion.

One of the important conclusions of the $x-y$ calculation is that it not only calculates the population of the first molecular layer, that directly in contact with the surface, but also the populations of all the higher "layers" for a flat surface,[22] provided other energetics are not present. It predicts that there is always more coverage, regardless of the surface geometry, in "layers" other than the first "layer", unless a physical barrier prevents this. The population of "layers" other than the "layer" in direct contact with the surface may be thought of as being dependent only upon the total amount the population of the first "layer" and physical barriers that may

---

[22]The collapse to yield a gas-liquid surface prematurely in pores is probably an exception.

prevent adsorption as expected on a flat unhindered surface. This is a very important point when one considers the isotherm for porous materials.

The first topic covered in this section is the question, "How thick is a 'layer?'" The layers are not centered exactly 1 diameter, 2 diameters, 3 etc., but rather they begin to spread out due to several factors, listed above. The assumptions are handled as follows:

> Assumption #1: The hard wall adsorbent will have no distribution in the normal direction. In analogy to NLDFT this assumption might be broken.
> Assumption #2: Only homogeneous surfaces will be considered here. Correction for heterogeneous surfaces requires the addition of a distribution due to the adsorbent.
> Assumption #3. The Lennard-Jones 6–12 potential (LJ 6–12) will be approximated by a quantum mechanical harmonic oscillator QMHO, which is commonly done, especially for the ground state.
> Assumption #4. The center of the molecules in layers other than the first layer will be fixed as an average of the tetrahedral position and the two-point position. A normal distribution between these two positions will assume they are $2 \times \sigma$ apart. This is a very questionable assumption and is used for lack of anything better at this time.

It turns out that the calculation is not very dependent upon these assumptions except at the low values of θ. This should make sense since it is the distributions between "layers" that is by far the most important determinant of the overall profile. Unfortunately, it is the "layers" close to the surface that are the most important in porosity calculations.

#3 The QMHO approximation for the LJ 6–12 potential

Since the ground state is to be calculated, a simplification of LJ 6–12 of a parabolic match is a good approximation. The LJ 6–2 potential is:

$$E_{LJ} = \varepsilon\left[\left(\frac{r_\sigma}{r}\right)^{12} - 2\left(\frac{r_\sigma}{r}\right)^{6}\right] \qquad (4.110)$$

To obtain the wave function for one molecule in the normal direction,[23] the mathematics is simplified by matching an HO potential to this 6–12 potential. (More details are in Appendix II.F) The HO potential is:

$$E_{HO} = \tfrac{1}{2} k_{HO}(r_{min} - r)^2 - E_{min} \qquad (4.111)$$

---

[23] Apologies for new symbolism, but there are other places where one could get confused without these distinctions.

By matching the minimums and the second derivatives one finds:

$$k_{HO} = \frac{72}{r_\sigma^2}, \quad E_{min} = \varepsilon \quad \text{and} \quad r_{min} = r_\sigma \tag{4.112}$$

The ground state energy is given by:

$$E_n = \hbar\omega\left(n + \frac{1}{2}\right) \Rightarrow E_0 = {}^1/_2 \hbar\omega \Rightarrow \omega = \frac{2E_0}{\hbar} \tag{4.113}$$

It includes the Hermite polynomials together with an exponential function and a normalizing function. However, in this case one need only consider the lowest state, a fact well known for low temperature spectroscopy. The population of the next state is easily calculated to be one part in 1.25×10^5 for argon at 80 K and can reasonably be ignored. The equation for the lowest state may be written as the simple Gausian-like function:

$$\psi = \left(\frac{m\omega}{n\hbar}\right)^{1/4} e^{\frac{m\omega(r - r_{min})^2}{2\hbar}} \tag{4.114}$$

where the QMHO potential $m\omega^2$ replaces the k_{HO} of the classical potential.

This is related to the normal distribution by the substitutions:

$$\sigma = \sqrt{\frac{\hbar}{2m\omega}} \Rightarrow \frac{m\omega}{\pi\hbar} = \frac{1}{2\pi\sigma^2} \tag{4.115}$$

and interpreting the probability function $\mathbf{P}(r) = \psi_0 {}^* \psi_0$ as a normal distribution[24] yields:

$$\mathbf{P}(r - r_{min}) = \frac{1}{\sigma\sqrt{2\pi}} \exp\left(-\frac{(r - r_{min})^2}{2\sigma^2}\right) \tag{4.116}$$

See Appendix II.G for a little more detail.

For the first monolayer, the $\mathbf{P}(r)$ from this equation is the overall probability of finding the molecule at the position r. However, for subsequent layers, the oscillators must be coupled to the underlying layers. Subsequence coupled harmonic oscillators would follow the same relationship, but with the $r = 0$ reference shifted to coincide with the first oscillator. Thus for the ground states this creates a concatenated string of σs of increasing size so:

$$\sigma_{n+1}^2 = \sigma_n^2 + \sigma_0^2 \quad n = 0, 1, 2, 3 \ldots \tag{4.117}$$

where the symbol σ_0 indicates the σ that would occur for the isolated molecule against a hard wall.

---

[24] Why go to the trouble of rewriting this as a normal distribution? The reason is because most computer programs have this normal distribution function shown here as well as its integral embedded. See your spreadsheet help.

#4 The packing problem uncertainty

For normal packing, the packing factor F_{1D} may be used to determine the geometrical distance between "layers" by the formula:

$$d_{\text{layers}} = 2F_{1D}r_\sigma \qquad (4.118)$$

On the other hand, the distance between the center of a layer and the center of a molecule sitting in the position of only two nearest neighbors from the layer is given by:

$$d_{2\text{point}} = \sqrt{3}r_\sigma \qquad (4.119)$$

An average value for an effective radius would be:

$$\langle r \rangle = \left(^1/_2 F_{1D} + ^1/_4 \sqrt{3} r_\sigma\right) r_\sigma = 0.841 r_\sigma \qquad (4.120)$$

The difference between these two distances, given here as the symbol Λ, is then:

$$\Lambda = \left(2F_{1D} - \sqrt{3}\right) r_\sigma \approx 0.099 r_\sigma \qquad (4.121)$$

Λ will then be treated as a standard deviation and thus adds to other σs and Λs by the variance addition rule to yield "σ_{total}." These are concatenated from layer to layer as described above.

In **Table 4.4** is a list of LJ potential parameters and the spacing parameters derived for the noble gasses and their counterpart QMHO and normal distribution parameters that might be derived from them.

These are assumptions that will help to reveal some of the characteristics that one might find in the generated isotherm and the accompanying family of standard curves. As will be seen, they make a large difference for the range of ultramicropores to the low end of micropores. For larger pores, the approximation of a very sharp distribution yields a good approximate answer.

Table 4.4 Parameters to determine spacing and distribution of "layers"

	E_{min} (J/molecule)	r_σ (nm)	$\langle r \rangle$ (nm)	$r_\sigma \times F_{1D}$ (nm)	k_{HO}/kg (s^{-1})	m (kg)
Ne	5.00×10^{-22}	0.274	0.230	0.224	0.48	3.35×10^{-26}
Ar	1.67×10^{-21}	0.341	0.287	0.278	1.04	6.63×10^{26}
Kr	2.25×10^{-21}	0.365	0.273	0.298	1.22	1.39×10^{-25}
Xe	3.20×10^{-21}	0.398	0.299	0.325	1.46	2.18×10^{-25}
	ω (s^{-1})	σ_1 (nm)	σ_1/r_σ	Λ (nm)	σ_{total} (nm)	$\sigma_{\text{total}}/\langle r_\sigma \rangle$
Ne	$3.78 \times 10^{+12}$	0.0204	0.074	0.0271	0.0339	0.147
Ar	$3.95 \times 10^{+12}$	0.0142	0.042	0.0337	0.0366	0.128
Kr	$2.96 \times 10^{+12}$	0.0113	0.031	0.0362	0.0379	0.139
Xe	$2.58 \times 10^{+12}$	0.0097	0.024	0.0394	0.0406	0.136

$F_{1D} = 0.8165$

$\sigma_1 = (h/4\pi m\omega)^{1/2}$, $\Lambda = r_\sigma \times (\sqrt{3} - 2F_{1D})$, $\sigma_{\text{total}} = [\sigma^2 + 1/4\Lambda^2]^{1/2}$

However, for small pores in the range of 1 to 2 monolayers, there is also another method that can be used to yield a good approximation. These distributions should be applied to the population of the individual layers as described in the next section.

Combining the normal direction distributions with the χ-n layer results

Although given in another chapter, a review to the χ-theory n "layering" is in order. There is no reason to assign a different probability for the fractional occupancy of the second layer than for the first layer,[25] nor for the third layer, etc. One might wonder how this could possibly be. This is the problem with classical mechanical thinking. In the quantum mechanical model, both the molecules and the forces are delocalized and unknown until the observation is made (or as some might say, that the molecule "shows up.") Of course, no layer may have a negative coverage. Thus, for whatever amount of adsorbate that is not in the first layer, the amount in the subsequent layer adsorbs as if the first layer plus the free surface is the new adsorbent. Thus a portion of the remaining adsorbate adsorbs in the second layer according to the χ theory. This leaves some adsorbate left over for the 3rd layer and the calculation is performed again. This is repeated for all subsequent layers, unless some other phenomenon interrupts the process.

These interrupting processes could include a physical barrier such as a pore wall, or some new energetic such as the formation of an gas-liquid interface. Excluding the exceptions mentioned, the probabilities may be written mathematically as:

$$\mathbf{P}(\theta_{n+1}) = \mathbf{P}\left(\theta - \sum_{i=1}^{n} \theta_i \middle| \theta_n \right) \geq 0 \quad (4.122)$$

From the derivation of the χ theory, the surface that is not covered, or "bare," by any amount of adsorbate, is given by:

$$A_{\text{bare}} = A_s \exp(-\Delta\chi) \quad (4.123)$$

The area covered by at least one monolayer, A_1, is the total area, A_s, minus the bare or:

$$A_1 = A_s - A_{\text{bare}} = A_s(1 - \exp(-\Delta\chi)) \quad (4.124)$$

The occupancy, or monolayer equivalence, of the first layer, θ_1, is obtained by A_s. Thus, by χ theory this is given by:

$$\theta_1 = 1 - e^{-\Delta\chi} \quad (4.125)$$

---

[25] The definition of adsorbate layer by χ theory is not dependent upon distance from the surface but rather how many intervening adsorbate molecules there are between it and the surface. However, when the underlying layers have a value of θ approaching 1, then the correspondence to geometry is much closer.

With no constraints on the build-up of layers, this can be written in terms of the monolayer equivalence coverage for a flat surface:

$$\theta_1 = 1 - e^{-\theta} \tag{4.126}$$

with the definition of the θs as monolayer equivalents, so that:

$$\theta = \frac{n_{ads}}{n_m}, \; \theta_1 = \frac{n_{ads,1}}{n_m}, \ldots, \theta_n = \frac{n_{ads,n}}{n_m} \tag{4.127}$$

and for $n > 1$ layers (if no other energies enter into consideration):

$$\theta_n = 1 - e^{-\beta_m} \tag{4.128}$$

with the βs defined as:

$$\beta_m = \Delta\chi - \sum_{i=1}^{m} \theta_i \tag{4.129}$$

for as high as m is allowed by geometry or another limiting factor (see for example the section on mesoporosity for a potential energy limit.)

Special limiting case, m = 1

For $m = 1$, i.e., a single monolayer equivalent, a special law is derived from this since one can by taking to exponential for both sides of the χ-equation:

$$\exp(-\Delta\chi) \equiv \exp\left(+ \ln\left[-\ln\left\{\frac{P}{P_{vap}}\right\}\right] - \ln\left(-\frac{\overline{E}_a}{RT}\right)\right) \tag{4.130}$$

and simplifying:

$$\exp(-\Delta\chi) \equiv \left(\left[-\ln\left\{\frac{P}{P_{vap}}\right\}\right]\left(-\frac{RT}{\overline{E}_a}\right)\right) \tag{4.131}$$

Substituting into Eq. (4.125) θ_1 for $1 - \exp(-\Delta\chi)$ to arrive at:

$$\theta_1 = 1 - \frac{RT}{\overline{E}_a} \ln\left(\frac{P}{P_{vap}}\right) \tag{4.132}$$

This equation indicates for the first monolayer, the dependence of the coverage is linear with respect to $\ln P$. This is a commonly observed "law" for ultramicroporosity (as defined for χ theory.)

Amount in each layer as a function of $\Delta\chi$

The series for Eqs. (4.128) and (4.129) can be written as:

$$\theta_n = 1 - \exp\left(-\Delta\chi + \sum_{n=1}^{n-1} \theta_i\right) \tag{4.133}$$

Eq. (4.133) provides a convenient method to calculate the number of adsorbate molecules that exist in each layer. An interesting aspect of this equation is that there is no direct dependence upon the energy of adsorption or pressure since $\Delta\chi=0$ on a flat surface. In Fig. 4.12 are results obtained from this calculation for layers 1 through 5.

It should be obvious that to obtain the isotherm, we need only add the results of all the layers in Fig. 4.12 at each $\Delta\chi$ values required. Mathematically, it can be demonstrated that the series represented by Eqs. (4.125)–(4.129) adds up to $\Delta\chi$. However, the more detailed distributions require the superposition of the uncertainties arrived at in the previous section to be superimposed on these layer distributions.

In Fig. 4.13 the individual layer uncertainties have been incorporated. This figure shows the density profile that one expects at a very large value of $\Delta\chi(P\to P_{vap})$ if a perfectly flat hard wall surface is assumed. In this figure, the literature values for argon were used to calculate the parameters listed in Eq. (4.121) (in Chapter 6, the LJ parameters that were used for the DFT were, for consistency, are also used for the χ model.) For the normal direction, E_a has little effect for $n>1$. For the first layer, the surface is assumed to be fixed—that is, a "hard-wall." The hard-wall assumption is, of course, unrealistic and makes the profile of the first layer unrealistically sharp. The results are very similar to calculations made using density functional theory or grand canonical Monti Carlo calculations (see DFT section in Chapter 7.)

For coverages other than very high, Fig. 4.14 (also using information from Eq. (4.121)) shows what the profiles might look like with increasing $\Delta\chi$. These values are as labeled for $\Delta\chi=2, 4, 6,$ and 8. This figure also uses Ar as a model. These profiles are for the ideal flat surface only; curved surfaces and especially pores are quite different, as indicated in Chapter 6.

The transform to the isotherm is obtained by integration over the layers from $distance=0$ to ∞, of the layers, shown in Fig. 4.14. Thus, for the value for the isotherm

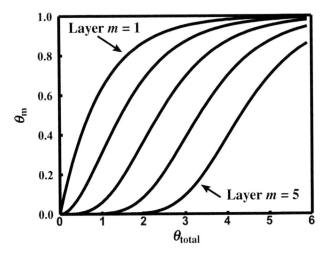

FIG. 4.12

The individual monolayer coverages for layers 1 through 5.

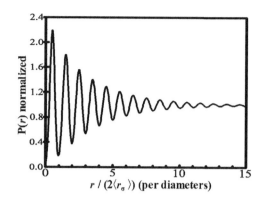

FIG. 4.13

Argon depth profiles against a hard-wall of an unlimited surface film.

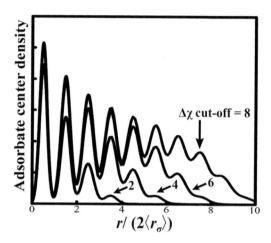

FIG. 4.14

The profiles of the adsorbate "layers" as a function of $\Delta\chi$.

for $\Delta\chi = 2$, the curve labeled "2" is integrated. This is easy to do with most programs since this integral is simply the "cumulative normal distribution" available in most computer programs. This would yield the isotherm as a function of the monolayer equivalents versus $\Delta\chi$. This is not normally done since the answer is directly available from the χ equation.

The addition by variance of the uncertainty in the layer distance and the vibrational σ yielded the initial σ, which added the uncertainty in the layer distance with the vibrational uncertainty. Thus for the isotherm of a flat surface:

$$\theta(\Delta\chi) = \int_{x=0}^{\infty} \sum_n \mathbf{P}_n(x - nd, \sigma_n) dx \tag{4.134}$$

where **P** will be the normal distribution related to the ground state with the arguments, and is increased from layer to layer by the sequence:

$$\sigma_n = \sqrt{\sigma_{n-1}^2 + \sigma_1^2} \qquad (4.135)$$

In the section on porosity in Chapter 6, Eq. (4.134) is used with cut-offs to χ-calculated universal standard curves for slit and cylindrical pores. The cut-off for the slit pores is a step function, whereas the cut-off for cylindrical pores is a linear decreasing function. The length of the pores, has no effect on this profile, but does naturally have an effect on the overall monolayer equivalence. There are also some listings of standard curves and directions on how to use them. Such an approach would be very difficult, and probably impossible, using the normal standard curve method, since each adsorbate-absorbent pair has its own standard curve. For the χ method this is not a problem.

The simulation shown in Fig. 4.13 could be simulated with an equation that has the form of a $\sin^2(r)$ since the probabilities are from the wave function is $\psi^*\psi$. This equation is:

$$\mathbf{P}(r) = A \sin^2(\pi r)\exp(-r/\lambda) + (1 - \exp(-r/\lambda)) \qquad (4.136)$$

where A and λ are for the moment parameters to fit Fig. 4.13.

This could make the calculation of the porosity much simpler once one obtains these parameters for the adsorbate. To illustrate this fit, Fig. 4.15 shows the least squares fit of Eq. (4.136) to the data up to five layers of Fig. 4.13. After five layers, the fit and the simulation are indistinguishable. The parameters for this are $A=2.15$ and $\lambda=3.528$. The standard deviation of the fit is 4.7%. One of the main problems with the fit is that the distribution for the first layer is "cut off" at the beginning of the figure, causing a misfit to the early profile data.

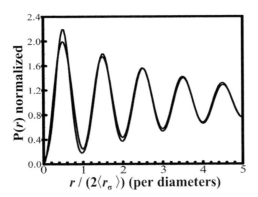

FIG. 4.15
Simulated profile fitted to an equation (inside line).

Simulations of the isotherm using n-layer

The above calculations can be used to simulate the isotherm. For the total isotherm, the populations calculated above are either:

- For slit pores, multiply by a unit step function that is 1 at the pore wall and continues as 1 to the midpoint between the slit walls, at which it drops to 0.
- For cylindrical pores, multiply by a linear function that is 1 at the pore wall and 0 at the center of the pore.

It is instructive to see what the profiles look like given the cut-offs mentioned. This is easily done in the spreadsheet as a result of the calculation. In Fig. 4.16 are the density profiles expected from a slit pore as a function of distance from the walls separated by 10 nm. In Figs. 4.17 and 4.18 are the profiles expected in a pore with a radius or 5 nm and a pore with a radius of 2 nm. Different adsorbent pressures in terms of $\Delta\chi$ are presented. For the cylindrical pores the first peak at a pressure corresponding to about $\Delta\chi=4$ is not much different from the of $\Delta\chi=10$. The closer the peak gets to saturate the first layer, approximately $\Delta\chi=4$, the less energy that is available to maintain a film that does not have an external sharp interface. Not surprisingly the dependence of the first peak follows the "log law" illustrated by Fig. 4.19.

The above treatment assumes that the gas-liquid interface is still ill-formed. Once the criterion for the trade-off in energies to formation of the gas-liquid interface, γ_{lg}, exceeds the incremental energy gain of $E(\theta)$, then the collapse of the film to a two interface film should occur. This should happen at about $\theta \approx 2.5$ for cylinders. This, of course, depends upon $\overline{E}_a, \gamma_{sl}$ and γ_{sg}.

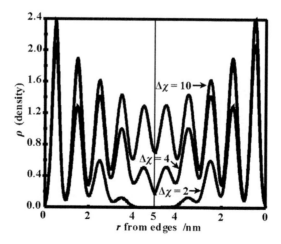

FIG. 4.16

Density profile in a slit pore walls separated by 10 nm at $3\Delta\chi s$.

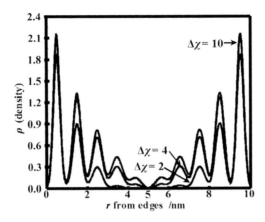

FIG. 4.17

Density profile in a cylindrical pore with $r=5$ nm at $3\Delta\chi s$.

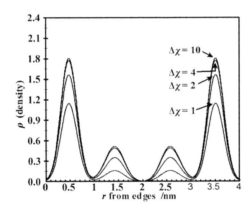

FIG. 4.18

Denisty profile in a cylindrical pore with $r=2$ nm at $3\Delta\chi s$.

Adsorption of binary adsorptives—More than one adsorbate

Binary adsorption in χ theory has not been thoroughly tested, mainly due to the lack of appropriate experimental data. Here two approximations are presented using only the χ theory. The BET theory and other theories have not proved to be of any use and DFT has not been attempted. Firstly, the approximation for the adsorption on nearly flat surfaces is discussed. Secondly, in Chapter 6, adsorption for porous adsorbents up to filled or nearly filled is presented. For both of these cases there is some information in the literature against which to test the derivations.

There are several ways of approaching the problem of binary adsorption. In this section two methods will be described using χ theory. The first one is to re-derive,

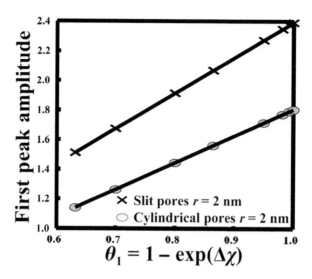

FIG. 4.19

Amplitude of the first peak as a function of the calculated.

starting by modifying the grand canonical partition function, and the second one is to take advantage of the "layer" population. The first approach is similar to the approach used with the binary Langmuir isotherms, except for χ theory, there are additional terms. In the case of a flat surface there is the addition E_a of the second adsorbate. For porous solids, there are obviously more parameters needed to express the shape, radius, and volume of the adsorbent.

The most useful application for these methods is for porous materials. First, however, the theoretical foundation is required for a plane surface. The modifications and applications for porous materials are presented in Chapter 6.

Method 1: Adjusting the grand canonical ensemble

The symbols in this section are the same as used previously except for the added subscripts 1 and 2 for labeling the adsorbates. In this and subsequent equations there are identical equations for adsorbate 2 with the indexes 1 and 2 switched. Thus, the quantities $\overline{E}_{a,1}$ for adsorbate 1 in relation to its $\overline{E}_a$ may be defined by:

$$\overline{E}_{a,1} \equiv \overline{E}_1 + {}^1/_2 RT + \mathbf{f}(T) \qquad (4.137)$$

Recall that $\mathbf{f}(T)$ for simple molecules should be zero and the ${}^1/_2 RT$ is small and will be ignored to simplify matters. In $\overline{E}_a$ of Eq. (4.30) is given here as E_1.

Another important restriction is the χ_1 and χ_2 must both fulfill the requirement:

$$\chi_n > \chi_{c,n} \quad n = 1 \text{ and } 2 \qquad (4.138)$$

With a large difference in χ_cs the adsorbate with the lowest χ_c, that is greatest energy, dominates the adsorption especially in the first layer adsorption. The second one fills in with the layers >1 as in the bulk solution. It appears that it if the two χ_cs become close, the situation becomes more complex.

Following the same prescription as before and noting that now molecules of type 2 may also form "teeth" in the particle in the box description, however the energy release for a type 2 will, in general, be different from a type 1. For the energy of adsorption, there are two considerations provided that the surface is homogeneous. One of the considerations is the energy of adsorption for the first molecule adsorbed on a clean surface, this is designated as E_1 and E_2 ($\overline{E}_1$ and $\overline{E}_2$ for molar quantities.) The second consideration is the amount of clean surface left over once adsorption begins. It will be assumed that in the summing of species 1, that the maximum term for species 2 with the amount N_2 is present and likewise for the summing of species 2 with the amount of N_1 present. Thus:[26]

$$E_N \equiv E_1 \int_{x=0}^{x=N_1-1} \left(1-\frac{a_1}{A_s}\right)^x dx \left(1-\frac{a_2}{A_s}\right)^{N_2} + E_2 \int_{y=0}^{y=N_1-1} \left(1-\frac{a_2}{A_s}\right)^y dy \left(1-\frac{a_1}{A_s}\right)^{N_1} \quad (4.139)$$

Added to this is the energy of interaction ($E_{in,1}+E_{in,2}$) between the adsorbed molecules. This replaces the previous term, which was simply $N\varepsilon$. Since this is a "big box" approximation, the energy between the molecules will be a weighted average, or for each adsorbate 1 and 2, this is:

$$E_{in,1} \equiv E_{in,2} = N_1 \frac{N_1 \varepsilon_{11} + N_2 \varepsilon_{12}}{N_1+N_2} + N_2 \frac{N_2 \varepsilon_{22} + N_1 \varepsilon_{21}}{N_1+N_2} \quad (4.140)$$

This is obviously the regular solution assumption, so one would expect that at high pressures the regular solution theory equation would be the result. This term could be replaced by other assumptions to yield different liquid solution functions that are more complicated than simply adding a Margules constant.

In constructing the grand canonical ensemble for the χ equations for one adsorbate, no accounting was needed for the sequence in which the molecules adsorbed, since they were all indistinguishable. In the case of two adsorbates, however, this is not the case. The number of ways one can arrive at a system with N_1 molecules of adsorbate 1 and N_2 molecules of adsorbed 2 is given by the (well known) expression:

$$\text{Number of sequences} = \frac{(N_1+N_2)!}{N_1! N_2!} \quad (4.141)$$

---

[26]When differentiating this expression, the difference between N_1 and N_1-1 will be assumed negligible due to the high value of N_1, and likewise for $N_2-1 \approx N_2$

From the above considerations, the grand canonical ensemble may be written:

$$\Xi = \sum \tilde{p}_1^{N_1} \times \tilde{p}_2^{N_2} \times \left[\frac{(N_1+N_2)!}{N_1!N_2!}\right]$$

$$\times \exp\left(-E_1 \int_{x=0}^{x=N_1-1}\left(1-\frac{a_1}{A_s}\right)^x dx \left(1-\frac{a_2}{A_s}\right)^{N_2} - E_2 \int_{y=0}^{y=N_2-1}\left(1-\frac{a_2}{A_s}\right)^y dy \left(1-\frac{a_2}{A_s}\right)^{N_1}\right)$$

$$\times \exp\left(-\left(\frac{1}{kT}\right)\left[N_1\frac{N_1\varepsilon_{11}+N_2\varepsilon_{12}}{N_1+N_2} + N_2\frac{N_2\varepsilon_{22}+N_1\varepsilon_{21}}{N_1+N_2}\right]\right)$$

(4.142)

The IUPAC symbols $\tilde{p}_1$ and $\tilde{p}_2$ indicate the fugacity of 1 and 2. Following the usual procedure and taking the partial differential with respect to N_1 of the ln of the maximum term of Ξ (designated as ln (Ξ^*), or the grand potential X_1) and switching to molar quantities:

$$RT\frac{\partial \ln(\Xi^*)}{\partial N_1} = RT\ln(\tilde{p}_1) - \overline{E}_1 e^{-\theta_1-\theta_2} + \Delta\overline{\varepsilon}X_2^2 - RT\ln X_1 - \overline{E}_2\frac{a_1}{a_2}\left(\exp^{-\theta_1-\theta_2} - \exp^{-\theta_1}\right)$$

(4.143)

where θ_1 and θ_2 have the same meaning with respect to components 1 and 2, respectively, as before:

$$\theta_1 = \frac{n_1}{n_{m,1}}, \quad \theta_2 = \frac{n_2}{n_{m,2}}$$

(4.144)

The symbols X_1 and X_2 are the mole fraction of component 1 and 2 in the adsorbate (liquid-like) phase and $\Delta\varepsilon$, (Margules 1st constant) is defined as:

$$\Delta\varepsilon = \varepsilon_{12} + \varepsilon_{21} - \varepsilon_{11} - \varepsilon_{22}$$

(4.145)

The left side of **Eq. (4.143)** is set to zero to yield the maximum line on the θ_2:

$$-\ln(\tilde{p}_1) = \left(\frac{\overline{E}_1}{RT} + \frac{\overline{E}_2 a_1}{RT a_2}\right)e^{-\theta_1-\theta_2} - \frac{\overline{E}_2 a_1}{RT a_2}e^{-\theta_1} + \frac{\overline{\varepsilon}}{RT} + \frac{\Delta\overline{\varepsilon}}{RT}X_2^2 - \ln X_1$$

(4.146)

X_1 and X_2 are the mole fractions of adsorbate 1 and 2, respectively.

If one writes the expression for component 2, solve for $\exp(-\theta_1-\theta_2)$ for both equations, and set them equal, this results in a contradiction when one looks at the experimental evidence. This is a troubling problem with this derivation. The explanation could be that the threshold pressure does not show up in this derivation as it should or it may be that the derivation is incorrect. This should be a topic of further research. Ignoring this caveat, the narrative continues. The n-layer derivation for binary adsorption seems to work as expected, so the total χ- theory at this point can not be entirely rejected.

In the above equations, the factor of ½RT for both of the adsorbates is not explicitly included, however, it is small and could be easily entered back in. The internal modes, $\mathbf{f}_1$ and $\mathbf{f}_2$ have likewise been left off. Using the low pressure relationship $\ln(\tilde{p}_1) = \ln(p_1)$ and the previous definition using the subscript "vap" to designate the vapor pressure of the pure adsorptive over its liquid with a flat surface:

$$-RT\ln\left(p_{vap,1}^*\right) = \overline{\varepsilon}_{11}$$

(4.147)

The symbol * is for the pure substance, that is, the vapor over the pure liquid at the temperature of interest (see the IUPAC Green Book 3rd edition 2nd printing [26]). Eq. (4.146) may be written in terms of $\Delta\chi_1$ as:

$$-\ln\left(\frac{p_1}{X_1 e^{-\Delta\bar{\varepsilon}X_2^2/RT} p_{vap,1}^*}\right) = \left(\frac{\bar{E}_1}{RT} + \frac{\bar{E}_2}{RT}\frac{a_1}{a_2}\right)e^{-\theta_1-\theta_2} - \frac{\bar{E}_2}{RT}\frac{a_1}{a_2}e^{-\theta_1} \quad (4.148)$$

Defining:

$$\Delta\chi_1 = \chi_1 - \chi_{c,1}(\theta_2), \chi_1 = -\ln\left\{--\ln\left(\frac{\tilde{p}_1}{X_1 e^{-\Delta\bar{\varepsilon}X_2^2/RT} p_{vap,1}^*}\right)\right\}, \chi_{c,1}(\theta_2)$$

$$= \ln\left\{\left(\frac{\bar{E}_2}{RT}\frac{a_1}{a_2}(e^{-\theta_2}-1) + \frac{\bar{E}_2}{RT}e^{1\theta_2}\right)\right\} \quad (4.149)$$

This yields an equation that has the same form as the original simple χ equations:

$$\theta_1 = \Delta\chi_1 \mathbf{U}(\Delta\chi_1) \quad (4.150)$$

Thus, if E_1 were the higher (absolute) energy, then even though χ_2 were below $\chi_{c,2}$ ($\Delta\chi_2 < 0$), there would still be some species 2 adsorbed on the surface due to the finite value for θ_1. On the other hand, if E_1 were the higher energy and χ_1 were below $\chi_{c,1}$ there must be adsorption for neither species 1 nor 2 even though there is an over pressure of adsorptives provided $\chi_2 < \chi_{c,2}$ Thus, there is still a phase transition to the surface liquid phase at either $p_{c,1}$ or $p_{c,2}$.

To simplify Eq. (4.146) a little, since the regular solution was assumed in the beginning, the vapor pressure expected over a regular solution[27] of:

$$p_{vap,1} := p_{vap,1}^* X_1 e^{-\bar{\varepsilon}X_2^2/RT} \quad (4.151)$$

where $p_{vap,1}$ is the partial pressure of adsorbate 1 over the bulk liquid mixture phase and $P_{vap,1}^*$ is the vapor pressure over the pure liquid adsorptive 1.

At low pressures the fugacity $\tilde{p} \approx p$. However, the partial pressure, using Y_1 and Y_2 for the gas mole fractions is related to the total pressure by:

$$p_1 = Y_1 P \quad (4.152)$$

Substituting:

$$-RT \ln\left(\frac{Y_1 P}{p_{vap,1}}\right) = \left(|\bar{E}_1| + |\bar{E}_2|\frac{a_1}{a_2}\right)e^{-\theta_1-\theta_2} - |\bar{E}_2|\frac{a_1}{a_2}e^{-\theta_1} \quad (4.153)$$

---

[27] This need not be restricted to regular solutions where the "correction function" F(X$_1$) is that given in this equation. There are several potential equations that are given for specific solutions where **F(X$_1$)** is a polynomial that would work as well. As applied to Arnold's work this term is small but noticeable in the "linear" plot that has a slight bow, especially at the 50% N$_2$, O$_2$ gas composition.

There are a few things to notice about Eq. (4.153) and its companion definition Eq. (4.151):

1. As $n_{ads,1}$ and $n_{ads,2}$ both approach ∞ in the same ratio, the regular solution theory relationship is approached (Since the $\ln(1)=0$, which would be the right side of this equation.) This fulfills one very important requirement for a valid adsorption theory, which is this limit should yield a reasonable bulk liquid answer.
2. As $n_{ads,2} \to 0$, and thus $p_1 \to P$, the equation approaches the simple χ theory equation for a single adsorbate.
3. As $n_{ads,2}$ approach ∞ and if $n_{ads,1}$ is small, the equation yields Raoult's law for solutions.
4. As both $n_{ads,1}$ and $n_{ads,2}$ approach 0 in the same ratio, the pressure approaches a threshold pressure and the threshold pressure should be the same as for the threshold pressure for the adsorbate that has the highest energy for the first molecule adsorbed.

Conclusion 3 follows from 1, and conclusions 1 and 4 can be tested from the data by Arnold [27].

To simultaneously obtain the maximum for 1 and 2, one would subtract the number 2 counterpart. Thus from Eq. (4.153) one obtains:

$$-RT \ln\left(\frac{p_1 P_{vap,2}}{p_2 P_{vap,1}}\right) = \left(\overline{E}_2 - \overline{E}_1 - \frac{\overline{E}_2 a_1}{a_2} + \frac{\overline{E}_1 a_2}{a_1}\right) e^{-\theta_1 - \theta_2} - \overline{E}_2 e^{-\theta_1} + \overline{E}_1 e^{-\theta_2} \quad (4.154)$$

There are two important approximations to this equation:

1. With θ_1 and $\theta_2 \to \infty$ together, this equation yields the ratio of adsorbates 1 and 2 in the Raoult' Law limit. Even before high values are reached, the coverages increase together using a constant gas mix, the slope of the ratios of θ_1 and θ_2 versus the $\Delta\chi$ of the total pressure follows Raoult's Law, but has an off-set. This can be demonstrated by the limits of this equation going to ∞ and to 0. That is if:

$$\theta_1 = k_a \theta_2 + k_b \quad \theta_1 > k_b \quad (4.155)$$

where θ_1 is the high energy adsorbate and therefore adsorbs to the extent of k_b before the threshold for θ_2 then:

$$\lim_{\theta_2 \to \infty}\left[\ln\left(\frac{p_1 P_{vap,2}}{p_2 P_{vap,1}}\right)\right] = 0 \Rightarrow \frac{p_1 P_{vap,2}}{p_2 P_{vap,1}} = 1 \quad (4.156)$$

and:

$$\lim_{\theta_2 \to 0}\left[\ln\left(\frac{p_1 P_{vap,2}}{p_2 P_{vap,1}}\right)\right] = \left(-\frac{\overline{E}_2 a_1}{RT a_2} + \frac{\overline{E}_1 a_2}{RT a_1}\right) e^{-k_b} = \text{a constant} > 0 \quad (4.157)$$

2. As an approximation, for a special case that assumes the molecular sizes are approximately equal, then setting $a_1 = a_2$ and noticing then that the left side of Eq. (4.154).

Adsorption of binary adsorptives—More than one adsorbate

$$\frac{\overline{E}_2}{RT}e^{-\theta_1} = \frac{\overline{E}_1}{RT}e^{-\theta_2} + \mathbf{F}(X_1). \tag{4.158}$$

is a function, **F**, that is fixed for a fixed mixture of gasses, where for the regular solution:

$$\mathbf{F}(X_1) = RT \ln\left(\frac{X_1}{X_2}\right) + \alpha\left(X_2^2 - X_1^2\right) \tag{4.159}$$

The component with the highest energy needs to be on the right side of these equations. To determine the off-set, set $\theta_2 = 0$ so:

$$\lim_{\theta_2 \to 0^+} \left(e^{-\theta_1}\right) = \frac{\overline{E}_1}{\overline{E}_1} + \frac{RT}{\overline{E}_2} \ln\left(\frac{X_1}{X_2}\right) + \alpha\left(X_1^2 - X_2^2\right) \tag{4.160}$$

Notice that the mole fractions remain constant so long as θ_2 is approached from the positive side. Also:

$$\lim_{\theta_2 \to 0^+} (\theta_1) = \ln\left(\frac{\overline{E}_1}{\overline{E}_1} + \frac{RT}{\overline{E}_2} \ln\left(\frac{X_1}{X_2}\right) + \alpha\left(X_1^2 - X_2^2\right)\right) \tag{4.161}$$

Notice that the χs are shown as functions of the P_{vap}s as described in Eq. (4.47). In the following narrative this relationship will be used for shift in the extrapolated on-sets of adsorption:

$$\Delta n_c \equiv n_{m,1} \Delta \theta_c \tag{4.162}$$

This offset then is the binary counterpart to the pure adsorbate threshold requirement. The consequence of this is the linear fit of #2 adsorbate versus #1 adsorbate will be the Raoult' Law line shifted by the amount indicated by Δn_c, if one assumes the ideal solution; if the regular solution or some other Margules equation applies, then Eq. (4.151) needs to be changed appropriately.

Demonstrations of evidence for the validity of the above equations are given here with the data by Arnold of the adsorption of N_2 and O_2 on anatase. To compare to experimental results that are in terms on n_{ads} one converts θ_1 to $n_{\text{ads},1}$, since the off-set is in N_2.

The n_ms from the pure gas isotherms are for nitrogen $n_m \approx 0.090$ mmol and for oxygen $n_m \approx 0.140$ mmol.

The data for the five binary mixtures are presented in Figs. 4.20–4.24. The experiment fulfills the requirements of the above derivation, that is, a simple non-porous adsorbent and two similar and simple adsorbates. These plots show Arnold's data displayed to demonstrate Eq. (4.156). The lines passing through the origin represent the ratios of the $n(O_2)$ to $n(N_2)$ for the liquid phase as measured by Dodge and Davis [28]. In Table 4.5 are the numerical comparisons. There is good agreement between the slopes, except for the 85% O_2_15% N_2, which is evident from Fig. 4.24. The explanation for this is unclear; however, the data scatter is quite large for this mixture.

The differences in the $\Delta\chi_c$s of N_2 and O_2 are needed to make a comparison for the off-set seen for the "start" of the O_2 adsorption. The χ_c for the pure N_2 and O_2 adsorption can be extracted from the pure N_2 and O_2 isotherms.

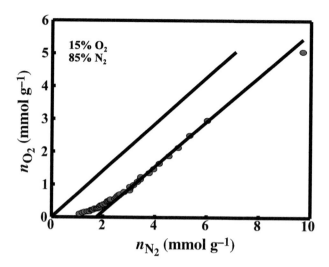

FIG. 4.20

Arnold's adsoptive mixture on anatase of $O_2:N_2 = 15\% : 85\%$.

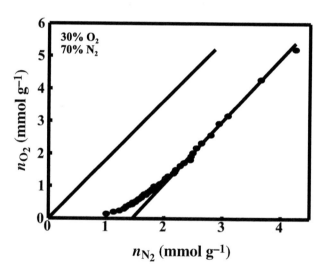

FIG. 4.21

Arnold's adsoptive mixture on anatase of $O_2:N_2 = 30\% : 70\%$.

There is some uncertainty in determining these numbers, apparently because of temperature control error, as mentioned in the section "Temperature Control of the Sample" in Chapter 2. However, the low χ values should yield values that are reasonably close as this section would indicate. So, here values of $\chi < 0$ are used. In comparison to this data, the χ plots for these pure gas isotherms are presented in Fig. 4.25.

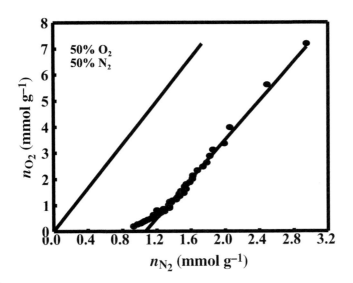

FIG. 4.22

Arnold's adsoptive mixture on anatase of $O_2:N_2 = 50\% : 50\%$.

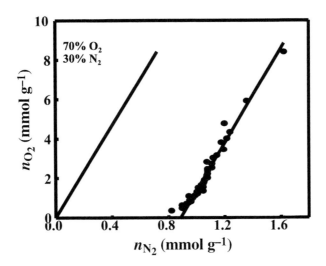

FIG. 4.23

Arnold's adsoptive mixture on anatase of $O_2:N_2 = 30\% : 70\%$.

The χ_c for the N_2 and O_2 pure adsorption were -2.97 and -2.55, respectively. The r^2 for the fits to these isotherms shown in Fig. 4.25 are 0.992 and 0.989 for the N_2 and O_2, respectively. This is not particularly good even though the low pressure region, which appears linear, was selected. The data diverts from the expected linear and is probably due to the sample temperature being about 1K above the

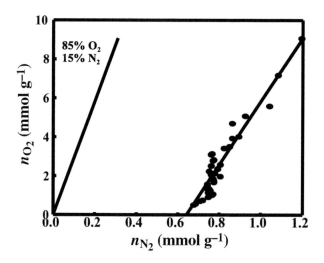

FIG. 4.24

Arnold's adsoptive mixture on anatase of $O_2:N_2=85\%:15\%$.

Table 4.5 Comparison of O_2 vs N_2 slope for adsorption versus bulk liquid

Mix	Slope fit[a]	Raoult Law slope	$\Delta\theta_c$	Δn_c (N_2)
15% O_2	0.79	0.82	1.74	0.157
30% O_2	2.22	2.10	1.45	0.131
50% O_2	4.95	5.49	1.06	0.095
70% O_2	12.4	11.7	0.89	0.080
85% O_2	16.3	29.0	0.64	0.058

[a]moles O_2/moles N_2.

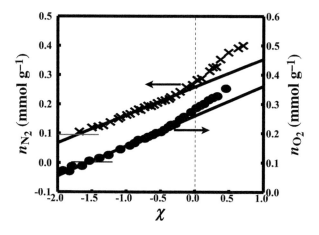

FIG. 4.25

χ-plots for the adsorption of pure N_2, ■, and O_2, ×, on anatase by Arnold.

measure temperature. This is a common problem with older data. More recent data does not deviate this much (see the section on standard curves.) The data was not "corrected" for this but rather only data below $\chi=0$ was used. The first O_2 data point marginally fails the Q test (90%) and is excluded, but it made an insignificant difference. The $\Delta\chi_c$s internally for each of the mixtures can then be calculated from total pressures; the percent of vapor pressure of each gas, and the two χ_cs for the pure gasses using Eq. (4.34) or (4.37). In Table 4.5 are the comparisons of the experimental slopes and the slopes expected from Raoult's law. The experimental off-sets in terms of n_{ads} and the calculated off-sets in terms of $\Delta\chi_c$ are also presented.

An important conclusion from this analysis and the off-sets illustrated in Fig. 4.25, is that the monolayer equivalent coverages, $n_{m,1}$ and $n_{m,2}$ are not what one obtains from the bulk liquid phase diagram, although the sum of them should add in such a way as to approximately satisfy the Lewis rule. This means that using the weighted average of the two $\overline{A}$ values one should obtain the about same surface area as the pure isotherm, which has been determined with the caveat that adsorbate molecules are very similar.

One might be tempted to use the Lewis rule to calculate how much binary adsorbate is in the pores of porous materials. This would be a mistake unless the pores were extremely large. The reason for this is that the first layer is usually different from the subsequent layers and a very large error is introduced. This is also demonstrated in Chapter 5 with some porous materials whose binary diagrams "bend" to exactly the opposite sides of the diagram from the ideal (Raoult's law) or the regular solution law calculation for the bulk liquid.

Method 2: Binary adsorbate using the *n*-layer analysis

Method 1 at this point leaves many questions unanswered and untested. Presented here is a second method that may or may not be a better solution. However, in Chapter 5 a simplified method of determining the binary phase diagram for porous materials is presented based on these concepts. It is a reasonable approximate method. The method assumes the following:

- The pressure of the gas mixture remains constant.
- One of the adsorbates energy of adsorption is much higher than the other and therefore dominates in the χ plot.

The first assumption is the normal way binary diagrams are presented. The second assumption seems to be justified by the results of Method 1. It is also the assumption needed for the use of the "heavy Henry" adsorbate as the controlling adsorbate in the carrier flow method (it should be obvious that the adsorbate with the higher energy, or lower χ_c should appear to be the "heavy Henry" adsorbate.) The question is how does one apply this information to porous materials. This seems obviously to be a more difficult problem.

In this section the focus is upon non-porous materials. This must be addressed before one can determine what to expect for porous materials, which follows from this derivation. The following definitions apply to the binary adsorption theory:

1) The first subscript indicates the layer a molecule belongs to. The layer will be indicated making it equal to the italic letter "*m*."
2) The second subscript indicates the species of the molecule, for example N_2 or O_2. If there is no $m=?$ then the value is for the total adsorbate 1 or 2.

Thus: $\theta_{m=1, 1}$ is a molecule in the first layer and from adsorbate 1, $\theta_{m\geq 2, 1}$ indicates a molecule in the 2nd and above layers and is adsorbate 1, and $\theta_{m=1, 2}$ is a molecule in the first layer and is adsorbate 2.

As can be seen from theory and from the data on non porous materials, the important variables that controls the adsorption are the $\Delta\chi$s. Below $\chi_{c, 2}$ but above $\chi_{c, 1}$ one would expect the surface to have only species 1 to be adsorbed. The presence of species 1 allows for some adsorption of 2 due to the solution attraction to layer 2. However, one would expect very little adsorption in the first layer.

In order to tell whether a quantity is a parameter, or a quantity determined by a previous experiment (and is thus not a parameter in the present case,) those quantities that are previously determined will be underlined. In other words in these equations the $\underline{X}$ values and others are predetermined by other experiments.

In contrast to first layer adsorption, the relationship between $\theta_{m\geq 2, 1}$ and $\theta_{m\geq 2, 2}$ should reasonably follow that in solutions:

$$\frac{\theta_{m\geq 2,1}}{\theta_{m\geq 2,2}} = \frac{\underline{X}_1(l)}{\underline{X}_2(l)} \tag{4.163}$$

The symbol $X_n(l)$ is the mole fraction one would expect in the bulk liquid state for the temperature conditions of the experiment. This follows from the Brunauer's [29] and others' argument that the polarization and other van der Waal (or non-ionic intermolecular) forces do not extend appreciably through one layer of molecules. This argument was used to discredit the deBoer-Zwikker proposed dipole-induced dipole proposal. Brunauer's argument, if correct, contradicts the existence of an "external potential" used by some other adsorption theories.[28] This approximation makes the unrealistic assumption that the interactions of the molecules between the first and second layer is the same as between all the other layers. Although it is unrealistic, the intermolecular force between adsorbates are quite a bit smaller that between the adsorbate and the adsorbent. Thus, the assumption might break down for weakly adsorbing systems.

The mole fractions of the liquid are normally known and given by the equations:

$$\underline{P}_1 = \underline{P}_{1,vap}\underline{X}_1\varphi(X_2)\underline{X}_1 + \underline{X}_2 = 1 \tag{4.164}$$

Where the known functions φ is given by:

- Ideal solution:

$$\varphi(X_2) = 1 \tag{4.165}$$

---

[28]It does not, however, invalidate the disjoining pressure theory since there is an alternate explanation determined in its derivation from χ theory.

- Regular solutions:

$$\varphi(X_2) = \exp\left(\Delta\varepsilon X_2^2\right) \quad (4.166)$$

- Generalized solutions:

$$\varphi(X_2) = \exp[\mathbf{P}(X_2)] \quad (4.167)$$

The function **P** could be a generalized Margules polynomial (usually starting with the second power. Also mole fractions are obviously not independent of each other.)

In the section on the occupation of layers, Eqs. (4.125), (4.128), and (4.129) yield the relationship of θ_1 to θ and $\Delta\chi$ that is given in the quantum mechanical derivation. The same relationship should hold for the sum of two adsorbates, or since:

$$\begin{aligned}\Theta &= \exp(-\Delta\chi_1\mathbf{U}(\Delta\chi_1) - \Delta\chi_2\mathbf{U}(\Delta\chi_2)) \\ \Rightarrow \theta_{m=1,1} + \theta_{m=1,2} &= 1 - \exp(-\Delta\chi_1\mathbf{U}(\Delta\chi_1) - \Delta\chi_2\mathbf{U}(\Delta\chi_2))\end{aligned} \quad (4.168)$$

where the symbol Θ is being used here to designate the adsorbent surface that does not have any adsorbate on top of it.

There are also the relationships:

$$\theta_{m\geq 2,1} = \theta_1 - \theta_{m=1,1} \quad \text{and} \quad \theta_{m\geq 2,2} = \theta_2 - \theta_{m=1,2} \quad (4.169)$$

where the symbols θ_1 and θ_2 designate the total adsorbates 1 and 2.

At this point there are four unknowns and three equations. Although one could relate the unknowns to each other, what is desired is a way to relate these to the pressure of the adsorptives.

To solve this as a function of pressure, there needs to be a relationship between one of the variables and the vapor pressure. For cases where $\chi_{,2c}$ is sufficiently lower than $\chi_{c,2}$ the effect of species 2 for the first layer could be ignored with a very small error. An example would be for Arnold's data for every ratio of pressures except those very rich in O_2. Between $\chi_{c,2}$ and $\chi_{c,1}(1=N_2)$ there is no O_2 adsorbed on the first layer. Therefore, at this point one could assume that $\theta_{m=1,2}=0$ and θ_1 is approximated by:

$$\theta_1 = \Delta\chi_1\mathbf{U}(\Delta\chi_1) \quad (4.170)$$

and the first layer coverage is therefore given by:

$$\theta_{m=1,1} = 1 - \exp(-\Delta\chi_1\mathbf{U}(\Delta\chi_1)) \quad (4.171)$$

as previously derived in Eq. (4.39) or (4.66). The value for $\theta_{m\geq 2}$, the total coverage in the layers greater that 1, is then:

$$\Theta_{m\geq 2,1} = \theta_1 - 1 + \exp(-\Delta\chi_1) \quad (4.172)$$

This seems to just be adding more unknowns to the mix. However, $m\geq 2$ consists of both adsorbate 1 and 2 in a ratio according to Eq. (4.163). Therefore, solving generally for $\theta_{m\geq 2, 1}$ and $\theta_{m\geq 2, 2}$:

$$\theta_{m\geq 2,1} = X_1(l)(\theta_1 - 1 + \exp(-\Delta\chi_1)) \quad (4.173)$$

and generalizing for 1 and 2:

$$\theta_{m\geq 2,1} = \underline{X}_1(l)[\Delta\chi_1\mathbf{U}(\Delta\chi_1) + \Delta\chi_2\mathbf{U}(\Delta\chi_2) - 1 + \exp(-\Delta\chi_1\mathbf{U}(\Delta\chi_1) - \Delta\chi_2\mathbf{U}(\Delta\chi_2))] \quad (4.174)$$

and

$$\theta_{m\geq 2,2} = \underline{X_2}(l)[\Delta\chi_1 U(\Delta\chi_1) + \Delta\chi_2 U(\Delta\chi_2) - 1 + \exp(-\Delta\chi_1 U(\Delta\chi_1) - \Delta\chi_2 U(\Delta\chi_2))] \quad (4.175)$$

When $\chi_2 \leq \chi_{c,2}$ the unit function, U, makes the expressions with $\Delta\chi_2$ drop out. Thus:

$$\chi_2 \leq \chi_{c,2} \Rightarrow \theta_{m\geq 2,2} = \underline{X_2}(l)[\Delta\chi_1 U(\Delta\chi_1) - 1 + \exp(-\Delta\chi_1 U(\Delta\chi_1))] \quad (4.176)$$

For $\chi_2 \leq \chi_{c,2}$ the first layer, from the basic equations of the χ theory one has:

$$\theta_{m=1,1} + \theta_{m=1,2} = 1 - \exp(-\Delta\chi_1 U(\Delta\chi_1) - \Delta\chi_2 U(\Delta\chi_2)) \quad (4.177)$$

It can be proven easily that this equation cannot be separated to make equations for 1 and 2 as written in Eq. (4.171) when two adsorbates are present and both $\Delta\chi_c$s are exceeded. However, when $\chi_1 \geq \chi_{c,1}$ and $\chi_2 \leq \chi_{c,2}$ then $\theta_2 = 0$ and Eq. (4.177) becomes Eq. (4.171). Thus, the first layer for #2 adsorbate "kicks in," much like a lower energy plain "kicks in" when its χ_c is exceeded. In this case, however, the subsequent layers are "pre-adsorbed" due to the presence of the #1 adsorbate.

This implies that at low coverages, the more energetic adsorbate controls the adsorption of the lesser adsorbate. This is the reason that the "heavy Henry" adsorbate dominates the adsorption of the "light Henry" adsorbate.

As you may have gathered, if there are two or more energy planes or energy distributions involved, each plane should be calculated separately, or if there are overlapping adsorption energies of adsorption.[29]

It seems obvious that for a separation process, it would be best to operate between $\chi_{c,1}$ and $\chi_{c,2}$.

Of course, this may not always be possible due the the possibility that $\chi_{c,1} \approx \chi_{c,2}$ or the amounts of adsorbates are simply too small. Furthermore, since there is some adsorption of adsorbate 2 for $m=2$ and greater, a material that is highly microporous or even ultramicroporous would separate the best on a thermodynamic basis. However, ultramicropores may have a kinetic disadvantage.

In Chapter 5, another approach is taken by starting over with the grand canonical partition function. That approach is even more messy, but some approximations can be extracted from it to obtain predictions of the binary isotherm from the pure adsorptive isotherms.

Is the χ plot compatible with the Freundlich and Dubinin isotherms?

The relationship to the Freundlich isotherms is important for two reasons. Firstly, there is the question as to whether the χ theory can predict isotherms such as the Freundlich (of which $r_F = 1$ Henry's law is a special case), Dubinin-Astakov, Dubinin-Radushkevich, and Tóth isotherms. All but the Tóth isotherm will be referred to as the Dubinin-Polanyi (DP) isotherm. Secondly, a reason is required

---

[29] I'll leave this and energy distributions to some graduate student to do. Happy figuring! Working on the details of porosity should also be a nice challenge. Here I'm sticking with the crude approximation.

for the observation that in most cases P appears to approach 0 as n_{ad} approaches 0. Even though there are cases where P approaches as finite value, the disproof of Henry's law is not convincing without an explanation as to why sometimes it, or rather the Freundlich isotherm, is observed.

The log-normal energy distribution has been expressed above in Eq. (4.96) which yields the isotherm in the χ representation as expressed in Eq. (4.97). The DP isotherms may all be expressed as:

$$n_{ad} = n_0 \exp\left(A\left[-\ln(P/p_s)\right]^k\right) \tag{4.178}$$

This formulation is the generalized form for all the DP isotherms. The details of each may be found in the literature [30] along with additional equivalency comparisons to χ theory. If $k=1$, this is the special case of the Freundlich isotherm. Define a quantity χ_0 as:

$$\chi_0 = \frac{\ln A}{k} \tag{4.179}$$

Then the χ representation of the low pressure isotherm is:

$$n_{ad} = n_0 \exp(-\exp(-k(\chi - \chi_0))) \tag{4.180}$$

The question is then whether this is the same as Eq. (4.97) in the low pressure range or not. In order to make a match, the second derivative of this expression should yield an expression that matches the energy distribution described by Eq. (4.96). The second derivative of Eq. (4.180) is:

$$\frac{\partial^2 n_{ad}}{\partial \chi^2} = n_0 k^2 \exp[-\exp(-k(\chi - \chi_0))][\exp(-2k(\chi - \chi_0)) - \exp(-(\chi - \chi_0))] \tag{4.181}$$

One of the important features of this equation to notice is that when $\chi = \chi_0$ the distribution is zero. If $\chi > \chi_0$ the distribution becomes negative. Noting that the fact that this second derivative yields the energy distribution is not dependent upon χ theory, one must therefore conclude that above χ_0 the relationships (that is, the DP etc.) cannot be literally correct. Luckily the amount of negative distribution above χ_0 is not too great. In order to match Eq. (4.181) with Eq. (4.96), the third and fourth derivatives (1st and 2nd of Eq. 4.181) are required to match the peak position and the curvature. Performing these operations yields the following relationships:

$$\chi_c = \chi_0 - \frac{2\ln(1+\sqrt{5}) - \ln 2}{k} \tag{4.182}$$

or:

$$\langle \chi_c \rangle \approx \chi_0 - \frac{0.962}{k} \tag{4.183}$$

and σ is related to the DP k parameter by:

$$\sigma = \frac{0.92423}{k} \tag{4.184}$$

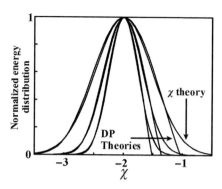

FIG. 4.26

A comparison of the χ theory energy distribution and Dubinin-Polanyi (DP) distribution. DP k values used were, starting from the outside, 1, 1.5, and 2.

In Fig. 4.26 are some examples of generated energy distribution curves for the DP isotherms and the χ theory. These are normalized by dividing by the constant at the beginning of the distributions. A value of -2.0 was picked for χ_c and χ_0 was calculated from Eq. (4.182). k values of 1, 1.5 and 2 were picked and the corresponding σs calculated from Eq. (4.184).

As may be discerned from the figure, the match between the two energy distributions is almost identical except where the DP distribution drops to zero at the high energy end.

The Freundlich isotherm is identical to the DP isotherm with $k=1$ and $\chi_c=0$. It is very unlikely for an adsorbate-adsorbent pair to have exactly this χ_c value. This value corresponds to an E_a at liquid nitrogen temperature of about $650\,\mathrm{J\,mol^{-1}}$. This is a very low value. For most ceramics the value is $\sim 10\,\mathrm{kJ\,mol^{-1}}$ range. Therefore, Freundlich isotherms with $r_F=1$ are extremely unlikely to be observed but higher powers, $r_F>1$ are likely.

The thermodynamics and spreading pressure

As noted early in this chapter, there is a definite relationship between the disjoining pressure theory of adsorption and the χ theory. In this section, some thermodynamic relationships for the spreading pressure are derived. It is questionable at this point how useful these relationships will be. They may be useful in extending the theory into the solution chemistry since these relationships are important in that area of research. A proposed short-cut theory for binary adsorption is also a possibility, using the "external potential" gradient, which yielded the disjoining pressure theory and DFT. In the following sections, an explanation is given as to why the disjoining pressure theory works and how it can be derived from χ theory.

It should first be noticed that any theories that claim both the continuity to the liquid state at high pressures and the consistency with "Henry's law" violates Gibbs' phase rule. "Henry's law" is in quotes because it is really not Henry's law as it applies to solutions. If it is assumed that the pressure and amount adsorbed approach zero simultaneously, then the relationship has the appearance of a Henry's law type behavior. The postulate that "Henry's law" must apply to any theory of adsorption can easily be disproved by finding only one system where this is not true. In Chapter 5 under "threshold pressure" several such examples are presented. One could also say that critical points violate the phase rule as well, so some researchers have made such an analogy. First, then, Gibbs' phase rule as it applies to surfaces will be reviewed.

Gibbs' phase rule in systems with surfaces

The origin of Gibbs' phase rule in thermodynamics is fairly easily deduced. It is not necessary to totally derive it here since it is available in almost any physical chemistry text (for example, see Ref. [31], starting on page 391.) Disregarding the surface as important the phase rule reads:

$$N = 2 + C - P \tag{4.185}$$

where N is the number of degrees of freedom, C is the number of chemical components, and P is the number of phases present.

The number 2 is a result of the terms in the free energy of "PV" and "TS." If one adds to this a surface area with a significant surface excess and addition term similar to these two, which is "γA_s," is to be considered. This adds one more degree of freedom to Eq. (4.185) or:

$$N = 3 + C - P \tag{4.186}$$

Consider, for example, the bulk case where there is only one phase and one component—for example, hydrogen in a container—then the temperature and pressure may be varied arbitrarily, providing of course that the container can be made larger or smaller. This is what is meant by 2 degrees of freedom. If both a liquid and gas are present, then the temperature and pressure are interdependent. This lowers N to 1.

Now consider the case where there is a surface. N now becomes 3 if only the gas phase is present. What is the 3rd degree of freedom? In a thought process, one could say this additional degree of freedom is the surface area, which could, therefore, be arbitrarily varied without requiring any adjustments in either T or P. If the adsorbed gas on the surface is contiguous with the gas phase (which is the basis for Freundlich isotherm with $r_F = 1$) then the surface excess must be zero for this to be true. Remember that this is about thermodynamics, which is related to molecular theories through statistical mechanics. This requires large numbers of molecules, and a few adsorbate molecules here and there would not be counted. On the other hand if a new phase forms on the surface—for simplicity call it the adsorbate—then the number of degrees decreases to 2. Now if the adsorbate phase changes, an adjustment must be made in either T or P (or both, but there is now a triplet relationship between n_{ads}, T, and P). Of course, normally it is T that is held constant to produce the isotherm. Thus, the adsorbed phase is contiguous with the bulk liquid phase and not the gas phase. This requires a phase transition at some pressure that is not zero.

One of the most strenuous objections to the χ theory, which for consistency should be applied also to the disjoining pressure theory, has been the prediction of a threshold pressure for adsorption. The above consideration not only allows a threshold pressure for the adsorbate phase to form, it requires it. This does not preclude the possibility of a surface gas phase, but some simple energy calculations demonstrate that if such existed, and given reasonable energies of adsorption, the amount adsorbed would be well below today's limit of detection. Assuming a very high energy of physical adsorption, $15\,kJ\,mol^{-1}$, and a very thick distance of 1 nm for this energy to operate, the number of moles that one would adsorb is about $2 \times 10^{-8}\,mol\,m^{-2}$. Even with a large surface area, this is still below most limits of detection. As an example, a realistic value for N_2 adsorption on silica at liquid N_2 temperature would be $1.5 \times 10^{-11}\,mol\,m^{-2}$.

It should be noticed at this point that most thermodynamic derivations of various quantities assume that, baring deformation of the adsorptive, the surface area, A_s remains constant and the chemical potential of the adsorptive, μ^S, does not change. This is a mistake. One of the reasons for concluding that the adsorbate is in patches is the fact that the interface areas, A_{sa} = the interface between the adsorptive and adsorbate, and A_{gs} = the interface area between the adsorptive and adsorbent, must vary. Thus, A_{sa} and A_{gs}, although not independent (if the surface cannot stretch or shrink,) must vary together:

$$A_s = A_{sa} + A_{gs} \tag{4.187}$$

Thus, both areas need not be specified in the thermodynamics and it is common to use only $A(ad)$ with the symbol A. Of course, γ_{gs} (gas solid) and γ_{as} (adsorbate-solid) are not the same. This means that as the adsorption increases, the areas change, since the total energy changes according to:

$$F_s + F_{ads} = \gamma_{gs} A_{gs} + \gamma_{sa} A_{sa} \tag{4.188}$$

The γ_{lg} (liquid-gas) is also important and is discussed later for mesopore "prefilling."

Another way the term γA can vary is by taking into account that the adsorbate "film" has a thickness and, as mentioned in the section on disjoining pressure, consists of two or three parts: the interface tensions and the disjoining pressure. It is the disjoining pressure, $\Pi(t)$, that is not taken into account in Eq. (4.188) (this has been explained in detail on in the section on Disjoining Pressure Theory.) Thus, regardless of the selection of the system type and the energy function, the term γA has a partial derivative.

The concept that distinguishes between the gas-solid and adsorbate-solid γs is implied in the derivation of the Freundlich isotherm, through the Gibbs-Duhen equation, and with the spreading pressure theory.

The Olivier monolayer criterion

Notice that a similar results for a maximum in the thermodynamic function for the total heat per gram adsorbent of adsorption. This is given as:

$$\mathbf{Q}(n_{ads}) = n_{ads}\overline{Q}(n_{ads}) \tag{4.189}$$

Dr. James Olivier noted in a presentation [32] at the Giardini Naxos FOA conference that this quantity reached a maximum at what he speculated was a monolayer. He was correct in this assumption as the following will demonstrate. Since the entropy change for adsorption is small according to the Dubinin criterion, then $\overline{Q}$ may be replaced by the $\overline{E}(n_{ads})$ and his unknown function will take the form of an exponential (given hindsight of Eq. (4.85) and Eq. (4.86), an exponential function is inserted here.) This functional form must be:

$$\mathbf{Q}(n_{ads}) = n_{ads}\overline{E}_a \exp(n_{ads}/n_m) \tag{4.190}$$

Differentiating this with respect to n_{ads} one obtains:

$$\frac{d\mathbf{Q}(n_{ads})}{dn_{ads}} = \overline{E}_a \exp(-n_{ads}/n_m) - \frac{n_{ads}\overline{E}_a}{n_m}\exp(-n_{ads}/n_m) \tag{4.191}$$

and setting this equal to zero to find the maximum and solving for n_{ads} one obtains at the max:

$$n_{ads} = n_m \tag{4.192}$$

So experimentally, the maximum does correspond to the monolayer equivalence. He did not use the exponential function to determine that this is correct, much less that the decay constant is n_m, he just reasoned that this maximum experimental point had to be n_m.

Summary of this chapter

In this chapter, the derivation of the χ theory is presented using quantum mechanical perturbation theory. Although the results of the perturbation theory were presented in the first edition of this book, the answer seemed to be self evident, and the derivation was not presented. The results of the perturbation theory were used as input into the grand canonical partition function and the χ equation for a single adsorbate and flat surface adsorbent was derived. This equation requires only a simple transformation of the normally used abscissa and no transformation of the amount adsorbed. Thus, the plot one gets is n_{ads} versus $-\ln[-\ln(P/P_{vap})]$. The slope of this plot yields the monolayer equivalence, n_m, and the abscissa intercept yield a value, χ_c, which is related to the energy of adsorption for the first adsorbed molecule. Eqs. (4.37)–(4.39) are the equations for this analysis.

χ theory also predicts the populations of the various adsorption layers. The equation for the population of adsorbate molecule in direct contact with the adsorbent surface is given in Eq. (4.41). This equation will be used to associate the log law with ultramicroporosity. Ultramicroporosity is defined in this book as adsorption that is physically restricted to one monolayer.

A discussion of the how the value of χ_c value might change due to change in temperature and the vapor pressure that it is referenced against is provided.

The problem of converting the monolayer equivalence into surface area is provided. This includes a discussion of what values to place on the adsorbate molecule area comparing the various possibilities from classical interpretation using densities and packing factors. A further complication is pointed out about the classical rolling over of the molecules using the Lennard-Jones potential. Although psychologically pleasant to have a number for surface area, all these uncertainties and others pointed out by Sing make the number for surface area somewhat disconnected from reality. A much better approach is to present the data foremostly in terms of monolayer equivalence, and possibly secondarily as surface area. Such an approach may be advised for porosity as well.

An alternative derivation of the isotherm form that is in agreement with χ theory by Churaev, Starke, and Adolphs is presented. This theoretical derivation is based solely upon thermodynamics and effectively one assumption, proved to be correct by the relationship to the quantum mechanical. It is essentially a model-less derivation and the one assumption, that excess surface work is an exponential function of the surface coverage with a decay constant of one monolayer equivalence, which, could have multiple model interpretations. The theory has been called the Disjoining Pressure Theory or the Excess Surface Work (ESW) theory. The theory yields identical equations with the identical two parameters to the χ theory. A model interpretation is given, which might provide the reader with some intuitive feel for the theory, but one should keep in mind that this modeling was not used in the derivation.

The combination of the Disjoining Pressure theory and the quantum mechanical χ theory has lead to several conclusions:

- A new thermodynamic definition of the boundary of the interface is defined by the Disjoining Pressure theory by setting a constant obtained from a differential equation as an integrating constant being set to zero. This is also makes the Disjoining Pressure theory and the χ theory identical along with the decay constant being equal to one monolayer equivalent.
- The thermodynamic Disjoining Pressure theory does not need a molecular model to be valid. However, for understanding purposes, a model is convenient to have.
- A misunderstanding of the Disjoining Pressure theory that it is a layer-by-layer model similar to the deBoer-Zwikker model has resulted in some DFT calculations to be incorrect.

The chapter continues with the question of heterogeneous surfaces with multiple $\overline{E}_a$s or possibly even an energy distribution. Due to the additivity of the χ plots, this question is easily handled.

For a distribution, the first derivative of the plot yields the monolayer equivalences and the second derivative yields the $\overline{E}_a$s or the distribution of the of the $\overline{E}$.

The expressions for the various "heats of adsorption" are provided and the expression for some of them in terms of the χ theory. The derivation from the BET equation is also provided for comparison.

The explanation of how χ theory calculates depth profiles is presented. This involves two calculations:

- the calculation of the depth profile for "layer" populations; and
- the quantum mechanical distribution for each "layer" using a reasonable potential, such as the Lennard-Jones 6–12 potential, and approximating it with the QM harmonic oscillator results.

The details of the derivation of the Disjoining Pressure theory is given and the theory developed by Olivier, which yields the same result for monolayer equivalence, is reviewed.

References

[1] J.H. deBoer, Proc. Roy. Acad. (Amsterdam) 31 (1928) 109.
[2] J.H. deBoer, C. Zwikker, Z. Phys. Chem. B3 (1929) 407.
[3] A.W. Adamson, Physical Chemistry of surfaces, second ed., Wiley, NY, 1967.
[4] S. Brunaur, Adsorption og Gases and Vapors, 1, Princeton University Press, Princeton, NJ, 1945.
[5] J. B. Condon, The Derivation of a Simple, Practical Equation for the Analysis of the Entire Surface Physical Adsorption isotherm, Y-2406 1988.
[6] E.L. Fuller Jr., J.B. Condon, Colloids Surf. 37 (1989) 171.
[7] C.W. Sherwin, Introduction to Quantum Mechanics, Holt, Rinhart and Winston, NY, 1960, p. 164. Section 7.1.
[8] E. L. Fuller, This Modification Was Inserted Into the Theory Very Early in Its Development.
[9] K. S. W. Sing, in "Adsorption by Powder and Porous Solids" by F. Rouquerol, J. Rouquerol, K. S. W. Sing, P. Llewellyn and G. Maurin, Elsevier Academic Press, Amsterdam, (2014) p. 237.
[10] N.V. Churaev, G. Starke, J. Adolphs, J. Colloid Interface Sci. 221 (2000) 246.
[11] B.V. Derjaguin, Progr. Colloid Polym. Sci. 74 (1987) 17.
[12] B.V. Derjaguin, N.V. Churaev, J. Colloid Interface Sci. 54 (1975) 157.
[13] J. Adolphs, M.J. Setzer, J. Colloid Interface Sci. 180 (1996) 70.
[14] J. Adolphs, M.J. Serzer, J. Colloid Interface Sci. 207 (1998) 349.
[15] P. Debye, E. Huckel, Phys. Z., 24, (1923) 185.
[16] P. Debye, Phys. Z. 25 (1924) 93.
[17] J.B. Condon, T. Schober, Solid State Ionics 77 (1995) 299–304.
[18] D.H. Everett, G.D. Parfitt, K.S.W. Sing, J. Chem. Biotechnol. 24 (1974) 199.
[19] R.B. Gammage, E.L. Fuller Jr., H.F. Holmes, J. Colloid Interface Sci. 34 (1970) 428. Digital data obtained directly from E. L. Fuller.
[20] K.S.W. Sing, S.J. Gregg, Adsorption, Surface Area and Porosity, second ed., Academic Press, London, (1991), p. 29.
[21] M.M. Dubinin, in: D.A. Cadenhead, J.F. Danielli, M.D. Rosenberg (Eds.), Progress in Membrane and Surface Science, vol. 9, Academic Press, New York, (1975) ISBN 0-12-571809-8, pp. 1–70.
[22] J.B. Condon, Microporous Mesoporous Mater. 53 (2002) 21.

[23] J.A. Morrison, J.M. Los, L.E. Drain, Trans. Faraday Soc. 47 (1951) 1023.
[24] T.L. Hill, J. Chem. Phys. 17 (1949) 520.
[25] T.L. Hill, Trans. Faraday Soc. 47 (1951) 376.
[26] I.U.P.A.C. Quantities, Units and Symbols in Physical Chemistry (The Green Book), third ed., RSC Publishing, Cambridge, UK, 2007.
[27] J.R. Arnold, J. Am. Chem. Soc. 71 (1949) 104.
[28] B.F. Dodge, H.N. Davis, J. Am. Chem. Soc, 49 (1927) 610–620.
[29] S. Brunauer, The Adsorption of Gases and Vapors, vol. 1, Princeton University Press, Princeton, NJ, 1945.
[30] J.B. Condon, Microporous Mesoporous Mater. 38 (2000) 377.
[31] A. W. Adamson, "A Textbook of Physical Chemistry" second ed., Academic Press, NY ISBN 0-12-044260-4.
[32] J. Oliver Presentation at the 9th International Conference of Fundamentals of Adsorption, Giardidi Naxos, Sicily, Italy, May 20-25, 2007, and personal communication.

CHAPTER 5

Comparison of the χ and ESW equations to measurements

Comparison to standard isotherms

In Chapter 3, a variety of standard plots were presented. It is instructive to plot these as χ-plots to see how well they obey the analytical expression. In the following, the χ-plot fits will be performed only on original data, where available. Creation of the standard plot by some fitting routine or simply using a manual spline fit is, inevitably, a distortion of the data. Indeed, the thoria and lunar soils standard plots were created using the insights of the χ-plot, so the generated standard plot, by definition, must fit the χ-plot. However, the question in this case is: statistically, how good is the fit to the original data? A similar question is encountered in analyzing heats of adsorption.

In the analysis that follows, the slope of the fit, the χ intercept (χ_c), the standard deviation of the fit, and the statistical r^2 will be presented, if appropriate. Occasionally, the original data has been reworked in such a way as to be a multiplicative of the original data. This is common for data that has been analyzed using the BET and reworked to yield "monolayer coverage" or "thickness."

The first two investigators presented here performed the experiment in the early stages of this type of research. Many of the precautions that are, or should be, routine today, such as sample temperature measurement and control, sample preparation, and low pressure adsorption, were not common at the time. Furthermore, the data is reported in terms of a quantity not in favor today, which is based on the BET. Although the original data might be back-calculated, there is still some uncertainty about some of the assumptions made. One of the assumptions probably is that N_2 has an effective area of 16.27 Å and a diameter of 4.3 Å.[1] This is not explicitly stated in all the publications, but to recover the original data this is judged to be the only choice. For data given as t-thicknesses, there is even more uncertainty regarding the conversion. Thus, the t-thicknesses as originally recorded are presented so as not to create even more uncertainty and confusion.

---

[1] These are the numbers that deBoer et al. apparently used throughout. The effective area is now the IUPAC convention, but what does the diameter mean? No mention is made for the packing factor in the direction normal to the surface.

Cranston and Inkley standard *t*-curve

The *t*-curve by Cranston and Inkley [1] is a fairly early standard curve. The data was an average curve for a variety of ceramic materials including alumina and silica. Given this, the statistics would seem to be meaningless, so they are not presented here. However, it is clear from Fig. 5.1 that the χ description is indeed a very good description of this standard plot. ($t = V_{ads}\sqrt{A}$ or $t = 1.547 V/A_{BET}$ for N_2, where V is the volume of the adsorbed gas as the gas state, not the adsorbed state.)

deBoer's standard *t*-plots

deBoer et al. performed many experiments from which a standard *t*-plot could be constructed. These included, most prominently, the standard *t*-curve on alumina by Lippens, Linsen, and deBoer [2]. The calculation of the "thickness" value depends upon the BET calculation. The stated conversion used was:

$$t = 3.53 \times 10^{-10}(V_{ads}/V_{m,BET})m \qquad (5.1)$$

where $V_{m,BET}$ is the STP gas volume of a monolayer according to the BET calculation.

Even though the actual value for the monolayer equivalence is in question, for the present discussion this does not matter. If the standard *t*-curve is plotted as a χ-plot, a noticeable curvature is detected. If, however, the original data in the form of *t*-thickness, available in the same series of papers by deBoer, Linsen, and Osinga [3], is plotted, it is not so obvious that this curvature is real.

In Fig. 5.2 is the original data used to construct the *t*-plot of N_2 on graphitized carbon black. The χ-plot appears as an excellent description for this data.

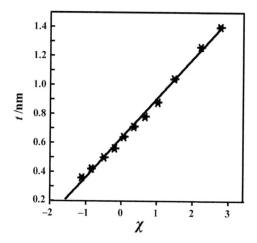

FIG. 5.1

Standard t curve constructed by Cranston and Inkley. The data point by the authors. The line is the least squares χ fit.

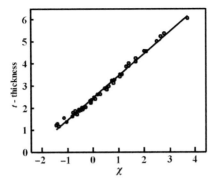

FIG. 5.2

Standard t-curve data by deBoer et al. The data are the circles and the line is the χ plot least squares fit.

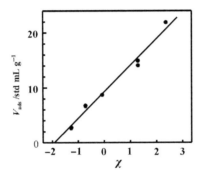

FIG. 5.3

Adsorption of I_2 on CaF_2 showing the χ plot according to deBoer.

The earliest plot of adsorbate versus $-\ln(-\ln(P/P_{vap}))$ was proposed by deBoer, which fit the adsorption of I_2 on CaF_2 [4]. Fig. 5.3 is an illustration of this data in χ-plot representation. It was recognized by deBoer at that time that the fit to the χ-plot was very good.

Another example is that used by deBoer and Zwikker [5] to develop the polarization model.[2] This example is of argon adsorption on tin II oxide as shown in Fig. 5.4. It appears that this sample had some microporosity; however, the fit is very good up to quite a high value of χ.

In addition to the well-known alumina adsorption, deBoer, Linsen, and Osinga created standard plots for $BaSO_4$, TiO_2, ZrO_2, MgO, SiO_2-aerosil, nickel antigorite,

---

[2] Empirically, this model, which plotted amount adsorbed versus $-\ln(\ln P/P_{vap})$, was a very good fit. However, after strong ridicule by Brunauer, this model, called the polarization model, was abandoned. It is unfortunate that this happened, since this held the field back from making advances in theory and quantitative analysis for at least 50 years.

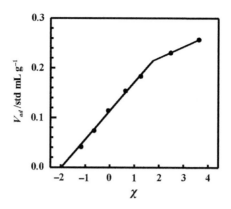

FIG. 5.4

Adsorption of Ar on SnO according to deBoer and Zwikker. The roundoff in the upper portion is probably due to some porosity.

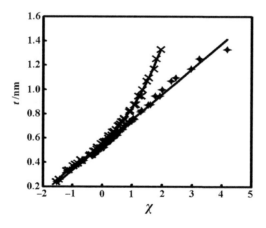

FIG. 5.5

N_2 on MgO aerosil according to deBoer et al.

Graphon 1 carbon, Graphon 2 carbon, and Sterling FT carbon. According to these authors, all but the carbon samples fit the standard t-curve well. Example χ-plots of some of these are presented in Figs. 5.5 and 5.6. All but the carbon samples had upswings at high pressures, indicating in all likelihood a 1 K error in measurement, as seen in many of the other data sets of the period. This is very likely since they assumed the temperature to be the temperature of the liquid nitrogen bath and there was no provisions to guard against radiative heating for the glass system. Thus, these plots are corrected for 1.0 K and the straight line is the least squares for these points.

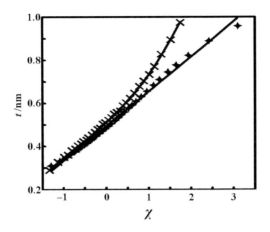

FIG. 5.6

N_2 on Ni antigorite according to deBoer et al.

Table 5.1 N_2 on MgO and Ni antigorite—A 3rd power χ fit

N_2 adsorbed on MgO			N_2 adsorbed on Ni antigorite		
Power 0	0.554	$\chi_c = -2.590$	Power 0	0.512	$\chi_c = -3.052$
Power 1	0.225	$t_m = 0.209$ nm	Power 1	0.181	$t_m = 0.162$ nm
Power 2	0.049	$r^2 = 0.992$	Power 2	0.028	$r^2 = 0.998$
Power 3	0.020		Power 3	0.012	

Correction ≈ 1 K.

Without the speculated temperature correction, the curves presented in Figs. 5.5 and 5.6 can be fitted to within a maximum absolute error of 2% with the following third order coefficients, given in Table 5.1. Increasing the power to the 4th power does not improve the fit.

Assuming a typical 1 K radiative heating a correction in Figs. 5.5 and 5.6 turns out to be nearly a straight line. These corrected data points are given as "+" in the figures. The analysis for the temperature corrected data is also given in Table 5.1.

A re-measurement of these data using a modern UHV system and either high resolution volumetric or high sensitivity gravimetric instruments, with good temperature control, is highly recommended.

The α-s standard plots

The most widely used standard plot is the α-s plot created by Sing et al., as recorded in a book on the subject, [6] for both N_2 and Ar adsorption. The original data by Bhanbhani, Cutting, Sing, and Turk [7] is presented in Fig. 5.7 for argon adsorption

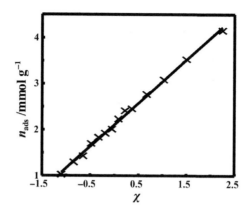

FIG. 5.7

Ar on SiO_2 data α-s plot as a χ plot. — is the least squares fit to χ plot.

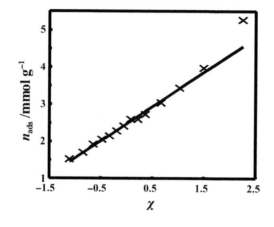

FIG. 5.8

N_2 on SiO_2 data α-s plot as a χ plot. — is the least squares fit to χ plot.

and in Fig. 5.8 for nitrogen adsorption. In Table 5.2 is a summary of the data that can be extracted from these α-s plots.

For the N_2 adsorption a second fit, indicated by the line in Fig. 5.8, was performed. As mentioned in the first chapter, occasionally the data at high pressures, for a variety of reasons, may not be reliable. The most likely deviation is in the positive direction, as seen here. Other silica data do not indicate this upswing. In Table 5.2, the last column provides the fit with the highest data point not counted. The original data, as shown here with the fit, is judged to be more reliable than the smoothed α-s tables. It appears that in the smoothing process some inaccuracies were introduced, since the r^2 changed from 0.992 to 0.957 in the process.

Table 5.2 Data from the α-s curves by Sing et al. of Ar and N$_2$ on SiO$_2$

α-s for Ar on SiO$_2$	α-s for N$_2$ on SiO$_2$	α-s for N$_2$ on SiO$_2$[a]
Slope = 21.1 mL	Slope = 23.7 mL	Slope = 20.8 mL
n_m = 0.942 mmol	n_m = 1.057	n_m = 0.53 mmol
χ_c = −2.23	χ_c = −2.36	χ_c = −2.67
σ = 0.36 mL	σ = 1.09 mL	σ = 0.53 mL
r^2 = 0.997	r^2 = 0.975	r^2 = 0.993

[a]Fit for line that leaves off the highest point.

Standard thoria plots

One of the advantages that thoria presents is that it is very stable with respect to high temperatures. Once a thoria produced powder is high fired to 1600°C, it is virtually physically stable. The surface chemistry is also stable with no change in stoichiometry. It is, therefore, an ideal powder with which to perform basic research.

Gammage, Fuller, and Holmes [8] (GFH) have performed extensive research on this material and have determined that for powders that are out-gassed at 1000°C there are several complicating features. Firstly, there is adsorption that is similar to chemisorption. This would be through high energy adsorption sites in small micropores. Secondly, there is some mesoporosity, and thirdly, a normal non-porous flat surface adsorption. If the material is exposed to water and then degassed at low temperatures, one observes only the flat surface area. The isotherm for the high temperature out-gas has been presented in Chapter 4. What is of special interest is the analysis of the low temperature out-gassed material. The thoria had previously been out-gassed at 1000°C and then exposed to water vapor. The subsequent high vacuum degas was at 25°C. This treatment apparently covered the high energy areas and filled the microporosity so that only the outer surface area is in this case being measured. In Figs. 5.9 and 5.10 and are the N$_2$ and Ar adsorption plots. In these figures, the data has been normalized to P/P_{vap} of 0.4, as one would do for an α-s plot.

Even the water adsorption isotherm reveals a good fit to the χ-plot. The plot in Fig. 5.11 is for water adsorption at 25°C on a powder that had been previously exposed to water seven times but had been out-gassed at 25°C for an extended period of time between exposures. For each exposure there was some additional irreversible adsorption. This would be an indication that the high energy planes and micropores were being masked for subsequent adsorption cycles. The fit to the linear χ-plot in Fig. 5.11 is quite good. In Table 5.3 are the statistics for the three thoria adsorption isotherms.

The units for the slope are given in Å but the conversion factor is uncertain. The data in the graphs is therefore $t_{ads}/t_{0.4} \equiv n_{ads}/n_{0.4}$ for an α-s plot using χ abscissa.

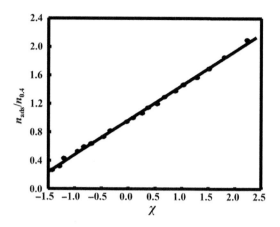

FIG. 5.9

N_2 adsorption on thoria normalized to 0.4 P/P_{vap}.

Data by GFH.

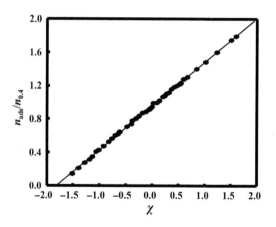

FIG. 5.10

Ar adsorption on thoria normalized to 0.4 P/P_{vap}.

Data by GFH.

Standard curves for lunar soils

Lunar soils have the interesting property that they are well out-gassed. The soils were collected on the moon and placed in a well-cleaned ultra-high vacuum aluminum alloy "moon box." The moon box was sealed on the moon with an indium seal. Upon arrival on Earth, the moon box was transferred to a very pure argon box and the soils transferred to smaller, well-sealed UHV spec containers for distribution. It is probably true that no sample, much less soil, has been handled in such clean and uncontaminating conditions. SEM indicated that the soils obtained were of surprisingly uniform composition.

Comparison to standard isotherms

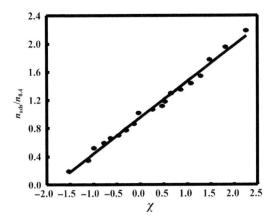

FIG. 5.11
Adsorption of water at 25°C after several prior adsorption cycles. Data is normalized to 0.4 P/P_{vap}.

Data by GFH.

Table 5.3 The statistics for the adsorption of gasses on 25°C out-gassed thoria

	N₂ adsorption	Ar adsorption	Water adsorption
t_m/Å	2.59	2.81	2.45
σ	0.03	0.01	0.06
χ_c	−1.993	−1.816	−1.855
r^2	0.998	0.999	0.990

Data by Gammage et al.

Gammage, Holmes, Fuller, and Glasson [9] were assigned the task of characterizing these samples. The standard NASA curves for lunar soils were obtained by Fuller and Agron. These are available in a US government report [10].

Several different isotherms were obtained. The χ-plots for these are in Figs. 5.12–5.15. One of the interesting features for the oxygen isotherm will be described in the section on the threshold pressure.

Apparently, due to the very clean and uniform conditions of the surface of these soils, the χ-plots are very linear (and it might have helped that a high precision UHV microbalance was used for the measurements.) In Table 5.4 are the statistics for the lunar soil χ-plot fits. The following data points were ignored for these fits: the first three data points for O_2, for an obvious reason, and the last three data points for N_2, which seemed to be experimentally out of line with the unlikely lose of adsorbate (more likely, this was due to a balance shift due to an earthquake. Oak Ridge region is prone to low-level but frequent earthquakes.)

Either from the graphs or from Table 5.4, it is obvious that the χ-plot is an excellent description. Notice that the energy of adsorption, E_a, for O_2 adsorption on the

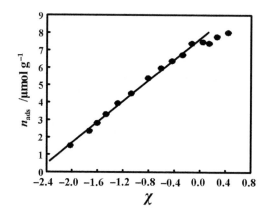

FIG. 5.12

N_2 adsorption on lunar soil.

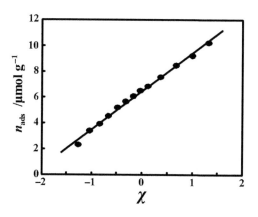

FIG. 5.13

Ar adsorption on lunar soil.

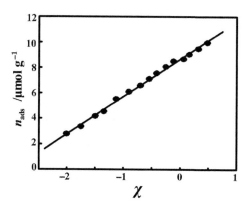

FIG. 5.14

CO adsorption on lunar soil.

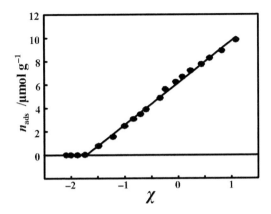

FIG. 5.15

O_2 adsorption on lunar soil.

Table 5.4 The statistics for the adsorption of gasses on lunar soil

	N_2 adsorption	Ar adsorption	CO adsorption	O_2 adsorption
$n_m/\mu mol\,g^{-1}$	2.99	2.96	2.94	3.60
σ	0.06	0.06	0.05	0.06
χ_c	−2.564	−2.186	−2.951	−1.718
r^2	0.9977	0.9976	0.9984	0.9983

The units for the slope and σ are $\mu mol\,g^{-1}$.

lunar soil is small enough that the threshold pressure is readily observed for this adsorbate/adsorbent pair. The other threshold pressures are below the limit of the pressure gauges, $\sim 1 \times 10^{-4}$ torr, for this particular experiment.

Isotherms by Nicolan and Teichner

Nicolan and Teichner [11] obtained several isotherms for various materials. They studied adsorption on non-porous silica and NiO. The χ-plots of the adsorption on silica are presented in Fig. 5.16 for N_2 and Fig. 5.17 for Ar. Although these indicate a nearly linear fit, the applicability is questionable since the lowest data point is more than a (postulated) monolayer equivalent of adsorbate. Furthermore, the range of the data is, compared to the α-s data, relatively rather short. The ordinate for these figures may be incorrect due to a lack of information and the assumed n_m may be incorrect. This, however, does not invalidate the conclusion for the fit the χ-plot.

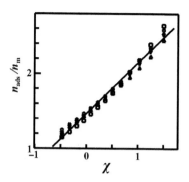

FIG. 5.16

N_2 adsorption on SiO_2 by Nicolan and Teichner.

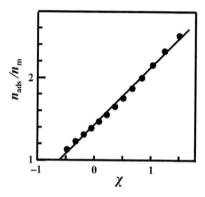

FIG. 5.17

Ar adsorption on SiO_2 by Nicolan and Teichner.

The data for the adsorption of N_2 on NiO is presented in Fig. 5.18. The information for the analysis for Fig. 5.18 is presented in Table 5.5. This is an obviously a good fit.

Isotherms by Bradley

In addition to his own work [12, 13] on Ar adsorption on sulfate salts, Bradley cited the work of McGavack and Patrick [14] on SO_2 adsorption on SiO_2, and water adsorption on CuO by Bray and Draper [15]. Although these data are quite old, there is no reason to suspect they are not accurate. Furthermore, they represent some rather unique isotherms, which here provide a broader perspective.

Figs. 5.19 and 5.20 are the isotherms of Ar on $Al_2(SO_4)_3$ and $CuSO_4$ by Bradley in the χ representation. These are composites from several close temperatures, the individual isotherms are more impressive, and the data for the #1 runs are shown as inserts. For the adsorption of Ar on $CuSO_4$ several measurements were made at slightly different temperatures in an attempt to extract the isosteric heat of

Comparison to standard isotherms 229

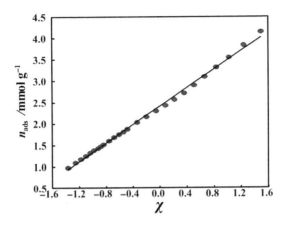

FIG. 5.18
N_2 adsorption on NiO by Nicolan and Teichner.

Table 5.5 Results of the χ plot fit for data by Nicolan and Teichner

N_2 adsorption on NiO Fig. 5.18
$\chi_c = -2.234$
$n_m = 1.06$ mmol g^{-1}
$20 \pm .02$ mmol g^{-1}
$r^2 = 0.997$

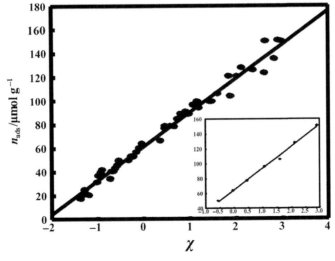

FIG. 5.19
Adsorption of Ar on $CuSO_4$ according to Bradley. Inset is data set #1.

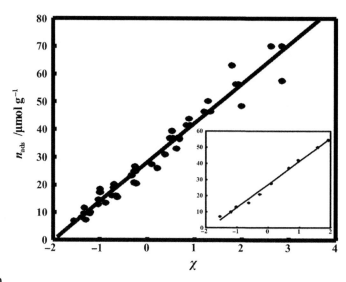

FIG. 5.20

Adsorption of Ar on $Al_2(SO_4)_3$ according to Bradley. Inset is data set #1

adsorption. One can see this in Fig. 5.20 by groupings of data with trends at the low adsorption end of the isotherm.

Bradley interpreted the data from McGavack and Patrick as consistent with the theoretical model for a standard. However, the original publication by McGavack and Patrick is much more extensive and more complex. Thus, a separate section is presented here to discuss their results.

Bradley followed an approach to explaining the isotherm base on the dipole-induced dipole similar to that by deBoer and Zwikker. Although the final equation is different from the deBoer-Zwikker equation, the form is the same:

$$\log_{10}\left[\log_{10}\left(\frac{P_{vap}}{P}\right)\right] = a\log_{10}(K_3) + \log_{10}\left(\frac{K_1}{T}\right) \quad 0 < K_3 < 1 \; K_1 > 0 \tag{5.2}$$

where K_3 and K_1 are constants.

This may be quickly associated with χ by the following:

$$-a\ln(K_3) \equiv \theta \therefore \frac{-1}{\ln(K_3)} \equiv n_m \; \& \; \frac{K_1}{2.303T} \equiv -\frac{\overline{E}_a}{RT} > 0 \tag{5.3}$$

Bradley provided tables of the constants obtained. These data are presented in Tables 5.6 and 5.7, using the units of specific that are normally used today. There is some uncertainty what P_{vap}s were used. Here a $\Delta_v\overline{H} = 6.447\,\text{kJ}\,\text{mol}^{-1}$ and $T_{BP} = 87.30\,\text{K}$ were used. An attempt is made here to determine the q_{st} from this publication. A pressure of 100 torr was selected. In this case, a data point at 100 torr was available for all the runs for the $CuSO_4$ adsorbent. This attempt was not very successful. The reason is obvious from the graph of $\ln(P/P_{vap})$ versus $1/T$ in Table 5.6 (Fig. 5.21). This plot should be in a fairly straight line in order to extract the isosteric heat but it is obviously not suitable.

Comparison to standard isotherms

Table 5.6 The analysis of the data by Bradley of Ar on crystalline $CuSO_4$

Quantity	Units						
T	K	84.5	85.1	85.8	86.5	87.2	87.9
K_1	K	377.5	360.5	336.5	270.8	293	286.3
K_3		0.796	0.794	0.796	0.813	0.803	0.796
$r^2(\chi)$		0.996	0.994	0.995	0.996	0.998	0.985
$n_m(B)$	$mmol\,g^{-1}$	0.027	0.027	0.027	0.030	0.029	0.027
$n_m(\chi)$	$mmol\,g^{-1}$	0.028	0.025	0.025	0.028	0.030	0.023
$\overline{E}_a(B)$	$kJ\,mol^{-1}$	6.91	6.60	6.16	4.96	5.37	5.24

(B) indicates this came from Bradley's table. (χ) indicate are recalculations by χ theory.

Table 5.7 The analysis of the data by Bradley of Ar on crystalline $Al_2(SO_4)_3$

Quantity	Units				
T	K	85	85.4	85.9	85.8
K_1	K	218.4	259.4	230	223.1
K_3		0.805	0.787	0.791	0.792
$r^2(\chi)$		0.996	0.994	0.995	0.996
$n_m(B)$	$mmol\,g^{-1}$	0.0135	0.0122	0.0125	0.0125
$n_m(\chi)$	$mmol\,g^{-1}$	0.0125	0.0129	0.0157	0.0149
$\overline{E}_a(B)$	$kJ\,mol^{-1}$	4.60	5.46	4.85	4.70

(B) indicates this came from Bradley's table. (χ) indicates recalculations by χ theory.

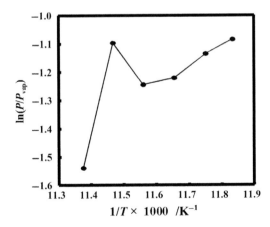

FIG. 5.21

Graph of Bradley-derived van-Hoff plot to determine q_{st}.

Table 5.6 illustrates the general problem of obtaining the isosteric heat. Notice that the numbers in the tables are quite reasonable and the individual (all not shown) and composite χ-plots are excellent. There is also good agreement between the constants that Bradley extracted and those obtained by χ analysis. However, even when one is confident that the data is very accurate and precise, it is not very likely to obtain good values for the isosteric heat. It is true that these data were obtained using what would be considered today as crude instrumentation; nevertheless, to obtain the type of precision needed to take a differential with the individual data points is quite challenging.

McGavack and Patrick [14]

The data by McGavack and Patrick may be represented quite well by a χ-plot, as may be discerned from Figs. 5.22 and 5.23. In Fig. 5.22, three plots of adsorption of SO_2 on silica gel are shown at 0°C with conditions as close as possible to each other, but with different samples. When this was done, very good agreement was seen. The χ-plots are very reproducible, with similar E_as of about 15.0 ($\sigma = \pm 0.4$) kJ mol^{-1}. Nearly all the isotherms produced good χ-plots. However, when conditions were varied—for example, temperature varied—there seem to be no correlations in the data. There was one very important conclusion regarding the contamination by either air or water. Contamination of the gas by 0.7 torr of air (for a 0°C isotherm) lead to a hysteresis due, apparently, as indicated by the χ-plots, to a change in observed E_a. The addition of a percentage of water to the SO_2 gas phase had a similar result, as demonstrated in Fig. 5.23.

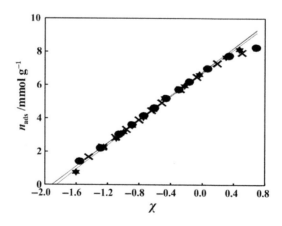

FIG. 5.22

The χ plot representation of the adsorption of SO_2 on SiO_2 gel according to McGavack and Patrick. Three successive runs with identical conditions but with different samples.

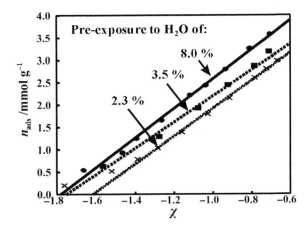

FIG. 5.23

The effect water contamination of the SO$_2$ at three levels. It appears that the higher the H$_2$O level the higher is the apparent E_a.

Data by Bray and Draper

The data by Bray and Draper of water on CuO and on a mix of 38.1% MnO$_2$ and 61.9% CuO show obvious evidence of porosity. It also appears that both systems may have heterogeneous surfaces to yield a distribution of E_as (for the data on CuO, analysis yields a n_m of the pore of ~0.29 mmol g^{-1} and for the external n_m of ~0.30 mmol g^{-1}, but a pore volume of ~12.4 mmol g^{-1}.) These do not appear to be a good choice for a standard.

Conclusion and some comments about carbon

From the discussion above, it should be quite clear that the χ-plot is at least a good empirical description for most simple isotherms. In constructing a standard isotherm, the fit to the χ-plot would be the overall best choice. Numerous other examples could be cited with a variety of adsorbates-adsorbent pairs and an analytical expression for a standard curve could then be constructed.

The χ-plot, however, is much more than just a standard curve. It frees one from the restrictions and uncertainties of the standard curve. As related in Chapter 3, it allows calculations of microporosity and mesoporosity without the use of a standard and all the uncertainties attached to this approach. Furthermore, it provides a value for the surface area that is founded upon some very sound principles and reasonable assumptions.

There are several cases where more than one energy of adsorption must be dealt with. One of these is carbon. Most carbon samples have the additional complicating feature of microporosity. Apparently some carbon samples, such as the Sterling FT and Vulcan 3G, do not have this complicating feature but still have more than one

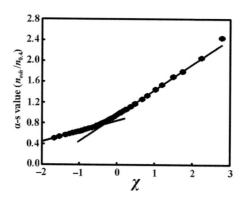

FIG. 5.24

The RMBM standard α-s carbon curve.

energy of adsorption. Indeed, one of these may be an in-register adsorption of either nitrogen or argon and has a very high adsorption energy. Representative of such adsorption is the RMBM (Rodrigues, Martin, Prado, and McEnaney) [16]) standard curve. Using the values of this standard curve and plotting them as a χ-plot, as in Fig. 5.24, one is able to see to two adsorption curves. The first one has a calculated energy of adsorption of about 45 kJ mol^{-1}, which is very high for delocalized adsorption. The second one has a reasonable physisorption energy of adsorption of about 4.5 kJ mol^{-1}. The individual carbon curves have similar double fits. In general the low-energy (higher pressure) line is about the same for all the curves, whereas the energy of the high-energy portion varies from about 30 to 100 kJ mol^{-1}. This is an obvious indication that something other than simple physisorption is present.

Summary with regards to standard curves

The χ-plot has demonstrated repeatedly that it is a good fit to well formulated standard plots on homogeneous nonporous adsorbents. For adsorbents that have more than one E_a, the reasonable explanation is that the sum of the two (or more) isotherms with differing energies yields the experimental data.

Knowing these facts, it becomes easy to construct a standard curve for any material with minimum of effort. This will become very useful for measurements of porosity, since the standard curve can be constructed from the low pressure parts of the isotherm, thus making unnecessary the task of finding a non-porous material that exactly matches the porous material; it is most likely that this latter matching is not possible, due simply to the fact that the porous and non-porous material are by nature not identical.

In the next section, the observation of a phenomenon that has been utilized by critics to discount the validity of the quantum mechanically based χ theory is demonstrated and the very important implications are examined.

The observation of χ_c

The implication of χ_c is one of the most controversial aspects of χ theory. The presence of this parameter, which is related to the energy of adsorption of the first adsorbed molecule, implies that below some pressure of adsorptive there exists no adsorbate on the surface (again, this is from a thermodynamic, i.e., large-numbers, point of view.) In Chapter 4, the argument was put forth that Gibbs' phase rule requires the presence of a threshold pressure. The theory that "Henry's law," in spite of the fact that it is hardly ever observed if one does not count the other Freundlich isotherms as "Henry's law," must be present is easily disproved by only one observation of the threshold pressure. This "Henry's law" phenomenon is not normally an observable, but is rather an extrapolation from the low pressure end of the observed isotherm. Of course, by definition this extrapolation is "observed."

It should be emphasized again at this point that "Henry's law" for adsorption is not derivable from, nor is it required by, thermodynamics. "Henry's law" for adsorption is a postulated equation of state just as, for example, the ideal gas law is for gasses. It is a result of the Langmuir isotherm; however, the Langmuir isotherm was formulated for chemisorption, in which case a new component is created in the process, which in turn changes the values in the Gibbs' phase rule. If the material on the surface is the same component as in the gas, then the Langmuir isotherm is not relevant. These thermodynamic arguments, however, do not seem to carry much weight; so in this section are presented some examples where there is clear evidence of a threshold pressure.

There are three reasons why the threshold pressure has not been recognized in the past. Firstly, researchers "knowing" that "Henry's law" should be obeyed have not looked for a threshold pressure. Indeed, there are many incidences in the literature where an extrapolation is performed on the data to include 0.0, and most computer programs for instruments likewise perform this extrapolation. Secondly, most adsorbents studied are ceramic materials, which have a fairly high energy of adsorption. The threshold pressure for these materials is typically below a P/P_s of 1×10^{-6}, below the normal measurement range. An extrapolation from 0.001 of P/P_{vap} to this value appears no different from an extrapolation through 0.0. In other words, precisely speaking the threshold pressure is usually insignificant, but important. The third reason is that many samples have heterogeneous surfaces or are contaminated with a variety of chemisorbed species, thus giving the appearance of a heterogeneous surface. With a heterogeneous surface, an energy distribution is obtained that obscures the threshold effect. The calculations in Chapter 4 demonstrate this.

Presented in this section is evidence by direct observation for the existence of χ_c and P_c, the threshold χ value and the threshold pressure phenomenon, respectively. Firstly, there is some indirect evidence for the presence of χ_c, which is the energy consideration.

Observations of the energy implications of χ_c

The value of χ_c is related to an energy, $\overline{E}_a$, by the equation:

$$\overline{E}_a = -RT e^{-\chi_c} \tag{5.4}$$

where E_a is interpreted to be the energy that the first adsorbate molecule for any particular patch of surface releases upon adsorption.

It is also related to the threshold pressure:

$$\overline{E}_a = RT \ln\left(\frac{P_c}{P_{vap}}\right) \tag{5.5}$$

A discussion of how this energy is related to the substrate and the adsorbing gas has been given by Fuller, Condon, Eager, and Jones [17]. Intuitively, one would expect this energy to be a function of both the gas and the solid. The expected trends for the value of $|E_a|$ would follow:

1. For adsorbing gases, the expected trends should follow the values of the dipole moment, polarity, etc. Thus, one expects for $|E_a|$:
   $H_2O > CO_2 > N_2 > O_2 > Ar > He$.
2. For solids, one expects the trend to follow the energy of a cleaved surface of the material (also follow the trend in surface dipole moments, etc.) Thus, one expects, for example: $ThO_2 > MgO >$ polystyrene $>$ polytetrafluoroethylene (Teflon®).

For a series of compounds, such as oxides, the trend in $|E_a|$ should follow closely the enthalpies of the compound formations. The reason for this is that the higher the $\Delta_f H^\ominus$s, the more polarized the oxide ions. Thus, for the following oxides the trends would be given as:

$$UO_2 \approx U_3O_8 > Y_2O_3 > Al_2O_3 > ThO_2 > SiO_2 > BeO > H_2O > CO$$

Experimental observation of such a trend in $|E_a|$ would be a strong indication that the threshold phenomenon is real. In Fig. 5.25 are the results of nitrogen adsorption on the above-mentioned oxides, most of which were reported by Fuller and Thompson [18] ($H_2O$is for water pre covered oxides and CO is for partially oxidized carbon.) The value for E_a of oxides is plotted as a function of the enthalpy of their formation. Since the threshold pressures for some of the oxides are too low to be measured directly, χ_c values are obtained from the χ theory equation. It is apparent that the correlation does exist as predicted. Although not claimed, due to the big question of stoichiometry to be used for the cleaved solid surface, this figure shows a linear relationship between the energies of the threshold and the enthalpies of formations. For the intersection at $\Delta_f \overline{H}^\ominus$s $= 0$, the value for E_a should be that expected for the liquefaction of N_2. The data point on this axis represents a surface whose energy is such that there is no preference for liquefying on the surface. A fit for the data yields a value of $8.6\,kJ\,mol^{-1}$, which is somewhat high but in qualitative agreement.

The observation of χ_c 237

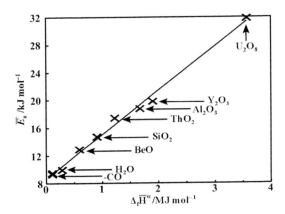

FIG. 5.25

$\overline{E}_a$ versus the enthalpy of formation of various oxides.

Another important point about Fig. 5.25 is, although it is the oxygen atoms that are being examined here, it is their polarizing power due to the link with the metal atoms that is determining the value of $\overline{E}_a$ and probably not the oxygen density, as some theories claim.

Direct observation of χ_c (threshold pressure)

In this section, to satisfy the disproof of "Henry's law," several instances of the observed χ_c are presented. This has indeed been reported in the literature by others. With the adsorption of water on NaCl reported by Peters and Ewing [19, 20], the threshold pressure is very clear—confirmed by both the isotherm and by infrared. In their investigation of the microporosity of Y-zeolites, for which very low pressure measurements were needed Guo, Han, Zou, Li, Yu, Qiu, and Xiao [21] reveal threshold pressures along with the reported oscillating adsorption. The oscillations are undoubtedly due to a variety of effects but one of these could be change in $\overline{E}_a$.

Gil, de laPuente, and Grange data

Gil, de laPuente, and Grange [22] (GPG) present data which seem to evidence a threshold pressure for N_2 adsorption. This observation was for nitrogen adsorption on microporous carbon. What is important about this data is that the threshold pressure is obvious even when looking at the data from the point of view of Henry's law. Fig. 5.26 illustrates this quite well. This plot illustrates that the threshold pressure is not an artifact of the transformation to the χ-plot. In this figure, which is the log-law plot given by the equations:

$$\theta_{m=1} = 1 - \exp(-\Delta\chi) \Rightarrow \theta \approx \theta_{m=1} = 1 - \frac{RT}{E_a} \ln\frac{P}{P_{vap}}, \quad E_a < 0 \qquad (5.6)$$

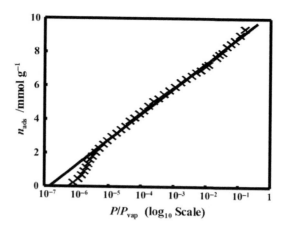

FIG. 5.26

Pc observation from data by GPG with microporous carbon.

This is the equation that is predicted for ultramicroporous material. In Fig. 5.26, the threshold pressure appears to be at about $1.0 \times 10^{-7} P/P_{vap}$. Since this equation indicates that the value for n_m is the extrapolated value to $P = P_{vap}$, it is given as 10.3 mmol g^{-1} In comparison, the χ plot indicates it to be about $7.6 \times 10^{-7} P/P_{vap}$. The problem in Fig. 5.26 is the dog-leg to yield a different value.

An important question is: what is the significance of the dog-leg? Every one of the isotherms that GPG measured demonstrated this feature. It is also obvious from the DR plots, as illustrated in Fig. 5.27. What causes this is unknown but here are some suggestions:

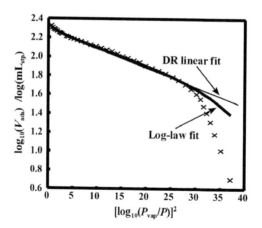

FIG. 5.27

DR plot of the GPG data showing the difference between either the DR fit or the log-law fit to the data.

- This is a result of the transition from the surface gas phase to the surface liquid phase. This would indicate that the dog-leg is nearly a vertical line in Fig. 5.26 for a homogeneous adsorbent and the slanting line is due to an energy distribution.
- This is due to the number molecules being so few that the value of the a is not correct.
- Even though they still act as a perturbation, the edges of small patches, or even isolated molecules, extent to the expected van der Waal radius rather than the molecular radius, as expected in larger patches. This was originally proposed in the 1st edition for all adsorbate molecules, but was dropped due to the realization and experimental evidence that the adsorption is mostly in patches.
- This is due simply to heterogeneity either of the surface energy or the pore shapes. This seems to be the least likely since this feature is fairly widely observed; although if the details of the dog-leg vary from adsorbate to adsorbate, this would be in the running.
- This is due to retained cover gas or calibration gas, such as He, as noted by Silvestre-Albero et al. This could be easily checked by following their advice on the sequencing the of experiment.
- Finally, this is due to something this author has not imagined—a good possibility.

Data by Nguyen and Do, N_2 on ultramicroporous carbon

Some adsorption isotherms on ultramicroporous do not have this dog-leg. For example, the adsorption of N_2 on Takeda ACF Carbon by Nguyen and Do [23], shown in Fig. 5.28, appears to be free of the dog-leg, as seen in Fig. 5.26. The analysis of this data indicates that that:

$$E_a = -9.78 \text{kJ mol}^{-1} \text{ within } 2\%$$

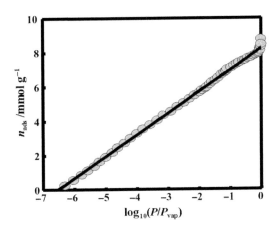

FIG. 5.28

N_2 on Takeda ACF carbon by Nguyen and Do. The "dog-leg" is not evident before the threshold pressure.

$$n_{ads} = 8.24 \text{ mmol g}^{-1}$$

$$2\sigma = 0.04 \text{ mmol g}^{-1}$$

$$r^2 = 0.999 \text{ for the fit to the log} - \text{law}$$

Observation by Thompson

In an attempt to observe the threshold pressure, Thompson selected a material that one might believe to have a low energy of adsorption. The direct observation of the threshold pressure is possible if the interaction energy between the surface and the adsorbed molecules is small. This can be easily illustrated with adsorption of N_2 or Ar on polytetrafluoroethylene (Teflon®) obtained by Thompson [24], which according to theory should have a very high threshold pressure.

From Thompson's data, the threshold pressure may clearly be seen at a pressure of about 0.01 atm (about 8 torr), well within (by a factor of at least 10^5) the capability of the most modern instrumentation. Direct observations of threshold pressures that are lower requires the use of more sensitive gravimetric techniques and high vacuum systems. Some of the modern high definition volumetric systems might also show this. This was also found experimentally by Thompson with adsorption data on diamond and alumina that had an ultrahigh vacuum surface cleaning. The results of Thompson's polytetrafluoroethylene experiments have not been reported in the open literature; however, it may be found in publically open government reports. Since it is not available in a more convenient place, the experiment will be discussed here in some detail.

The powder used was a Teflon® Dupont resin obtained from Aldrich Chemical Company (polytetrafluoroethylene lot #6). The measurements were made in an UHV system with a balance with a sensitivity of 10^{-9}g. The measurements on this material were performed over an extended period of time in both the adsorption and desorption mode. There was absolutely no indication that the isotherms exhibited any type of meta-stable condition or that the phenomenon reported herein is related to kinetics. The kinetics of both adsorption and desorption were indeed measured. The adsorption measurements and the desorption measurements were in agreement after the kinetic stage. What are shown here are only the stable thermodynamically valid portions of the measurements.

The results of the adsorption Ar on Teflon®, shown in Fig. 5.29, are in the untransformed form to illustrate the shape of the isotherm. The data for this figure are given in Table 5.8 to show the precision and accuracy that are obtainable with the instrumentation described. In this form, even with a high threshold pressure, the presence of a threshold pressure for most experiments, especially the volumetric type, would be missed. The zero mass recordings, however, are very obvious with the instruments described. This value is well within any conceivable error by a factor of 10^5. The flat portion of the pressure curve is more evident in the χ-plot. This plot

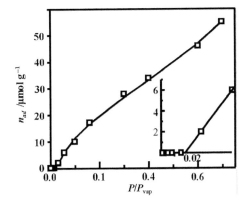

FIG. 5.29

Ar adsorption on polytetrafluoroethylene (Teflon®) with the normal P/P_{vap} axis by Thompson. In this experiment the threshold pressure is evident even without transforming to a χ plot.

Table 5.8 Data for the adsorption of Ar on polytetrafluoroethylene (Teflon®)

P/P_{vap}	$m_{ads}(g\,g^{-1})$	P/P_{vap}	$m_{ads}(g\,g^{-1})$
0.000003	0.0×10^{-8}	0.053560	2.40×10^{-7}
0.000023	0.0×10^{-8}	0.099731	4.00×10^{-7}
0.000129	0.0×10^{-8}	0.159684	6.80×10^{-7}
0.001273	0.0×10^{-8}	0.299779	1.12×10^{-6}
0.004805	0.0×10^{-8}	0.399902	1.36×10^{-6}
0.008105	0.0×10^{-8}	0.599674	1.84×10^{-6}
0.015051	0.0×10^{-8}	0.698356	2.20×10^{-6}

is shown in Fig. 5.30. In this figure, the presence of the threshold pressure becomes very obvious. This is very strong confirming evidence for the validity of the χ theory with respect to the threshold phenomenon. In light of this first experiment that reported a threshold pressure, the theories based upon "Henry's law" are disproved.

A variety of isotherms were obtained and the experiment repeated several times. In Fig. 5.31 are some data for three different types of experiments. For the low coverages, a slight rounding off of the χ equation plot is apparent as seen in Fig. 5.31. However, the threshold pressure still exists well above the limit of detection. This rounding phenomenon may be attributed to the heterogeneous nature of the surface energy. The threshold pressure with this rounding is also seen with some other common standard isotherms.

Similar threshold behavior is also apparent for both well cleaned diamond and alumina surfaces only at lower pressures. The results of these experiments are available in the open literature from a conference proceedings [25]. Thompson performed

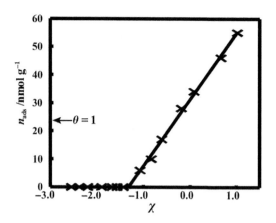

FIG. 5.30

χ plot of Ar adsorption on polytetrafluoroethylene (Teflon®) by Thompson. The χ plot makes the threshold pressure more obvious.

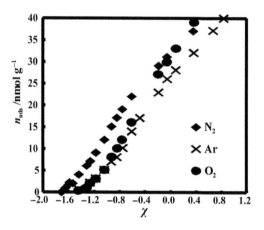

FIG. 5.31

Various adsorption isotherms on polytetrafluoroethylene (Teflon®) with N_2, Ar, and O_2.

several experiments on these materials to test the hypothesis that a uniform surface may be created by a good ultrahigh vacuum cleaning, thus simulating the possible conditions that the lunar soils had. Heating in hydrogen at a high temperature and degassing under an ultrahigh vacuum created the right conditions to observe a threshold pressure for the argon adsorption isotherm.

Some of the details of the experiment are as follows. The diamond powder was 1 μm powder obtained from Amplex Corporation. This powder was degassed and heated in H_2 to obtain a clean surface. It is well known that heating in H_2 up to 1000°C can eliminate the graphitic carbon that often contaminates diamond surfaces, but there should also be other chemically bonded contaminates. The alumina powder was NBS 8571, which was cleaned in a similar manner. Entirely different isotherms for both materials are obtained if the out gassing step is performed in a different fashion. According the Smirnov, Semchinova, Abyzov, and Uffmann [26], such a difference in surface structure with diamond may be due to the variation of the radicals on the surface. On the other hand, alumina may become slightly substoichiometric on the surface.

Figs. 5.32 and 5.33 contain the results of the adsorption isotherms in the χ-plot form on these materials. A very important observation was made with these materials: when the surfaces were contaminated, the threshold was not as apparent. Indeed, for diamond the adsorption isotherm followed the in-register χ theory analysis. This is probably due to the contamination creating a number of high-energy adsorption sites on the surface, thus masking the threshold effect. The hydrogen treated alumina evidences a threshold pressure; whereas, normally alumina χ curves, of which there is an abundance in the literature some of which are presented in this book, do not go to low enough pressures to observe this. The hydrogen treatment, which could yield a substoichiometric surface, apparently creates a lower energy of adsorption for nitrogen on alumina.

In Fig. 5.15, the χ-plot for the adsorption of oxygen on lunar soils was presented. It should be noted that the adsorption of oxygen below a χ-value of -1.72 was nonexistent. This was indeed observed for this material and was not an error in measurement. Thus, well-cleaned soil from the moon exhibits the threshold phenomenon with oxygen that are at a relative high value of P/P_{vap}, i.e., about $P/P_{vap}=0.0038$. Whether the other adsorbates would have exhibited such a clear threshold is unknown since the value of χ_c was below the detection limit.

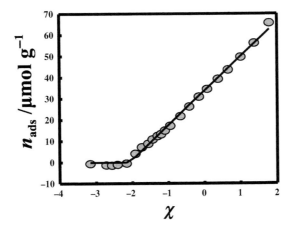

FIG. 5.32

χ plot of argon adsorption on H_2 cleaned diamond by Thompson.

244 CHAPTER 5 Comparison of the χ and ESW equations to measurements

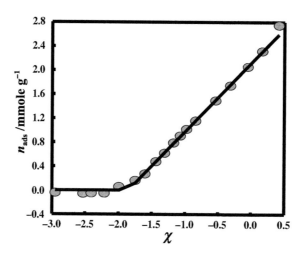

FIG. 5.33

Nitrogen adsorption on H_2 cleaned alumina.

Data by Thompson.

In both Figs. 5.32 and 5.33, the value dropped below 0 by a very small amount. In the case of alumina it was about 2 mg and in the case of diamond about 10 μg. This is probably due to a very slight error in buoyancy correction.

Direct observation with high resolution data

Some very important data has been published by Silvestre-Albero, Silvestre-Albero, Llewellyn, and Rodrigues-Reinoso [27] using high resolution isotherm measurements. They investigated N_2 adsorption on carbon samples, LMA233 and DD52, at 77.4 K. The measurements were highly resolved into the ultrahigh vacuum range, and they reveal the following important information:

- Their data provides confirmation of χ_c. In Figs. 5.34 and 5.35, one can see from the log-law plot a clear indication of the threshold pressure.

In addition to providing the evidence for the threshold pressure, this article is important for other reasons:

- It demonstrates the importance of High-resolution isotherms, which extend into at least the high vacuum range.
- It demonstrates that helium calibrating gas, used before the adsorption isotherm with the adsorbate gas is used, distorts the isotherm. The *calibration with He should be performed after the isotherm*. This is probably a surprise to many investigators.
- The amount versus the log scale of P/P_{vap} is observed to monolayer restricted porosity, thus confirming one aspect of the χ theory. The plots are **log**(P)

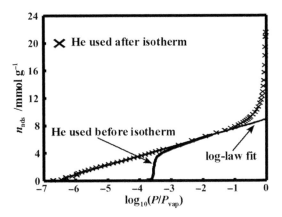

FIG. 5.34

High-resolution N_2 adsorption isotherms at 77.4 K an activated carbon LMA233 on a logarithmic scale.

(Reproduced with permission from the American Chemical Society—see reference [27].)

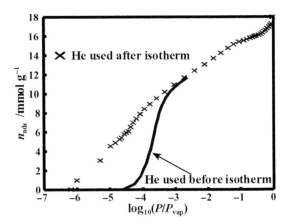

FIG. 5.35

High-resolution N_2 adsorption isotherms at 77.4 K an activated carbon DD52 on a logarithmic scale.

(Reproduced with permission from the American Chemical Society—see reference [27].)

versus amount adsorbed. This is what is expected according to χ theory if the pores are of the size that only a single monolayer can be adsorbed. (See Eq. 5.6)
- The validity of isotherms that depend on the existence of Henry's law are disproved.

This article is probably one of the most important articles published in the last ten years regarding physical adsorption. A few figures from this publication are shown here as Figs. 5.34 and 5.35 for adsorption of N_2 on carbon LMA223 and activated carbon DD52. Both of these are thought to have microporosity, but clearly have either ultramicroporosity or nearly so. There appears to be a residual dog-leg in The same could be said about Figs. 5.26 and 5.28.

Summary and conclusion concerning χ_c

As mentioned previously, in order to disprove the universality of "Henry's law" one needs to present only one example of a threshold pressure. Several examples have been presented above so the disproof is complete.

Along with the observation of the threshold pressure, the indirect evidence of the energy implications was also presented. The prediction of both the threshold pressure and energy implications is very strong supporting evidence supporting the χ theory. The predictions of the isosteric heats of adsorption, calculations of porosity adsorption, calculations of porosity, measurements of multiple plane adsorption (with its additive nature), and calculations of binary adsorbate mixture are not only supporting evidence but are quite useful. It is certainly an improvement over the BET, which is theoretically weak, creates an anomaly and predicts very little correctly.

Multiplane adsorption

The terminology "plane" and "multiplane" here are used in the sense that there are distinct areas with differing E_as. These may indeed be different crystallographic planes, but adsorption experiments cannot determine this. The different E_as may be due to other factors such as, for example, microporosity.[3] In the case of a distribution, it may be due to a multiplicity of chemical species on the surface or contamination.

Examples of two plane adsorption

An example of a multiplane adsorption has already been presented in Fig. 5.24. This, however, is a compilation of isotherms for carbon adsorbent. Examining just one isotherm for carbon, for example N_2 adsorption on Sterling FT carbon in Fig. 5.36, the break in the isotherm is still obvious, if not more so. In this figure, there are two lines drawn on the right axis corresponding to a monolayer of the total surface—the upper line—and a monolayer of the high energy planes only—the lower line. It seems unlikely that the adsorption on the high energy planes is by physisorption only, since the extrapolation to χ_c yields an energy of about $150\,kJ\,mol^{-1}$, which experience indicates is about 10 times too high for physisorption. There is probably one of two possibilities that would show up if lower pressure measurements were available. Firstly, there could be some chemisorption or in-register adsorption taking place on about 1/5th of the surface or secondly, and more likely as observed on

---

[3]More about this in Chapter 6.

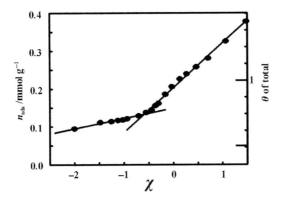

FIG. 5.36

χ plot of IUPAC lab "H" adsorption of N_2 on Sterling FT carbon.

other carbon samples [28], there is a considerable amount of ultramicroporosity present and the observed adsorption for the first fit is only the external area of these particular portions. The filling of the ultramicroporosity would have already been accomplished before the observation of the first data point.

Another example of multiplane adsorption is the 1000°C fired thoria powder by Fuller mentioned in Chapter 3. This sample, however, evidences some mesoporosity and will be a good example to analyze in the next chapter.

The Freundlich, Dubinin-Polanyi, and Tóth isotherms

The comparison to isotherms when there is a distribution comes back to the "Henry's law" question. Why is it that sometimes one observes the Freundlich isotherm and thus at least the appearance that the pressure and adsorbate amount simultaneously approach zero? As demonstrated in Chapter 4, a log-normal distribution in E_a yields the Dubinin-Polanyi (DP) set of isotherms, of which the Freundlich isotherm is a subset. The Tóth isotherm is similar but mathematically not in this class. The question becomes, are the generated isotherms, and not just the energy distributions, similar.

For these isotherms, especially the Dubinin-Polanyi and Tóth isotherms, not only must the distribution in E_a be considered but also the distribution in the micropore sizes. The reason for this is that these two distributions are close enough to overlap somewhat, thus interacting to change the values of the parameters.

As a review, the general form for the Freundlich-Dubinin-Polanyi equation is:

$$n_{ads} = A \exp\left(-r_F \ln\left(\frac{P_{vap}}{P}\right)^{r_{DP}}\right) \tag{5.7}$$

where A, r_F and r_{DP} are the parameters.

This may also be written in terms of χ:

$$n_{ads} = A \exp(-r_F \exp(-r_{DP}\chi)) \tag{5.8}$$

With $r_{DP}=1$, one obtains the Freundlich isotherms and if in addition $r_F=1$, then one obtains "Henry's law." With $r_{DP}=2$, one obtains the Dubinin-Raduchkevich [29](-Kaganer) equation. Other values of r_{DP} yield the Dubinin-Astakhov [30] equation. The Dubinin-Polanyi equations were originally used to analyze porous carbon, for which the porosity is slit-like. Thus the simple formulation of the χ theory, that is initial adsorption followed by a cutoff of adsorption with a simple normal distribution for both, is appropriate. One need not be concerned about the possibility that geometrical changes will change the effective surface area, as might be the case with cylindrical pores. Thus, the energy distribution together with the cutoff of the pores will consist of two normal distributions:

$$\frac{\partial^2 n_{ads}}{\partial \chi^2} = \frac{n_m}{\sigma\sqrt{2\pi}}\left[\exp\left(-\frac{(\chi-\langle\chi_c\rangle)^2}{2\sigma_c^2}\right) - F\exp\left(-\frac{(\chi-\langle\chi_p\rangle)^2}{2\sigma_p^2}\right)\right] \quad (5.9)$$

where the subscripts "c" and "p" correspond to the energy distribution and the pore distribution, respectively.

The parameter F is the fraction of the surface area that is inside the slit pores. For the purposes here it will be assumed that F is 1. How to determine this is clarified later. For very porous carbon samples, this could be close to 1. To see the correspondence between Eqs. (5.8) and (5.9) one needs to double integrate Eq. (5.9) and set the values of each at $\chi=\infty$ to be equal. One then needs to find the maxima and minima in Eq. (5.9) and the maxima and minima in the second derivative of Eq. (5.8) (given in Chapter 4) and set the magnitude and curvature of each to be equal. This is mathematically a little messy, but possible. In Chapter 4, this was performed with only the energy distribution for the χ equation and the match between these demonstrated. Here the porosity is introduced.

Figs. 5.37–5.41 illustrate the relationships between some common isotherms that fit porous materials and the χ-plot. They have all been normalized to a final pore volume of 1 in order to make the comparisons clear.

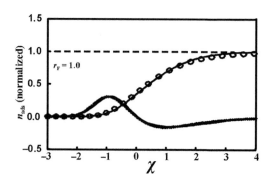

FIG. 5.37

Comparing DP isotherm with χ-plot with $\sigma=1$ and $r_F=1$ (Henry's Law).

Multiplane adsorption 249

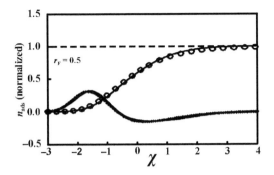

FIG. 5.38

Comparing DP isotherm with χ-plot with $\sigma=1$ and $r_F=0.5$ (Freundlich).

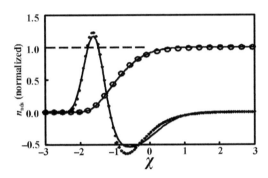

FIG. 5.39

Comparing DR isotherm with χ-plot.

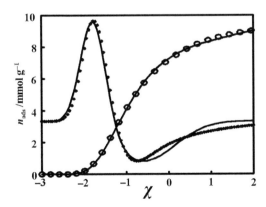

FIG. 5.40

Comparing Tóth T-equation with Ar χ-plot.

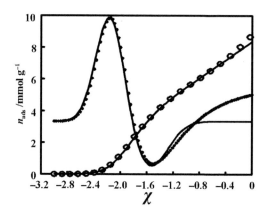

FIG. 5.41

Comparing Tóth T-equation with N_2 χ-plot.

Examples of the matches are shown in Figs. 5.37–5.39. In these figures the dotted line, barely distinguishable from the solid line, is the second derivative of the DP equations and the circles are the overall DP isotherm. The solid lines are the results expected from the χ theory with a distribution for both the E_a and pore size. Fig. 5.37 is the DP isotherm, which is identical to the "Henry's law" isotherm, $r_F = 1$, whereas Fig. 5.38 is a more realistic Freundlich isotherm with $r_F = 0.5$. The only difference is a shift on the χ axis. In Fig. 5.39 is the case of the Dubinin-Raduchkevich equation. The second derivative match is not perfect in Fig. 5.39, but it is evident that it is good enough to match the overall isotherm.

The Tóth isotherms [31], referred to as the T-equation [32], were presented in Chapter 3 as a good representation for many isotherms. This should not be surprising since it includes five fitting parameters (n_m, K, m, k, and $P_{r,e}$). $P_{r,e}$ is a low relative pressure value and can be ignored with a small amount of distortion. Figs. 5.40 and 5.41 show two examples of a comparison with the T-equation fit for nitrogen and argon adsorbed on SiO_2. In these figures the second derivative for the Tóth T-equation was obtained digitally and is slightly offset due to this. It should be noted that the T-equation second derivative has a tendency to exceed zero at high values. This does not make sense from an energy point of view. In terms of χ theory, an upward bend in the isotherm is due to either additional lower energy planes adsorbing or capillary filling has commenced. It is unlikely that SiO_2 would have these low energy planes and capillary filling would be more rapid than shown here.

As was the case for the DP isotherms, Eq. (5.9) was used to simulate the energy and pore size distributions. For the nitrogen adsorption about 55% of the adsorbate is in micropores whereas for argon 90% is in micropores, according to the χ analysis. The Tóth analysis indicated nearly identical surface areas using either N_2 and Ar, whereas the χ analysis indicates the surface area with the N_2 was less than with

Ar. These conclusions from the χ analysis are quite possible since the N_2 molecule is about 10% larger than the Ar atom. In addition to fewer molecules being packed into the micropores, some of the micropore volume available for Ar adsorption may not be available to N_2.

The above comparison to some well-known isotherms is not strong support for χ theory since porosity must be assumed without any other indicators. It does, however, demonstrate that the theory is consistent with the literature.

Summary of comparisons to some other isotherms

With the addition of energy distributions, the χ theory is consistent with the Freundlich, Dubinin- Polanyi, and Tóth isotherms, which seem to demonstrate "Henry's law." Although this does not lend any strong evidence for χ theory, it also demonstrates that the observation of these isotherms does not preclude χ theory.

Do isotherms for multiplane adsorption have anomalies?

This question is a critical question for all theories. A mathematical anomaly is a deadly problem for a theory. The mathematics itself unequivocally disproves a theory as written. A theory that leads to an anomaly has to be either modified—as was, for example, early string theory—or discarded. Even using it as an approximation, such as classical mechanics, is not appropriate beyond the empirically tested use. However, extrapolation into the unknown is often done for the lack of anything else. The test of whether the isotherm can handle more than one energy of adsorption is a test for consistency.

It is not unusual for a sample adsorbent to have different types of surfaces that have difference energies of adsorption. More than one energy of adsorption can obviously happen if a powder is a mix of two or more powders. This may also happen if the adsorbent has two or more crystallographic planes, or if the sample is porous where the energy of adsorption is not the same inside the pores as it is on the outer surface. There is also the question of heterogeneous surfaces where a distribution of E_as exist. The question then is, "Can the isotherm yield the correct overall surface area and possibly be able to give the individual surface areas for the differing energies?" The answer most researchers would give for this question is: "Of course one can obtain the overall surface areas, but separating the individual areas may be difficult." Unfortunately, from a statistical standpoint this is usually impossible. The following paragraphs will illustrate that the equations themselves can be tested by looking for mathematical anomalies. Of course, just because that expression that is used does not lead to an anomaly does not prove the theoretical bases of the expression, but it certainly can disprove the proposed equations.

There are two anomaly avoidance criteria that any isotherm must fulfill to be mathematically correct. These anomalies will show up for the particular way the data is analyzed. These are, given adsorption on two differing energy surfaces, either different planes or different powders mixed, at the same time then:

$$\mathbf{G}_1\left(\frac{P}{P_{vap}}, T, k_1...k_n\right) + \mathbf{G}_2\left(\frac{P}{P_{vap}}, T, k_1...k_n\right) = \mathbf{G}_{total}\left(\frac{P}{P_{vap}}, T, k_1...k_n\right) \quad (5.10)$$

where the subscripts on the ks are to distinguish the various parameters that are used. For example, for the BET equation $k_1 = C_{BET}$:

$$n_{m,1\ or\ 2\ or\ total} = \frac{A_{s,1\ or\ 2\ or\ total}}{A_m} \quad (5.11)$$

The first equation is simply that the amount adsorbed, regardless of the adsorption pressure, on area 1 added to area 2 must be equal to the total adsorbed (assuming these are the only adsorbing planes; the expansion to more than 2 should be obvious.) The second is then one of the criteria, which is simply the addition of the two surfaces to yield the total. The third one is a companion requirement to the first criterion that any portion of the isotherm equation that is ignored in demonstrating the first criterion that is "left over" must fulfill this requirement. This is often associated with the adsorption energy. In practice, this latter assumption possibly could break down as the pressure approaches the vapor pressure or if the bed porosity changes due to mixing of the two powders. Here, however, this is a mathematical construction to determine if the equations themselves have any anomalies, and such practical considerations are not of relevance.

Example 1

The χ and the disjoining pressure equation
To check this theory, recall the χ/Π equation, Eq. (5.12):

$$n_{ads} = n_m \Delta\chi U(\Delta\chi) \quad \Delta\chi = \chi - \chi_c \quad (5.12)$$

where:

$$\chi = -\ln\left(-\ln\frac{P}{P_{vap}}\right) \text{ and } \chi_c = \ln\left(-\frac{\overline{E}_a}{RT}\right) \quad \overline{E}_a < 0 \quad (5.13)$$

Above $\chi_{c,1}$ and below $\chi_{c,2}$ there is only one if $\chi_{c,1} < \chi_{c,2}$ adsorbate due to the step functions. So the criteria would apply only when $\chi > \chi_{c,2}$. In that case the equations simplify to:

$$n_{ads,1} = n_{m,1} \max\left(0(\chi - \chi_{c,1})\right) \quad n_{ads,2} = n_{m,2} \max\left(0(\chi - \chi_{c,2})\right)$$
$$n_{ads,total} = n_{m,t} \max\left(0(\chi - \chi_{c,t})\right) \quad (5.14)$$

(From here out the **max** will not be written, but $\Delta\chi \geq 0$ is implied.) Using Eq. (5.14) then:

$$n_{m,total}(\chi - \chi_{c,t}) \equiv n_{m,1}(\chi - \chi_{c,1}) + n_{m,2}(\chi - \chi_{c,2})$$
$$n_{m,total}\chi \equiv n_{m,1}\chi + n_{m,2}\chi - n_{m,1}\chi_{c,1} - n_{m,2}\chi_{c,2} + n_{m,t}\chi_{c,t} \quad (5.15)$$

Since Eq. (5.15) is true regardless of the value of χ, then the variable and constant portions may be separated:

$$n_{m,1}\chi + n_{m,2}\chi = n_{m,total}\chi$$
$$n_{m,1}\chi_{c,1} + n_{m,2}\chi_{c,2} + n_{m,total}\chi_{c,t} \quad (5.16)$$

Substituting into the first part of this equation using Eq. (5.11) one obtains:

$$\frac{A_1 \chi}{A_m} + \frac{A_2 \chi}{A_m} = \frac{A_{total}\chi}{A_m} \quad (5.17)$$

Thus:

$$A_1 + A_2 = A_{total} \quad (5.18)$$

which is the first criterion. For the second criterion, one uses the second part of Eq. (5.16). It is not obvious simply looking at this that the two sides of this equation are equal due to the addition sign. The substitution of the meaning of χ_c is needed to demonstrate this. This is provided by Eq. (5.13). Substituting in:

$$n_{m,1}\left[-\ln\left(\frac{-E_{a,1}}{RT}\right)\right] + n_{m,2}\left[-\ln\left(\frac{-E_{a,2}}{RT}\right)\right] = n_{m,total}\left[-\ln\left(\frac{-E_{a,total}}{RT}\right)\right] \quad (5.19)$$

Thus, substituting, cancelling common terms, and changing signs:

$$A_1\left[+\ln\left(\frac{-E_{a,1}}{RT}\right)\right] + A_2\left[+\ln\left(\frac{-E_{a,2}}{RT}\right)\right] = A_{total}\left[+\ln\left(\frac{-E_{a,t}}{RT}\right)\right] \quad (5.20)$$

Bring the As inside the $\ln$ terms and combining the sum of the $\ln$ terms on the left side:

$$E_{a,1}^{A_1} E_{a,2}^{A_2} = E_{a,total}^{A_{total}} \quad (5.21)$$

It is obvious that if $E_{a,1} = E_{a,2}$ then the above equation is correct. For $E_{a,1} \neq E_{a,2}$ there is little useful meaning for $E_{a,t}$. It is actually a reflection of the x-axis intercept for the total χ curve, that is, the high χ values. At any rate, this equation presents no conflict and there is no internal mathematical anomaly.

This makes sense since for the plot of both adsorbents, the abscissa is the same and the ordinate is a sum of the experimental weights, and one can add weights without breaking any law of physics.

Similar arguments may be made to demonstrate that the Disjoining Pressure theory does not exhibit any anomaly. This should not be surprising since the two theories are essentially the same.

Example 2

The BET method

Using the BET equation, as given in Chapter 1, for one energy of adsorption is:

$$n_{ads} = \frac{n_m CP}{(P_{vap} - P)[1 + (C-1)(P/P_{vap})]} \quad (5.22)$$

For two planes, subscripts "1" and "2," and the total, subscript "t," and conservation of matter requires adding $n_{ads,1}$ and $n_{ads,2}$ yields $n_{ads,t}$ the following equation is obtained:

$$\frac{n_{m,1} C_1 P}{(P_{vap} - P)[1 + (C_1 - 1)(P/P_{vap})]} + \frac{n_{m,2} C_2 P}{(P_{vap} - P)[1 + (C_2 - 1)(P/P_{vap})]}$$

$$= \frac{n_{m,total} C_{total} P}{(P_{vap} - P)[1 + (C_t - 1)(P/P_{vap})]} \quad (5.23)$$

By conservation of surface area then requires:

$$\frac{C_1}{[1+(C_1-1)(P/P_{vap})]} \equiv \frac{C_2}{[1+(C_2-1)(P/P_{vap})]} \equiv \frac{C_{total}}{[1+(C_{total}-1)(P/P_{vap})]} \quad (5.24)$$

These identities are true if and only if:

$$C_1 = C_2 = C_t \quad (5.25)$$

This, however, is not generally the case. The only case where this is true is if $C_1 = C_2$ obviously. But even when the Cs are nearly equal, this is still strictly not true. Thus, the BET equation has a built-in mathematical anomaly. Although the BET equation might fit the shape of the isotherm in the range of P/P_{vap} from 0.5 to 3.5, it cannot possibly be an accurate picture of the physisorption, at least if two adsorption energies are present.

The argument that the BET contains a built in anomaly has been repeatedly rejected. The first presentation of the proof was not the above sequence but rather ended up saying that the BET did not conserve either mass or (exclusive or) surface area. The usual argument that reviewers put forth is similar to one reviewer's comment, "it is ridiculous." [sic: antecedent of "it" is not clear.] Nevertheless, it would seem to be difficult to argue against a clear mathematical proof.

The question then is about the many other isotherms, the Freundlich, the Tóth, etc. It can be demonstrated that any isotherm, like the Langmuir and BET, that has a pressure term added in the denominator and is rearranged for analysis in such a way to have something other than n_{ads} as the ordinate will fail this test. Most of them are treated in this fashion, which makes the conclusion questionable.

This does not mean that such an isotherm, for example the Langmuir isotherm, is not useful but the basis for surface area and energy calculation is not simple. Thus, for chemisorption and the Langmuir isotherm behavior, the surface area, or monolayer equivalent, is not being measured but rather the number of chemical bonding sites. Therefore, other factors, such as crystallographic arrangements and locations of these sites, are needed for practical purposes.

It should be noted that all the "Henry's law" compatible isotherms have the same problem with an anomaly as does the BET. This can be proven by taking the Henry's law limit:

$$\lim_{P \to 0}(n_A + n_B) = \frac{P}{K_A}A_A + \frac{P}{K_B}A_B = \frac{P}{K_{total}}A_{total}, \quad K_A \neq K_B \quad (5.26)$$

and the conditions:

$$A_A + A_B = A_{total} \quad A_A \neq f(A_B) \quad (5.27)$$

The stated condition that A_A is not a function of A_B simply means that their ratios cannot be adjusted arbitrarily to compensate for the differing Ks. But these two equations cannot be equal with the stated conditions, as demonstrated by cancelling the P and multiplying by K_{total} substituting the second equation for A_{total}:

$$\frac{K_{total}A_A}{K_A} + \frac{K_{total}A_B}{K_B} = (A_A + A_B) \quad (5.28)$$

Since A_A and A_B are independent, then this implies:

$$\frac{K_{total}}{K_A} = \frac{K_{total}}{K_B} = 1 \Rightarrow K_{total} = K_A = K_B \quad (5.29)$$

in contradiction to the first equation condition (of course if $K_A = K_B$ this is the degenerate case and the areas are indistinguishable).

Contrasting the Henry's law isotherms to the χ and disjoining pressure isotherm

Henry's law isotherms, as noted above, have a problem both mathematically and theoretically. However, this does not mean they are totally useless. Such a law could be considered simply an empirical equation to fit the isotherm over a certain range. In this respect, they are no better than a polynomial fit. However, the best polynomial fit would be to a transform that should yield a straight line with some deviations.

This latter possibility is what χ and Disjoining Pressure (ESW) isotherms offer. These are theoretically sound and for simple adsorbents a linear fit. Deviations from this linear fit have clear meanings, and calculations of these begin with the underlying transform.

Conclusion concerning multiple energies

χ theory is capable of very simply explaining the results obtained from isotherms when an adsorbent has more than one energy of adsorption and does not seem to follow the standard isotherm model. It is straightforward and consistent with the entire theoretical framework of the χ theory.

The equation is simply the addition of two, or more, isotherms in terms of amount adsorbed, and since the amount adsorbed is directly proportional to the surface areas, the surface areas are likewise added. The energies may be added according to the χ equation to yield a quantity that has meaning but is not too useful except to calculate the break-point in the χ-plot. At any rate this latter addition does not generate a contradiction.

The BET equation is useless for calculating isotherms with more than one energy plane. Furthermore, along with other "Henry's law" isotherms, it has the built-in flaw of having the same anomaly. This is due to the attempt to do chemically contradictory calculations—a chemical equilibrium expression to form new components and the obvious requirement to keep the adsorbate as the same component.

There are no other theories that can make the distinctions that the χ-theory (or ESW) makes, including modeling methods such as NLDFT of Monte Carlo techniques. With the χ-theory, however, the determinations of multiple plane energies and energy distributions is straight forward and, even better, simple.

Heats of adsorption

The relevant equations for the heat of adsorption are the simple χ-theory equation, Eqs. (5.12) and (5.13) for the isotherm and that portion of the equation relating to the energy heats of adsorption, Eqs. (4.101), (4.105) and (4.106). Since there are several heats of adsorption, depending on the measurement and conditions, q_{lg} is a typical example:

$$\bar{q}_{la} = (\bar{E}_a - \tfrac{1}{2}RT)e^{-\theta} := \mathbf{E}_a(\theta) \tag{5.30}$$

for which the function $\mathbf{E}(\theta)$ is defined. The isosteric heat of adsorption differs by $\Delta_l^g \overline{H}(\theta)$, as given in Eq. (4.1.2). There are several other equation in the section on heats of adsorption related to how the measurements made and the reporting of the data might need to be recalculated to yield the proper form for the total energy function used here, but all of them contain the function $\mathbf{E}_a$.

Heats of adsorption is an area where the χ theory is clearly superior to other theories of adsorption. The Dubinin concept of adsorption potential and the postulated "thermodynamic criterion" implies that the adsorption potential does not vary with temperature. This Dubinin thermodynamic criterion is very widely observed, and there are rare exceptions especially for simple adsorbates. This can be easily derived from χ theory. The adsorption potential is simply the Gibbs' free energy going from the bulk liquid state to the physisorbed state and the χ theory predicts that this value should be ½RT. This is very small compared to most heats of adsorption, thus yielding the thermodynamic criterion.

Many researchers will disagree with the above statements, pointing out that the $\Delta \overline{S}$ is not ≈ 0 as demonstrated below. However, these instances are from publications that rely upon the BET equation to yield the heat of adsorption and have not checked with the calorimetric data, if it exists, to see that the BET does not yield the correct results.

A few examples of the predictions of χ theory to yield the various defined heats of adsorption are presented here. Other examples and more details can be found in an article on the subject [33]. In order to make a parameterless prediction of the isosteric or integral heat of adsorption, an adsorption isotherm is first obtained. If there are no complicating features, such as simultaneous chemisorption or microporosity, then these heats of adsorption can be predicted without any further information. This is what is referred to as a *parameterless* prediction or fit. That is, all the constants needed to make a calculation are available from some other measurements.

One of the problems encountered in the literature is that the data has been transformed and presented in such a way that it is difficult, perhaps impossible, to unscramble the presentation to obtain the original data. Luckily, some can be obtained directly, as is the case with data by Pace, Dennis, and Berg [34, 35], from original sources, such as Ph.D. dissertations by Dennis [36] and Berg [37], or mathematically unwinding it, as is the case with information supplied by Harkins and Jura [38].

Data by Harkins and Jura—The absolute method

In Fig. 5.42 ("energy curve") is the molar integral heat of adsorption of water on anatase as obtained from the data by Harkins and Jura (HJ). The solid line shown in this figure is the calculation obtained from the isotherm shown in Fig. 5.43. Notice that this is a two-energy surface, but the energy of interest is the high energy portion of the plot, that is, to the left of the break.

Heats of adsorption 257

Table 5.9 shows the calculated parameters from the isotherm data as a χ-plot and the energy of adsorption data that HJ calculated. The fit to the HJ energy curve was nonlinear and also yielded the monolayer equivalents. There is excellent agreement between the HJ calorimetry calculation and χ theory for both the energy E_a and the n_{ads}.

Only the high energy part of the isotherm is relevant, in that the break occurs above $n_{ads} > 0.39\,\text{mmol}\,g^{-1}$ ($\chi >$) and there are only three data points well above this

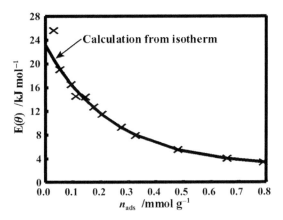

FIG. 5.42

The energy of adsorption with amount adsorbed.

Data by HJ.

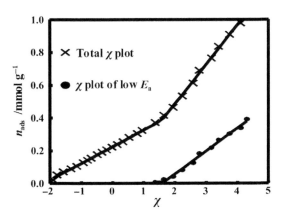

FIG. 5.43

χ-plot of the H$_2$O adsorption isotherm by HJ to Fig. 5.42.

Table 5.9 Parameters extracted from data by Harkins and Jura

From	χ_c or (E_a/kJ mol^{-1})
Isotherm	23.2±2σ=0.9
χ_c	−2.235±2σ=0.039
"Energy curve"	22.8±2σ=0.3
	n_{ads}/mmol g^{-1}
Isotherm	0.2415±2σ=0.0094
Energy curve	0.2410±2σ=0.0020

level for the energy curve. The calculation is made on the basis of the 18 data points that would properly extrapolate to the high energy E_a. The other data points are not relevant.

All the calorimetric data points were used except for the first one in order to calculate the E_a. Including the first point, which seems out of line, a value of 24.7 kJ mol^{-1} $2\sigma_{max}=1.9$ kJ mol^{-1} is obtained. The fit for this point has a value of 20.5 kJ mol^{-1} (20.75 kJ mol^{-1} in the χ-plot) and the data point has a value of 25.6 kJ mol^{-1}. This is a difference of 4.1 kJ mol^{-1} compared to $2\sigma_{max}=1.9$ kJ mol^{-1}. This implies that the data point should be discarded by the M test.

The reason that first data point could be incorrect is that the material might have had some adsorbed water before it fell into the water. This would make the moles adsorbate low and thus the division by the number of millimoles of water for this particular point is recorded as lower than believed. All the other increments past this one would be referencing the same initial coverage and therefore would be, incrementally, correct.

Data by T. Berg, Kr on Anatase—Simultaneous measurements

Another example is the adsorption of Kr on anatase that was performed by Berg [37]. In this case the isotherm in Fig. 5.44 is used to obtain the entire dependence of the molar heats of adsorption as a function of the amount adsorbed. In Fig. 5.45 is the result of the calculation obtained for this, along with the heat of adsorption data by Berg.

This is one of the few examples where the isotherm and the heats of adsorption were measured simultaneously on the same sample. Therefore, the isotherm corresponds exactly to the heats of adsorption. The calculation of Fig. 5.45 is referred to then as a parameterless (this means there are no adjustable constants) calculation for the heats of adsorption, since no unknown physical quantities are adjusted for the "fit" to the data. Furthermore, there is no ambiguity about whether a point should or should not be included in the analysis. On the other hand, one of the data points by Harkins and Jura that seemed to be very far from correct was ignored. The

Heats of adsorption

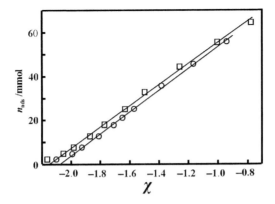

FIG. 5.44

χ plot of data by Berg. Parameters obtain here are constants to calculate the lines in Fig. 5.45.

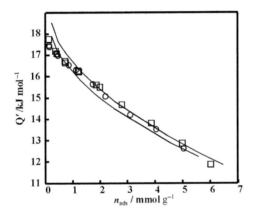

FIG. 5.45

The E of adsorption with amount adsorbed from the data by Berg. Lines are from the χ-plot analysis.

calculation of the heats of adsorption uses the isotherm data of Fig. 5.44 and the χ-theory to calculate the lines drawn in Fig. 5.45.

The data is precise and accurate enough that the subtle translational loss going from a 3 dimensional fluid to a 2D fluid can be detected. The line drawn includes the χ predicted by χ theory. This effect should decay as the first layer approaches one monolayer. At 140 K this is a significant difference (about $0.6\,kJ\,mol^{-1}$) in this figure. The uncertainty lines drawn are for one standard deviation as determined in the χ-plot. One point to notice is that even with this treatment the calculated uncertainty increases going from the adsorption isotherm to the heats of adsorption.

CHAPTER 5 Comparison of the χ and ESW equations to measurements

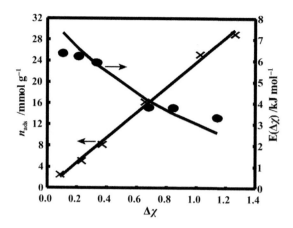

FIG. 5.46

χ plot and heat of adsorption Kr adsorbed on rutile at 126K by Dennis.

Similar data was obtained by Dennis [36] for Kr on rutile. The problem with this data is that it is too sparse to really draw a conclusion, It was performed at 121 K, which is above the freezing point, so the χ treatment should apply, but with 6 points full range and calculating the differential heat from integral heat introduces some uncertainty. The main objective of the study, however, was to measure the heat capacity of the adsorbed Kr and there is plenty of data for that purpose.

The data from the work by Dennis is presented in Fig. 5.46. The isotherm data was available as the original values from the experiment, whereas the heats of adsorption were given in a reworked form. There is some uncertainty in back-calculating the heat to obtain E as a function of n_{ads}. The line drawn for $E(\Delta\chi)$ in Fig. 5.46 is the *parameter-less* line calculated using the data from the isotherm.

Data by Kington, Beebe, Polley, and Smith on anatase

Another example with oxygen and nitrogen adsorption on anatase by Kington et al. [39] is shown in Fig. 5.47. The data points using "×" are for nitrogen and those using ovals are for oxygen.

The fit to this data gives an energy of 20.7 and 15.5 kJ mol^{-1} for nitrogen and oxygen, respectively. The monolayer equivalency is 0.107 and 0.295 mmol g^{-1}, respectively. The data was analyzed using the χ theory fit to the q_{dif}. This involves the transform of the n_{ads} to $(1-e^{-\theta})$ for the least squares fit. The isotherms were not very useful due to the small number of data points taken and the lack of range. This is a common problem using the BET to obtain the surface area, since it is assumed that the question has for a long time been settled, and investigators have been in the practice of not reporting the raw data, but rather just the BET surface

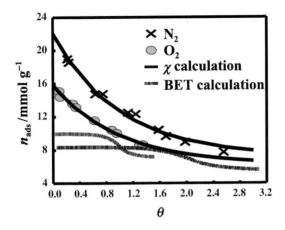

FIG. 5.47

Heats of adsorption by Kington et al. The lines are the least squares fits.

area and sometimes the C constant.[4] Therefore, there is considerable uncertainty for value found for the n_m, however fortunately, the value of E_a is not greatly dependent upon this value.

For more examples, readers should consult the cited article [33].

Summary of heats of adsorption

Unlike most other adsorption theories, χ theory and Disjoining Pressure theory (ESW) do an excellent job of calculating the heat of adsorption. It must be emphasized here that this can be done using the two parameters from the isotherm to calculate the heat of adsorption without any additional variables (variable constants used to fit a set of data are referred to as "parameters," whereas, constants determined by other means and not variable to make a fit are referred to as "constants.") Thus, the heat of adsorption calculated by the χ and Disjoining Pressure theories are *parameterless*. This is something no other theory is capable of to date.

Adsorption of more than one adsorbate

The theoretical foundation for the interpretation of binary adsorption by χ theory was presented in Chapter 4. A few examples are in order to illustrate these predictions.

---

[4]Sadly including this author. This means that thousands of data points were in vain and conclusions about surface areas for many of the materials investigated are probably incorrect. The raw data may still exist in some DOE vault that I do not have access to.

Adsorption on non-porous surfaces

The only experiment of binary adsorption on non-porous materials, at least to this author's knowledge, where the adsorbates are different enough to have differing E_a is that by Arnold [40]. Arnold Studied the co-adsorption of N_2 and O_2 on anatase. An advantage of using N_2 and O_2 is that their $\overline{V}$, and therefore their $\overline{A}$, are very similar. Raoult's Law does a reasonably good job of calculating the liquid mixture, with a Margules constant of $\alpha=0.181$ for a regular solution fit. The α needs to be used, however, to fit the data more precisely.

In Fig. 5.48 are the χ-plots for the adsorption of the pure N_2 and O_2 with an extrapolated χ_cs of -2.665 and -2.477, respectively. There is considerable uncertainty in these numbers as one would gather from the scatter in the lower pressure deviation, as seen in Fig. 5.48. The resultant binary plot with the ratios of the pressures staying constant at 50.2% O_2 and 49.8% N_2 is shown in Fig. 5.49 along with the calculation. Fig. 5.49 should show a relative approach to Raoult's Law as the amount adsorbed increases. The relative approach to Raoult's Law is evident by the parallel slope comparison (that is, the offset of ~0.096 mmol becomes trivial with additional adsorption.) In the low pressure range the formula given in Chapter 4 is:

$$\frac{\overline{E}_2}{RT}e^{-\theta_1} = \frac{\overline{E}_1}{RT}e^{-\theta_2} + \mathbf{F}(X_1). \tag{5.31}$$

To inspect the low pressure end of this figure, a plot of $\exp(-\theta_1)$ versus $\exp(-\theta_1)$ should be linear as the θs approach 0. In Fig. 5.50 is the result of one of these plots. In Chapter 4, plots were presented for the plots of θ_1 versus θ_2, which demonstrated that the relationship approached a straight line with the slope approximating the literature value for Raoult's Law except for the plot of 85% O_2 versus 15% N_2. Notice that in Fig. 5.50 the low pressure data where the linear fit seems to work are to the upper right. The function $\mathbf{F}$ is dependent upon the mole fractions of the gasses (and thus the mole fractions in the liquid phase) and is nearly constant for any particular gas mixture.

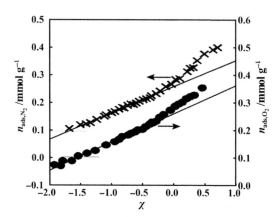

FIG. 5.48

χ plot of the pure nitrogen and pure oxygen adsorption on anatase by Arnold.

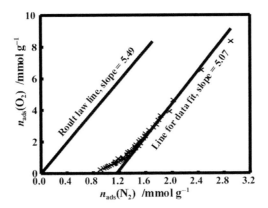

FIG. 5.49

The moles of nitrogen adsorbed versus moles of oxygen adsorbed for a 50% mix of gasses.

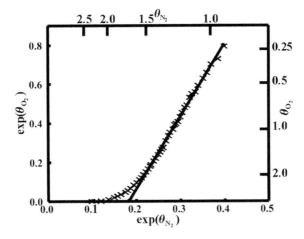

FIG. 5.50

a plot of **exp(−θ)** for both θs illustrating the linearity expected at low pressure. 50% O_2+50% N_2 data on anatase by Arnold.

Thus, the data at low pressure should follow Eq. (5.31) and at high pressures it should follow the slope of Raoult's law. Furthermore, the adsorbate that has the highest E_a should have a linear χ-plot for a nonporous adsorbent. This is due to the fact that as individual pure gasses the high E_a adsorbs first with very little of the second adsorbate present. This is the effect that has been referred to as the "heavy" component dominating the adsorption for the carrier gas systems. Thus, what causes this "heavy" phenomenon is the difference in E_a of the adsorbates.

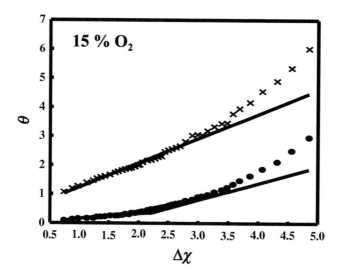

FIG. 5.51

Adsorption of a gas mix of 15% O₂ 75% N₂ on anatase by Arnold.

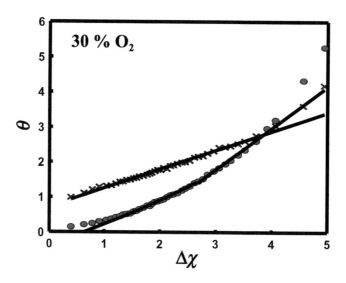

FIG. 5.52

Adsorption of a gas mix of 30% O₂ 70% N₂ on anatase by Arnold.

For some cases, the plots of the individual adsorbate may cross. With a sufficient mole fraction, the low energy isotherm will start out lower than the high energy adsorbate, and cross over to be higher at high pressures. In Figs. 5.51–5.55, this is illustrated with Arnold's data.

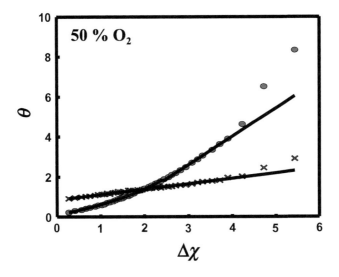

FIG. 5.53

Adsorption of a gas mix of 50% O_2 50% N_2 on anatase by Arnold.

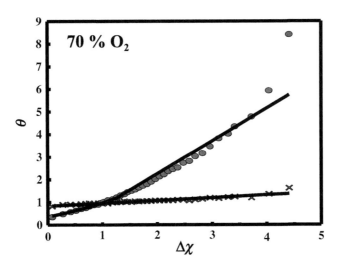

FIG. 5.54

Adsorption of a gas mix of 70% O_2 30% N_2 on anatase by Arnold.

These figures show an agreement with the theoretical prediction, except for the data for 85% O_2^- 15% N_2 graph, where the high pressure data using the expected Raoult's Law does a very poor job. Of course, using the fitted slope from the data does work well, but this is, obviously, to be expected. The question, then, is; why

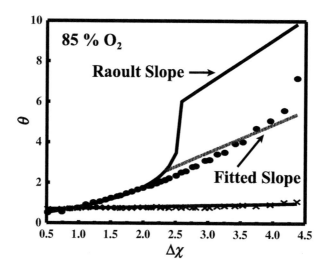

FIG. 5.55

Adsorption of a gas mix of 85% O_2 15% N_2 on anatase by Arnold.

does this seem to fall out of the predictions made by the theory presented so far? There are a number of possible answers.

The reason for the deviation may be that the approach of the threshold pressures is not sufficient to allow the N_2 adsorption to "dominate," but is beginning to look like identical adsorbates in the low pressure region. In other words, due to the shift in χ_c, as mentioned in the section on "Adsorption of Binary Adsorptives" and "Method 1," the effective χ_cs are about the same, although the amount of O_2 adsorbate is greater; something that Arnold noticed for this set of data (although he didn't quite say it that way.)

Another factor to consider for the 85% O_2 data is that the slope of the nitrogen plot is very shallow. It is only 8.5% of the slope of pure nitrogen, whereas the O_2 is 81% of the slope of its pure isotherm. The χ_c value is ~ -9, which indicates that the exponential portion is very short and probably below the recorded data. Another problem is that the N_2 data from this has an extreme amount of scatter with an r^2 value of 0.75. This is in contrast to other values in this series, which ran between 0.956 and 0.999.

The experimental setup, described in Chapter 2 in the section "Arnold's Vacuum Flow System," is an excellent arrangement for making the measurements with the possibility that non-equilibrium condition being the problem as being very unlikely.

Finally, this may be an indication that qualitative explanations presented here are insufficient for calculating the phenomenon of binary adsorption. This seems to be a good area in which some careful experimental and theoretical research would be fruitful.

Binary adsorption on nonporous adsorbents. Summary and conclusion

There is little data to make definitive conclusions about binary adsorption. There is much literature data on separations, etc., but what is needed is the actual adsorption isotherms using instrumentation design for that purpose. The instrumentation that Arnold used was designed to assure adsorption under equilibrium with the adsorbent properly controlled. It does not seem that one can obtain such instrumentation commercially, and so in Chapter 2 a detailed description and operating procedure is given, or one can refer to the original paper.

Using the data that Arnold presents, the χ theory predictions are tested. This data by Arnold was a series of mix gas isotherms of O_2 and N_2. The advantage of this set is the molecular sizes are roughly the same. This is also convenient for χ theory approximations in the theory to yield some simple equations. Using the approximations for the low pressure region and the high pressure regions, attempts are made to predict the isotherm using the isotherm parameters of the pure O_2 and N_2. The results were quite good for all the isotherms in the low pressure region; however, in the high pressure region, where Raoult's Law is an important part of the calculation, the slopes of the O_2 isotherm for the 85% O_2 gas mix was quite different. The explanation for this is unknown. The other four binary isotherms were fitted quite well in this region.

Binary adsorption in micropores

Binary adsorption in micropores is simpler to determine than one would expect once the information about adsorption of one component is understood. Most of what is new is quite classical, and could be explained by the "heavy" component idea that has been around for some time. In terms of χ theory, this means that the plots of the binary adsorption phase diagrams will may be plotted using a graph of $\Delta\chi$ of the adsorbate with the highest $|E_a|$ and from that pick off values of component 1 and 2 and plot them in a traditional binary diagram.

As seen in previous sections, the binary isotherms can be predicted fairly well using only the pure isotherms. The most likely pressure range that will be used for separations, as an example, would be the low pressure range, where the isotherms predict the possibility of the cross-over in the isotherm. There would seem to be little advantage with an adsorbent to select a pressure range where Raoult's Law dominates.

Many process, however, use higher pressures, so how can one take advantage of the low pressure range enhancement. Recall that in the low pressure range the coverage in direct contact with the surface, θ_1, follows the equation:

$$\theta_1 = 1 - \exp(-\Delta\chi) \tag{5.32}$$

Thus, for example if one were to expose the flat surface in such a way as to have a coverage of one monolayer equivalence ($\Delta\chi = 1$), then the molecules in direct contact with the adsorbate have the coverage $\theta_1 = 0.63$. It is desirable to have as many molecules as possible to make contact with the surface, so at a pressure such that

$\Delta\chi=3$, $\theta_1=0.95$. However, $\theta_{m>1}$, that is molecules in the 2nd, 3rd, etc. layers is 2.05. Most these molecules will be tending toward Raoult's Law values. Notice in the previous section that the Raoult's Law cross-over is at about $\Delta\chi=2\,\text{to}\,2.5$. The more layers become filled, the worse the situation becomes. This has been intuitive in engineering, the smaller the pores, the more surface is exposed, but also the higher the pressure, the greater are the kinetic problems due to the thicker liquid film.

With the binary mixes, there is the added problem with getting the pressure in an optimum position on the isotherm. Inspecting the 50% mixture above, one can see that the separations at $\Delta\chi=1$ and $\Delta\chi=3$ are opposite. So if one wanted to get a separation characteristic of $\Delta\chi=1$ at a pressure characteristic of $\Delta\chi=3$, using anatase, there is no hope if a flat surface is used. However, what about the possibility of using a high pressure but physically preventing more than one monolayer from adsorbing—in other words, eliminating "layers" 2 and above? Then success may be achieved. Engineers also have to worry about kinetics. To begin with, some simplifying rules are sometimes followed. These are for binary adsorption. For simplification, one might expect that Lewis' rule [41] should apply:

1. The pores are filled or nearly filled.
2. The adsorbate molecules are approximately the same size.
3. The adsorbate molecules have simple geometries.
4. The intermolecular forces are simple van der Waal forces.

Lewis' rule assumption

Lewis' rule assumes that (1) the densities of the adsorbates are the same as the densities of the liquid phase adsorptive, and (2) the volumes of the adsorbates add to yield the pore volume. Both assumptions could be incorrect, but for mixing liquid phases, assumption (2) is usually fairly good. These assumptions yield for <u>filled</u> pores:

$$n_{\text{ads},1}\overline{V}_1 + n_{\text{ads},2}\overline{V}_2 = V_p \tag{5.33}$$

or:

$$\frac{n_{\text{ads},1}}{n_{p,1}} + \frac{n_{\text{ads},2}}{n_{p,2}} = 1 \tag{5.34}$$

where the n_ps are the determined number of moles adsorbed to fill the pores for each adsorbate alone.

Assumption 1) could be incorrect, as observed by Dubinin, Zhukovskaya, and Murdmaa [42] and calculated by χ theory [43]. An intuitive explanation for this phenomenon is that the first "layer" is not fully dense, therefore subsequent layers also cannot be fully dense. Since the areal density is not the same as the liquid density, then the molar volumes also cannot be the same as the molar volume of the liquid. (The BDDT equation also predicts this.)

Fig. 5.56 shows a calculation of the molar volume as a function of $\Delta\chi$, or monolayer equivalent coverage for a flat surface, as calculated from χ theory. In this

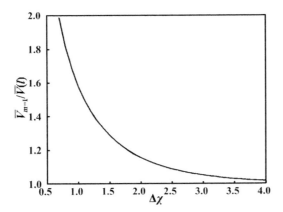

FIG. 5.56

Relative molar volume as a function of coverage in terms of monolayer equivalence on a flat surface or $\Delta\chi$.

calculation, it is assumed that the density in the normal direction from the surface is not affected, but only the areal density. From the figure, it is apparent that, by a surface coverage of 3 monolayers (flat surface calculation), the molar volume of the adsorbate directly in contact with the adsorbent is nearly the same as the liquid. In any case, micropores with a radius or width less than a monolayer thickness would not allow adsorption within the pore. Therefore, the minimum measurement possible for a filled pore is at 1 monolayer and it is more likely to be greater than this amount. Therefore, the correction for the change in molar volume would not seem to be an issue for adsorption in pores.

For pores, the layers greater than those in contact with the adsorbent will not have as much material as a flat surface. This would introduce an entropy term, but the energy of adsorption would be undisturbed. The pore restriction would cut back on the amount of adsorbate as a fraction according to the geometry and the n-layer analysis. This will be discussed further in Chapter 6. For pores that are restricted to 1 monolayer, referred to here as ultramicropores, there is a special form of the χ theory that may be used. This equation is given by the log-law as mention in Chapter 4. The caveat for using this equation alone is that the ultramicropore surface area is much greater than the external surface area. Otherwise, the external surface adsorption would be added to the ultramicropore adsorption according to the additive principle of the χ theory.

The analysis of binary adsorption in micropores depends somewhat upon the analysis of adsorption of the pure adsorptives. The ideal situation would be to analyze the adsorption of the pure adsorptives and from this information predict the adsorption of the binary adsorptives. The analysis using the pure adsorbates isotherms is given in Chapter 6 in some detail. Some of the results of the analysis will be used here to demonstrate a few points.

Although preferred, it is not necessary to do a thorough investigation of the pure adsorbates, if one is willing to make a few measurements for the binary system. The following analysis will demonstrate this.

Binary adsorption at a constant pressure

Assuming that Lewis' rule applies regardless of the pressure and that the value for $n_{p,1}$ and $n_{p,2}$ are specified only by the value one expects for the pure adsorbate (1 or 2) at the specified pressure, then Eq. (5.34) could be symbolized as:

$$\frac{n_{ads,1}}{\mathbf{n}_{p,1}(P)} + \frac{n_{ads,2}}{\mathbf{n}_{p,2}(P)} = 1 \tag{5.35}$$

where P designates the total pressure and $\mathbf{n}_{p,1}(P)$ and $\mathbf{n}_{p,2}(P)$ are now functions (thus, written as bold), which follow Lewis' rule. Here the external amount adsorbed may be included in $\mathbf{n}_{p,1}(P)$ and $\mathbf{n}_{p,2}(P)$, so that even at pressures where the pores are completely filled there might be a slight pressure dependence. Obviously, the simplest case, both theoretical and experimental, is to hold P constant and just vary the composition. For such a case, $\mathbf{n}_{p,1}(P)$ and $\mathbf{n}_{p,2}(P)$ revert to being input parameters, which, if required, are relatively easy to obtain from the pure adsorbate isotherms.

It is clear that within the space of the pores it is not possible for both adsorbate 1 and adsorbate 2 to follow the χ equation or the standard curve. If adsorbate 1 has a much higher $|E_a|$ than adsorbate 2 then the adsorption of 1 will predominate and adsorbate 2 will fill out the remaining space according to Lewis' rule. Therefore, the value of χ_c for adsorbate 1 will remain unchanged, whereas χ_c for adsorbate 2 will change due to the pre-adsorption of 1. For whatever total pressure is used, then $n_{p,1}$ will equal $n_{ads,1}$ at that pressure. Picking a particular pressure for a standard (in many cases 1 atm at which the experiment is performed) and since n_{ads} is linear with χ, this yields two equations:

$$n^*_{ads,1} = m\chi^*_1 + b$$
$$0 = m\chi_{c,1} + b \tag{5.36}$$

where the symbol "*" indicates at the pressure picked for the experiment.

From these equations m and b may be obtained. Thus:

$$n_{ads,1} = \frac{n^*_{ads,1}(\chi_1 - \chi_{c,1})}{\chi^*_1 - \chi_{c,1}} \tag{5.37}$$

The quantities $n^*_{ad,1}$ and $\chi_{c,1}$ may be obtained from the isotherm of the pure adsorbate 1. $n_{ads,2}$ is therefore:

$$n_{ads,2} = n^*_{ads,2}\left(1 - \frac{n^*_{ads,1}(\chi_1 - \chi_{c,1})}{\chi^*_1 - \chi_{c,1}}\right) \tag{5.38}$$

where $n^*_{ad,2}$ may be obtained from the isotherm of pure adsorbate 2.

Notice that $n_{ads,2}$ is not linear with χ_2 but rather linear with χ_1. One need not have the information from the pure adsorbates to obtain the parameters for Eqs. (5.37) and (5.38). One may instead use some data from the binary adsorption isotherm at the

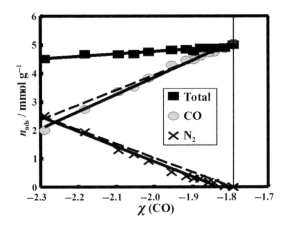

FIG. 5.57

Adsorption of CO-N$_2$ mix on 5A zeolite at 1 atm.

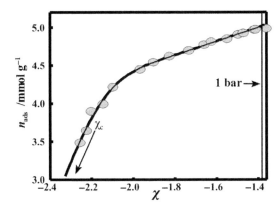

FIG. 5.58

Isotherm of CO on 5A zeolite to extract the constants χ_c and χ(1 bar).

pressure of interest. This is particularly advantageous for obtaining $n^*_{ads,1}$, and $n^*_{ads,2,2}$, since these quantities would normally be obtained in such a measurement. A few additional data points are needed to obtain $\chi_{c,1}$.

Comparison to experiments

An example of some data where both the pure adsorption isotherms were obtained over a broad pressure range and the binary phase diagrams at 1 atm pressure were also measured are those by Danner and Wenzel [44] and analyzed here. The measurements were made for the various combinations of CO, N$_2$, and O$_2$ on 5A and 10X zeolites.

Table 5.10 Analysis of the parameters for binary adsorption versus the pure adsorbates

	Adsorbate	$\chi_c{}^a$	$n_{ads, 1\ atm}$
Binary 5A	$\underline{N_2}$-O_2	−2.399	4.52
	$\underline{N_2}$-CO	−2.642	5.02
	$\underline{CO}$-O_2	−3.051	5.00
Pure 5A	CO	−2.751	5.02
	N_2	−2.446	4.59
	O_2	−2.751	4.94
Binary 10X	$\underline{N_2}$-O_2	−2.225	4.52
	$\underline{N_2}$-CO	−2.238	5.71
	$\underline{CO}$-O_2	−2.554	5.57
Pure 10X	CO	−2.559	4.72
	N_2	−2.323	5.59
	O_2	−1.873	5.08

[a] χ_c for the pure adsorbates is the value from the micropore analysis discussed in Chapter 6.

A summary of the obtained parameters is given in Table 5.10. The analyses of the adsorption isotherms for the pure adsorbate are given in Chapter 6. The simple, flat-surface χ theory is not appropriate for analysis in micropores and the expansion on the theory is present in Chapter 6. The order of the $|E_a|$s are:

$$CO > N_2 > O_2$$

Thus, for the combination CO-N_2 and CO-O_2 the χ_c for CO should be used, and for N_2-O_2 combination the χ_c for N_2 is proper (as underlined in the second column). Thus, the χ_c for O_2 is not relevant.

The $\chi_{c,1}$ values from the pure adsorbate experiments and the binary experiments are in fair agreement except for two cases. The adsorption of CO-O_2 on 5A zeolite is particularly variant and the adsorption of N_2-CO on 10X zeolite is nearly as bad a fit.

Figs. 5.57 and 5.59 present two cases of the plots of the n_{ads}s versus the appropriate χ_1. The gas mixture was constant at 1 bar. In these figures:

- the solid lines are the χ fits from the binary experiment;
- the small vertical line is the data for the pure CO adsorption at 1 bar; and
- the dotted line is the prediction from the measurements with the pure adsorbates.

In order to make the theoretical prediction, the individual isotherm of the CO and the N_2 adsorption were analyzed. The analysis required in this case is for microporous adsorbent and is described in Chapter 4 and in more detail in Chapter 6. Fig. 5.58 is the isotherm for CO. The vertical line is the 1 bar pressure, which is needed for the vertical line in Fig. 5.57. The χ_c value is needed for the dotted lines in Fig. 5.57.

Fig. 5.57 is for the 5A with the binary mix of N_2–CO, which according to Table 5.10 was fairly well predicted by the pure adsorption isotherms. The difference

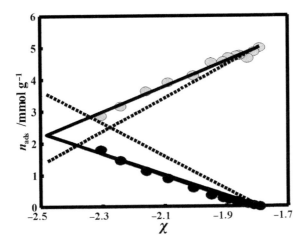

FIG. 5.59

Adsorption of CO-O_2 mix on 5A zeolite at 1 atm.

between the $\chi_{c,1}$ from the pure to binary measurement was 0.13. Fig. 5.59 is for the adsorption of CO and O_2 on 5A, which according to the data of Table 5.10 was the set with the worst agreement between the binary adsorption and that expected from the pure adsorbate isotherms. The difference in $\chi_{c,1}$ for this latter set was 0.30. The reason for the difference for this latter data set could be experimental. The calculated value for E_a for the binary adsorption is about 25.4 kJ mol^{-1}, which seems to be quite high. One normally does not observe E_as for these adsorbates on silica materials greater than 20 kJ mol^{-1}.

A common method of presenting the adsorption data for binary mixes is the gas-adsorbate phase diagram. This is a plot of partial pressure versus the amount adsorbed at constant total pressure. The data and fits shown in Figs. 5.57 and 5.59 may be redrawn to form such phase diagrams. These are shown in Figs. 5.60 and 5.61. In these figures:

- the solid lines are the χ fits to binary measurements; and
- the dashed lines are predictions from the pure adsorption.

If one were to use Henry's Law to determine the position of these lines, they would actually be on the opposite side of the diagram from where the lines are shown. Thus, the difference between a liquid-gas diagram and adsorbate-gas diagram is very obvious in these cases.

Except for the two cases mentioned, the predictions from the pure adsorbate isotherms would be, for most practical purposes, acceptable. The advantage of being able to predict the phase diagrams from the pure isotherms is that if one wished to do a screening study the number of isotherms for N adsorbates and M adsorbent is NM, whereas for the various combinations it is MN(N − 1), which for a large number of adsorbents could be considerably more work. For screening

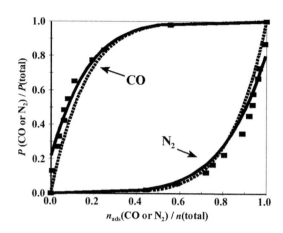

FIG. 5.60

Phase diagram of N_2-CO in 5A zeolite.

Data by Danner and Wenzel.

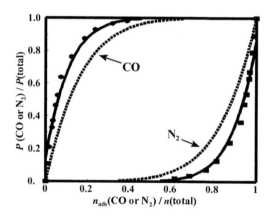

FIG. 5.61

Phase diagram of CO-O_2 in 5A zeolite.

Data by Danner and Wenzel.

studies, it is clear that measuring the pure adsorbates is the best strategy. There are several other sets of experimental data available in the literature. For most, the χ formulation works quite well.

Conclusions regarding binary adsorption

With the possible exception of density functional theory, χ theory is the only theory which is capable of making some predictions regarding binary adsorption. Density functional theory, in principle, should be able to calculate the binary adsorption for all types of pores, given all the atomic details. The latter proviso is the principal problem with density functional theory, which at present is not capable of dealing with unknown surfaces and unknown geometries. Furthermore, the number of matching kernels, roughly speaking - models, will become overwhelming when one introduces multiple adsorbates and the calculation might be too difficult for today's computers. The relatively simple χ theory determines some of the properties from the experimental data and then goes on to make predictions.

There is no doubt that much more research is needed in the area of binary adsorption, both theoretical and experimental. The binary adsorption in micropores depends upon the development of the theory of adsorption in micropores, which, as noted in Chapter 6, itself could benefit from further development.

Statistical comparisons of other isotherms to the χ-plot

A statistical comparison of χ theory with the BET or the DP isotherm fits is not completely possible due to the fact that for the latter two a best fit range is required in order obtain the parameters. This requires some judgement as to what this range is. The normal recommendation for the BET is to select the range in $P/P\text{vap}$ from 0.05 to 0.35. However, this can also vary, as noted previously, depending upon the energetics of the adsorption. For ceramic materials, this range is usually OK. The DP range, however, is best determined by an examination of the transformed plot, i.e., $\ln(n_{ads})$ versus $\ln(P/P_{vap})$. In Fig. 5.62 is a typical example of the three fits to the data used for the construction of the α-s plot (the original digital data used for this purpose is from a publication by Payne, Sing, and Turk [45], considered a very good standard curve.) It should be noted that the DP isotherms were originally not intended for non-porous materials, although the extension by Kagener would indicate this. The DP formulations are best for fitting the data at the high coverage end of the isotherm for microporous adsorbates. In Fig. 5.62 it is obvious that the deviation is very great for the BET equation at the higher pressures. The DP

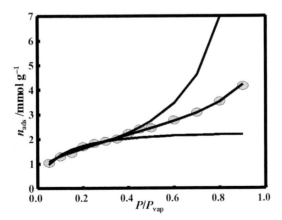

FIG. 5.62

A graphical comparison of the BET, χ, and DP theories respectively from top to bottom. Data points are those of the silica αs standard.

Table 5.11 Statistics comparing the BET, DP, and χ theories

	χ	BET	D-P
N$_2$ adsorption			
Range:	0.05–0.80	0.05–0.35	0.05–0.60
Sigma:	1.31	1.53	1.86
F-test full[a]:	0.6088	0.0024	0.0221
F-test in range:	0.9903	0.8339	0.8511
Ar adsorption			
Range:	0.05–0.90	0.05–0.35	0.05–0.70
Sigma:	1.21	1.31	1.66
F-test full:	0.9953	0.0087	0.4785
F-test in range:	0.9953	0.9136	0.7120

α-s plot for silica.
[a] Full range was 0.05–0.90. χ range extends to UHV.

formulation deviates somewhat in the low direction for these high pressures. In Table 5.11 are the statistics for the fit to the data used to construct the α-s curves. Not surprisingly, the F-test for the full range of the isotherms for the BET and DP isotherms are very poor. Even over the range that was judged best for these fits, the F-test would indicate a slightly better fit for the χ theory. As noted before, the last data point for the N$_2$ adsorption is probably too high and is ignored in this analysis.

For another example, the adsorption of N$_2$ and Ar on the 25°C out-gassed thoria are presented in Table 5.12. The advantages for these data have been presented in a previous section, which is the stability and uniformity of this powder with this treatment; but in addition to these advantages, the measurements were performed in a very accurate and controlled gravimetric system and many data points were collected. There is still, however, the question of range selection. The fewer data points

Table 5.12 Statistics comparing the BET, DP and χ theories

	χ	BET	DP
N$_2$ ad			
Range:	0.016–0.90	0.05–0.35	0.016–0.42
Sigma:	0.0213	0.0180	0.0209
F-test full:	0.9971	2.3×10^{-5}	0.105
F-test in range:	0.9971	0.9938	0.9939
Ar ad			
Range:	0.011–0.82	0.05–0.35	0.011–0.35
Sigma:	0.00998	0.01249	0.00781
F-test full:	0.9985	2.5×10^{-4}	0.130
F-test in range:	0.9985	0.9835	0.9835

Adsorption on thoria.

selected for the BET and DP fit, the better the statistics should be. After all, if one were to select two data points one would obtain a perfect fit. A best effort for selecting the DP range was used and the BET range was selected as the normal recommended range. Even so, the fit for the χ theory is still a little bit better.

These statistics are so close, at least in the selected ranges, that a definitive distinction is not possible. It is, however, possible to create a large number of equations that would fit the data very well. There are more than a hundred isotherms listed in the literature from which one could choose.

General conclusions

In this chapter, the terminology χ theory has been used, but one must remember that many of the applications could use a good standard curve. Furthermore, as previously demonstrated, χ theory and disjoining pressure (Excess Surface Word or ESW) theory are basically the same with the modification specified to calculate the surface area. The advantage of χ theory over other standard curve methods is that the standard is internal; that is, the energy of adsorption is calculated directly from the specific adsorbent sample being investigated rather than from a simulated sample. Using a simulated sample could be a source of considerable error.

If one prefers to reject the theoretical basis of χ theory, then the formulation as a standard curve is still very useful. As noted in the comparison to standard curves, the χ function is a very good analytical form for most standard curves. Having an analytical form for the standard curve is extremely handy for both practical measurements and theoretical development.

The prediction of the heats of adsorption from the adsorption isotherm without the introduction of any parameters is very difficult to explain away. This provides an explanation for the Dubinin "thermodynamic criterion," which was an assumption for which previously there was little theoretical basis. Its implication is that the entropy difference between the bulk liquid state and the adsorbate state is very close to zero, implying that the adsorbate state is nearly identical to the liquid state. The only difference seems to be that there is a the lack of an initial absorbate-gas film interface which forms sometime between the adsorption of 2 or 3 monolayer equivalencies.

References

[1] R.W. Cranston, F.A. Inkley, Adv. Catal. 9 (1957) 143.
[2] B.C. Lippens, B.G. Linsen, J.H. deBoer, J. Catal. 3 (1964) 32.
[3] J.H. deBoer, B.G. Linsen, T.J. Osinga, J. Catal. 4 (1965) 643.
[4] J.H. deBoer, Proc. R. Acad. (Amsterdam) 31 (1928) 109.
[5] J.H. deBoer, C.Z. Zwikker, Z. Phy. Chem. 28 (1929) 407.
[6] K.S.W. Sing, in: D.H. Everett, R.H. Ottewill (Eds.), Surface Area Determination, Butterworths, London, 1970, p. 25.
[7] M.R. Bhanbhani, R.A. Cutting, K.S.W. Sing, D.H. Turk, J. Colloid Interface Sci. 82 (1981) 534.

[8] R.B. Gammage, E.L. Fuller, H.F. Holmes, J. Colloid Interface Sci. 34 (1970) 428.
[9] R.B. Gammage, H.F. Holmes, E.L. Fuller, D.R. Glasson, J. Colloid Interface Sci. 47 (1974) 350.
[10] Fuller E. L., Jr., Agron P. A. (1976) The Reactions of Atmospheric Vapors With Lunar Soil, U. S. Government Report ORNL-5129 (UC-34b).
[11] G.A. Nicolan, S.J. Teichner, J. Colloid Surf. Sci. 34 (1972) 172.
[12] R.S. Bradley, J. Chem. Soc. (1936) 1467.
[13] R.S. Bradley, J. Chem. Soc. 1926 (1936).
[14] J. McGavack Jr., W.A. Patrick, J. Am. Chem. Soc. (1920) 42946.
[15] W.C. Bray, H.D. Draper, Proc. Natl. Acad. Sci. U.S.A. 12 (1926) 297.
[16] F. Rodrigues-Reinoso, J.M. Martin-Martin, C. Prado-Burguete, B. McEnaney, J. Phys. Chem. (1987) 91515.
[17] E.L. Fuller Jr., J.B. Condon, M.H. Eager, L.L. Jones, Sorption Analysis in Material Science: Selected Oxides, DOE Report Y-DK-264, US Government Printing Office, Washington, DC, 1981.
[18] E.L. Fuller Jr., K.A. Thompson, Langmuir 3 (1987) 713.
[19] S.J. Peters, G.E. Ewing, J. Phys. Chem. B 101 (1997) 10880.
[20] S.J. Peters, G.E. Ewing, Langmuir 13 (1997) 6345.
[21] X. Guo, H. Yu, Y. Zou, D. Li, J. Yu, S. Qiu, F.-S. Xiao, Micropororous Mesoporous Mater. 42 (2001) 325.
[22] A. Gil, B. de la Puente, P. Grange, Microporous Mater. 12 (1997) 51.
[23] C. Nguyen, D.D. Do, Carbon 39 (2001) 1327–1336.
[24] Thompson K. A. (1982) Mostly unpublished work—personal communication—publication rejected by *Langmuir* by editor William Steel on the basis that it disagreed with his book.
[25] K.A. Thompson, E.L. Fuller Jr., J.B. Condon, in: Further Evidence Supporting the Autoshielding Physisorption Equation, 17th DOE Surface Studies Conference, US Government Printing Office, Washington, DC, 1989.
[26] E.P. Smirnov, O.K. Semchinova, A.M. Abyzov, D. Uffmann, Carbon 35 (1997) 31.
[27] J. Silvestre-Albero, A.M. Silvestre-Albero, P.L. Llewellyn, F. Rodrigues-Reinoso, J. Phys. Chem. 117 (2013) 16885–16889.
[28] J.D. Lopez-Gonzalez, F.G. Carpenter, V.R. Deitz, J. Res. Natl. Bur. Stand. 55 (1955) 11.
[29] M.M. Dubinin, E.D. Zaverina, L.V. Radushkevich, Zhur. Fiz. Khim. 21 (1947) 1351.
[30] M.M. Dubinin, V.A. Astahov, Izv. Akad. Nauk SSSR, Ser Khim. 1971 (1971) 11.
[31] J. Tóth, Adv. Colloid Interf. Sci. 55 (1955) 1.
[32] J. Tóth, Colloids Surf. A 49 (1990) 57.
[33] J.B. Condon, Micropororous Mesopororous Mater. 33 (2002) 21.
[34] E.I. Pace, K.S. Dennis, W.T. Berg, J. Chem. Phys. 23 (1955) 2166.
[35] E.I. Pace, W.T. Berg, A.R. Siebert, J. Am. Chem. Soc. 78 (1956) 1531.
[36] Dennis K. S.. Heat Capacities From 15-125K and Entropies of Keypton Adsorbed on Rutile. PhD Thesis, Western Reserve University (now Case western Reserve University) Cleveland, OH, USA, 1954.
[37] Berg W. T.. Heat Capacities From 15-140 K and Entropies of Keypton Adsorbed on Anatase. PhD Thesis, Western Reserve University (now Case western Reserve University) Cleveland, OH, USA, 1955.
[38] W.D. Harkins, G.J. Jura, J. Am. Chem. Soc. 66 (1944) 919.
[39] G.L. Kington, R.A. Beebe, M.H. Polley, W.R. Smith, J. Am. Chem. Soc. 72 (1950) 1775.

[40] J.R. Arnold, J. Am. Chem. Soc. 71 (1949) 104.
[41] K. Denbigh, The Principles of Chemical Equilibrium, 3rd ed., Cambridge University Press, Cambridge, UK, 1971, 152.
[42] M.M. Dubinin, E.G. Zhukovskaya, K.O. Murdmaa, Ivza. Acad. Nauk SSSR, Ser. Khim. 1966 (1966) 620.
[43] J.B. Condon, Microporous Mesoporous Mater. 38 (2000) 359.
[44] R.P. Danner, L.A. Wenzel, AIChE J. 32 (1986) 1263.
[45] D.A. Payne, K.S.W. Sing, D.H. Turk, J. Colloid Interface Sci. 43 (1973) 287.

Further reading

[46] J.B. Condon, Langmuir 17 (2001) 3423.

CHAPTER 6

Porosity calculations

Introduction to porosity calculations

This chapter is the most speculative in this book. The analysis of porosity, especially microporosity and mesoporosity, follows the logic used for the standard curve method. There is no question that porosity needs, firstly, a better database to test and, secondly, better methods of calculations on the basis of quantum mechanical modeling. (This is also the case for binary adsorption.) As such, these are now fertile areas for more research.

Those familiar with pore calculations by density functional theory would immediately notice that the quantum mechanical base calculation does not yield the detail that DFT does. On the other hand, various DFT (NLDFT, QSDFT) theories keep being modified when things do not seem to fit, whereas χ theory and ESW theory have not changed since their conception, but rather expanded the implications, and such fine detail seems to be missing and unnessary. The problem for χ and ESW theory is that at this point, the slight errors in measurements, as small as a part per thousand, yield wildly differing results in the second derivative. It is tricky enough to take the first derivative from data points, much less the second. This brings up the question: "Is the fine detail seen in DFTs real, or is it a phantom also of small non-systematic error?" Also on the quantum mechanical side: "With better data, will the distributions become something other than the PMD?" and "Is the smoothing of the data, which is implicit with least squares fits, too severe?"

On a more basic level, the questions become: "How can one test these theories when an independent measurements do not seem to exist at this time?" Barring this, both theories beg the question: "Can one use the results of either of the theories to predict the behavior of adsorbent-adsorbate or multiple adsorbate systems?" and "Are they reliable enough for practical engineering problems?"

With respect to porosity, in Chapter 3, three approximate methods were presented. One method was based on QM modeling only, and the other two were based on the concept of the standard curve. These three were for calculations of ultramicropores, micropores, and mesopores. The definitions of these three pore types were modified from the IUPAC definitions to be more universal and in recognition that the IUPAC isotherm classification depends upon the adsorbent but not on the adsorptive. In this book, ultramicropores are defined as being small enough to restrict the adsorption to only one molecular layer. This is analyzed using the "log law" derived from quantum mechanics to apply to the first "layer." Micropores are large enough to accommodate more than one "layer," but not large enough to reach the point of the collapse of the less than dense multilayers to a dense liquid. The characteristic of mesopores is this collapse, referred to as "prefilling." According to QM, this collapse of the adsorbate is from the distributed surface film to something approaching the bulk liquid. Later on in this chapter, a few calculations will be provided to illustrate these points.

In this chapter, alternative methods that are more sophisticated are presented. In this area of investigation, many of the outstanding questions presented in the first addition of this book are resolved. Some areas, however, need further investigation. More experimentation would be welcomed and is needed. The most useful formulations are those that are not dependent only upon the specifics of the adsorbent. As mentioned previously, the reason for this is often that the details of the surface of the adsorbent are unknown in spite of the accepted notion that the surfaces of chemically similar adsorbents are identical.

Two of the methods utilize the insights that χ theory provides. One method is relatively easy and has already been presented in Chapter 3 in an approximate form. The problem with the approximation is that it does not take into account the possibility that a distribution exists for $\overline{E}_a$ or pore radius. Introducing these variables, one of which usually has insufficient experimental data for calculations, increases the complexity somewhat. This latter method utilizes the full equation that describes adsorption with porosity. At this time, the most promising technique employs a least squares minimization that uses the full χ description equation. The disadvantage of using the χ equations is that, at this time, an assumption about the form of the distributions must be made. This form is, for lack of a better choice, the PMF (Probability Mass Function, which is conveniently provided in most spreadsheets). Another technique would be to use any arbitrary function to match the χ curve in order to obtain the first and second derivatives. This eliminates the problem of assuming a form for the distributions. This works well for micropores, but mesopores present challenges since even a very high power polynomial will yield an unrealistic oscillatory fit. Investigations of these techniques seem to be a good area of research for some ambitious graduate student to undertake.

As mentioned previously, philosophical problem exists for the definition of the physical quantities "surface area," "pore volume," and "pore radius." What is meant by these terms? At first, this seems to be simple. However, consider that the physical quantity being measured and that the measuring device, namely the adsorbate molecules, have approximately the same size, and the answer to this question then becomes a little more difficult. Add to this the possible molecular-sized roughness for the adsorbent and the problem becomes more complex. This problem is the well-known fractal problem. That is, the measurement made depends upon the ruler being used. Thus, one should not expect to obtain the same answer for these physical quantities using different adsorbates. Furthermore, it should not be surprising that techniques other than physisorption, such as X-ray analysis or NMR, might also yield different results. The theoretical problem is intended not necessarily to obtain a perfect match between measurements, but to correlate these measurements and possibly bring them into predictable agreement. The practical consequence is that, given a certain set of physical quantities, all obtained by measuring with the same adsorptive, one should be able to reproduce the same correlated physical behavior from sample to sample based upon these physical quantities. Thus, the effort to pursue reproducible, reliable, and possibly accurate measurements of the basic physical quantities is not a waste. One could argue that such agreement is not important and that the correlations between the extracted parameters and the physical-chemical behavior—for example, catalytic activity—are all that is important. This, of course, can be and is done, but this then becomes an art rather than science, and one is unlikely to be able to make the predictions of which the scientific use of physical quantities and theories is capable.

Another fertile area of research, and one of the biggest problems in the area of mesopore analysis, is the hysteresis effect.[1] Several complications have been postulated for the phenomenon. Very often the two branches of the hysteresis loop are roughly parallel. This would indicate that the slope of the isotherm is determined by either an energy or a size distribution.

Reasons are proposed for hysteresis, some of which are presented in this chapter. Kinetic reasons have been investigated and, excluding metastability, have usually been discounted. Normally, it is postulated that the adsorption is the thermodynamically stable isotherm, whereas the desorption is metastable.[2] What causes this metastability? One possible answer is a stressed condition, in this case of the adsorbate film, in one direction that does not exist in the other. For example, for a meniscus that is metastable in the reverse direction, the film stress will eventually exceed a breaking point. This causes a rapid change to a different adsorbate configuration. Other examples might include partial chemisorption and plastic deformation of the adsorbent. The chemisorption hypothesis seems unlikely since chemisorption should make at least a semipermanent change in the both branches in contradiction to the experiment. The deformation hypothesis would make the adsorption branch metastable, which is a possibility for some adsorbents. The solution to nonspecific hysteresis may be found in density function theory calculations, of which several have been quite insightful.

For ultramicroporous calculations, only one method is presented: the χ theory method. This method is very simple, based on a long-term observed adsorption law known as the log law. Other reliable analyses are not available.

For microporous calculations there are three classes of theories described here. They are (1) the BDDT theory, (2) the theories classified here as Dubinin-Polanyi (DP), which include several similar isotherm forms, and (3) the χ theory for which there are two

---

[1] A quick review from Chapter 1: hysteresis is the phenomenon whereby at a certain position in the isotherm, adsorption occurs at a higher pressure than desorption for the same amount of adsorbate. Thus, there are two branches for the isotherm. At least one of these branches must be metastable.

[2] This is the opposite to the usual chemical reactions of solids with gas, where the stress is in the formation due to internal volume increase with the release of loop dislocation punching. Thus, the formation branch is metastable.

advanced methods. The BDDT theory is not found often in recent literature; however, the DP theories are widely used. The DP equations yield an answer for the pore volume that explains its wide use. The χ theory, although available in the literature, is hardly ever used. It has, however, the advantage of yielding pore radius, the pore volume, and perhaps a parameter that indicates the average fractal geometrical factor.

Two methods are proposed using the χ theory to analyze microporosity and mesoporosity. The first method is the simplest. This involves taking the first and second derivatives of the isotherm data. This requires the use of an arbitrary function, usually a polynomial, that is capable of doing a good job of fitting the data in a least squares routine. Two least squares routines are presented in Appendices IIA and IIB for this purpose. The second derivative yields a "spectrum" that needs some interpretation with distributions for the adsorption energy and pore size. This might also be used for mesoporosity if a proper fitting function can be found, perhaps a piecewise use of polynomials.

The second technique is mathematically a more difficult method. This uses a least squares fitting to the equation that describes both of these phenomena at the same time. This method of fitting is a minimization of the standard deviation, but it needs initial estimates and is prone to false minima. The other problem is that the number of parameters is very large and the question is whether these parameters really have physical meaning.

There are several explanations for the hysteresis phenomena. Some possibilities are presented, but no judgement is made as to which one, if any, is close to correct. This is an area that needs more research, but first the fundamentals need to be settled.

As one can see, many questions are left unanswered that are addressed in this chapter. Nevertheless, the narration continues *as if* the questions are all settled, otherwise there would be constant reminders that "I'm not too sure about this."

Ultramicropore analysis

From the n-layer analysis of χ theory, the fraction of the adsorbent which has adsorbate in direct contact is given by the equation (given in Chapter 1 as Eq. 1.27):

$$\theta_{m=1} = 1 - \exp(-\Delta\chi) \tag{6.1}$$

Recall that $\theta_{m=1}$ is the coverage for the adsorbate in direct contact with the adsorbent. (For simplicity, the "U$\Delta\chi$" is left off. $\Delta\chi \geq 0$ always.) Back-substituting the meaning of $\Delta\chi$, one gets:

$$\theta_{m=1} = 1 + \frac{RT}{\overline{E}_a} \ln\left(\frac{P}{P_{\text{vap}}}\right) \tag{6.2}$$

or, as normally plotted with $\log_{10}$ and noticing the $\theta_{m=1} = n_{\text{ads},\,m=1}/n_{\text{m}}$:

$$\frac{n_{\text{ads},m=1}}{n_{\text{m}}} = 1 + \frac{RT}{(2.303\ldots)\overline{E}_a} \log\left(\frac{P}{P_{\text{vap}}}\right) \tag{6.3}$$

If the plot of moles adsorbed, n_{ads}, versus $\log(P/P_{vap})$ is a straight line from $n_{ads}=0$ to a fairly high value of P, then there is ultramicroporosity. If it is close to a straight line but has an upward bow, then there is some adsorption in layer $m=2$ and a different analysis might be required. Notice that $n_m = n_{ads,\ m=1}$ on the ordinate axis intercept, $P=P_{vap}$, of the extrapolated line.

For a monolayer confinement, the energy distribution and pore size are given by the first and second derivatives:

$$\frac{\partial \theta_{m=1}}{\partial \Delta \chi} = +\exp(-\Delta \chi) \quad \text{and} \quad \frac{\partial^2 \theta_{m=1}}{\partial \Delta \chi^2} = -\exp(-\Delta \chi) \qquad (6.4)$$

There are also some external surfaces, so there will be a positive deviation from the straight line of the plot near $P=1$. This needs to be ignored for the extrapolation for the monolayer calculation. This up-swing is due either to the external surface, which is much smaller than the pore surface area, or to additional mesoporosity. Both need to be handled with the normal χ plot analysis. These values from the χ plot would then need to be subtracted from the "log law" values, or conversely the fitted values of the "log law" need to be subtracted from the χ plot analysis. In the "log law" plot, the details of the isotherm as $\ln P \to 0$ become obscured; therefore, inspection of the normal plot and the χ plot should be performed to identify if these other parameters are needed.

The value for $\overline{E}_a$ can be obtained from the slope of the plot, or more simply from the abscissa intercept in a fashion similar to obtaining the χ plot, in this case:

$$\overline{E}_a = -\frac{RT}{2.303} \log\left(\frac{P_c}{P_{vap}}\right) \qquad (6.5)$$

To illustrate the technique, the outstanding data by Silvestre-Albero et al. [1] (SSLR) are analyzed here. Figs. 6.1–6.3 show the data by SSLR and the attempted linear fits at low pressures to the log-law transform.

Fig. 6.1, LMA233 carbon, has a long straight line fit to the "log law." The conclusion from this is that there are pores for which only a monolayer can adsorb. According to SSLR, this is confirmed by the QSDFT (Quenched Solid DFT) fit, which (without interference from He pre-adsorption) peaks at one monolayer. However, mesopores are obvious from the large up-swing at the higher pressures and a hysteresis loop. This is confirmed by the QSDFT with a peak center of about 3.7 monolayer thickness, followed by a hysteresis loop visible in the normal isotherm presentation. (Statistically, the fit to the linear portion is not as good as it seems in this graph.)

Fig. 6.2, DD52 carbon, has a slight bow which indicates that there is porosity with more than one monolayer that is adsorbed. (Alternatively, there may be some residual adsorbate left in the pore, but, given the history of careful experimentation by the authors, this seems unlikely.) There is no hysteresis loop or large up-swing at high pressure indicating an absence of mesoporosity. Here this isotherm is analyzed using the slopes and the intercepts from the χ plot and from the "log law." This should yield nearly the same result, with the proviso that the "log law" has a bow in it, so perfect agreement is not expected.

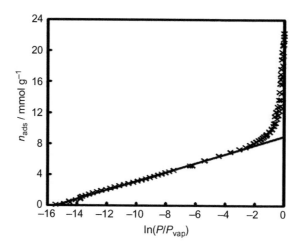

FIG. 6.1

N$_2$ isotherm on LMA233 activated carbon by SSLR.

Reproduced with the permission from the American Chemical Society.

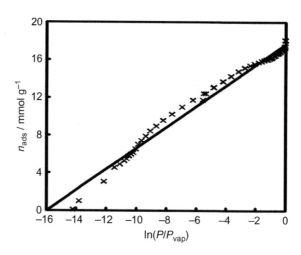

FIG. 6.2

N$_2$ isotherm on DD52 activated carbon by SSLR.

Reproduced with the permission from the American Chemical Society

The plot of Fig. 6.3 indicates that SCA-15 silica has both ultramicropores and mesopores. This is confirmed by the clear hysteresis in the normal isotherm and the large upswing in the "log law" plot.

From straight-line fits to the linear portions of the "log law" and χ plots, the graphics fit yields the following:

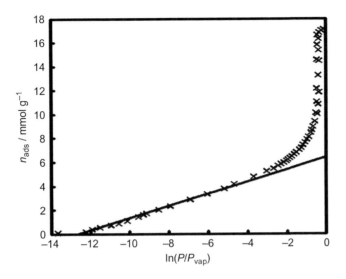

FIG. 6.3

N$_2$ isotherm on SBA-15 silica by SSLR.

Reproduced with the permission from the American Chemical Society.

(Fig. 6.1) LMA233 carbon:

Ultramicropores for "log law" $n_m = 8.91$ mmol g^{-1} $\pm 2\sigma = 0.06$ mmol g^{-1}

$$\overline{E}_a = -10.0 \text{ kJ mol}^{-1} \pm 2\sigma = 0.3 \text{ kJ mol}^{-1}$$

$$r^2 = 0.997$$

Mesopores from χ plot:

$$n_{ex} = 1.13 \text{ mmol g}^{-1} \pm 2\sigma = 0.42 \text{ mmol g}^{-1}$$

$$n_p = 17.7 \text{ mmol g}^{-1} \pm 2\sigma = 1.4 \text{ mmol g}^{-1}$$

$$r^2 = 0.934*$$

(*too few data points to obtain a decent r^2.)

(Fig. 6.2) DD52 carbon:

Ultramicropores for "log law"

$$n_p = 17.7 \text{ mmol g}^{-1} \pm 2\sigma = 0.2 \text{ mmol g}^{-1}$$

$$\overline{E}_a = -10.3\,\text{kJ}\,\text{mol}^{-1} \pm 2\sigma = 0.4\,\text{kJ}\,\text{mol}^{-1}$$

$$r^2 = 0.995$$

Micropores from χ plot:

$$n_{ads} = 18.2\,\text{mmol}\,\text{g}^{-1} \pm 2\sigma = 0.7\,\text{mmol}\,\text{g}^{-1}$$

$$\overline{E}_a = -9.37\,\text{kJ}\,\text{mol}^{-1} \pm 2\sigma = 0.7\,\text{kJ}\,\text{mol}^{-1}$$

$$n_{ex} = 0.49\,\text{mmol}\,\text{g}^{-1} \pm 2\sigma = 0.08\,\text{mmol}\,\text{g}^{-1}$$

$$n_p = 15.0\,\text{mmol}\,\text{g}^{-1} \pm 2\sigma = 0.2\,\text{mmol}\,\text{g}^{-1}$$

$$r^2 = 0.946*$$

(*too much curvature in higher pressure to obtain decent r^2.)

(Fig. 6.3) SBA-15 silica:

Ultramicropores for "log law"

$$n_{ads} = 6.71\,\text{mmol}\,\text{g}^{-1} \pm 2\sigma = 0.02\,\text{mmol}\,\text{g}^{-1}$$

$$\overline{E}_a = -8.11\,\text{kJ}\,\text{mol}^{-1} \pm 2\sigma = 0.5\,\text{kJ}\,\text{mol}^{-1}$$

$$r^2 = 0.995$$

Mesopores from χ plot:

$$n_{ex} = 0.18\,\text{mmol}\,\text{g}^{-1} \pm 2\sigma = 0.01\,\text{mmol}\,\text{g}^{-1}$$

$$n_p = -16.76\,\text{mmol}\,\text{g}^{-1} \pm 2\sigma = 0.004\,\text{mmol}\,\text{g}^{-1}$$

$$r^2 = 0.989$$

It would be instructive to look at simulations to see what the "log law" looks like for something other than one monolayer restriction, and how it looks for cylindrical pores versus slits. The following is a simulation using n-layer χ theory. This is an appropriate simulation, since if one were to use χ theory to determine surface area and porosity, then the χ theory simulations would be more appropriate than using Monte Carlo of DFT.

These simulations assume reasonable parameters seen in actual data. Each layer has a distribution base of the QMHO (quantum mechanical harmonic oscillator) using only the ground state. The HO normally used is a parabolic fit to the LJ-6-12 (Lennard-Jones 6-12) potential. (See appendix IIH.) Figs. 6.4–6.7 show the results of some typical calculations. Here the following parameters are assumed:

- The first $\sigma = 0.209$ nm, typical of a $\bar{E}_a$ about $8\,\text{kJ}\,\text{mol}^{-1}$.
- Subsequent $\sigma s = 0.303$ nm, typical of Ar, which includes uncertainty in siting configuration (σs are concatenated using summation of the squares).

For the slit pores, one obtains exactly what would be expected, as Fig. 6.4 indicates. The "log law" is precisely followed as the distance between the layers approach a three-monolayer thickness that is enough room for a monolayer equivalence can adsorb on both sides of the pore. (Recall how this works: at one monolayer equivalence there is roughly ~66% in the first layer and ~34% in the second layer. Thus, if the second layer is blocked, then the "log law" is followed but the amount is then truncated.) At this point, there does not seem to be a way to distinguish a two-atom slit and a three-atom slit from the adsorption isotherm. One possibility is to use various adsorptives.

For slit pores, the "log law" is precisely linear up to the point that the three monolayer equivalence is adsorbed ($r = 1.5\phi$). For cylindrical pores, the observation is quite different. The "log law" is quite far from the simulation. Exceeding three monolayers is still the point at which the "log law" breaks down. However, the above derivation of the "log law" does not take into account the linear cutoff due to the

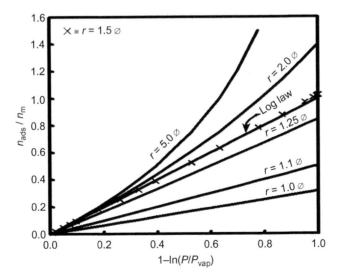

FIG. 6.4

The amount adsorbed as monolayer equivalents versus the "log law" for various slit sizes.

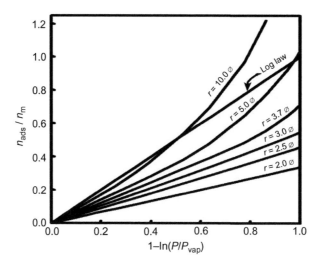

FIG. 6.5

The amount adsorbed as monolayer equivalents versus the "log law" for various cylindrical pores.

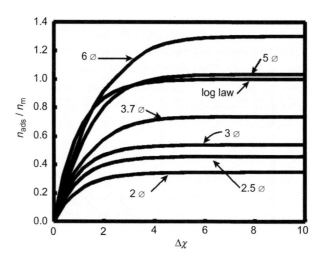

FIG. 6.6

The amount adsorbed as monolayer equivalents versus $\Delta\chi$ for various slit sizes.

decreasing circumference with decreasing radius—thus a factor of ½ is needed for a correction, which is not included in the calculations shown.

For adsorption in microporous materials, that is, adsorbent/adsorbate pairs, the distinction between slits and pores may not be possible, at least with one adsorbate. However, for mesoporous materials there seems to be a way to figure out whether the pores are slits or cylinders (or possibly other fractal factors).

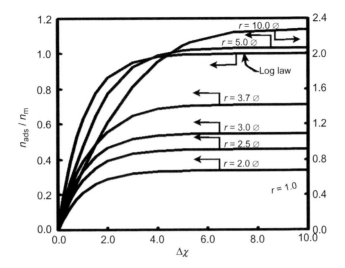

FIG. 6.7

The amount adsorbed as monolayer equivalents versus $\Delta\chi$ for various cylindrical pores.

Micropore analysis

Microporosity is defined by the IUPAC as pore sizes (diameter or slit width) of 2 nm or less. Although this is the official definition, the practical definition would be in terms of the isotherm produced. The type of isotherm that is produced is usually a type I isotherm, although this could be misleading. The χ feature associated with microporosity is feature 2 in the absence of feature 3. This would be a negative curvature in the χ plot without any preceding high-pressure positive curvature.

All micropore analyses make the simple assumption that the adsorption is limited by the size of the pores, specifically the pore volume. Indeed, for the Dubinin-Radushdkevich and Dubinin-Astakhov equations, the pore volume is the only practical physical quantity claimed obtainable.

The BDDT equation

One attempt to account for the adsorption in micropores was to modify the BET equation by limiting the number of adsorbed layers. The resultant equation is the Brunauer et al. [2] (BDDT) equation. With N_{BET} being the number of layers allowed, this is given as:

$$\frac{n_{ads}}{n_m} = \frac{C_{BET}P\left[P_{vap} - (N_{BET}+1)(P)^{N_{BET}} + N_{BET}(P)^{N_{BET}+1}\right]}{(P_{vap}-P)\left[P_{vap} + (C-1)(P) - C(P)^{N_{BET}+1}\right]} \tag{6.6}$$

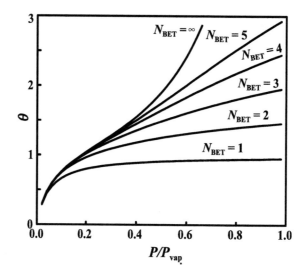

FIG. 6.8

The BDDT equation for various values of N_{BET}. The C constant used for this was 20.

Fig. 6.8 illustrates the shape of this isotherm for several values of N_{BET}. An obvious question is, "If only integer values of N_{BET} can exist, how could one obtain a fit to the isotherm that is not an integer?" There are two possible answers to this. Firstly, there is no reason to assume that the adsorbate molecule stack exactly in a row (although this seems to be implied by BET). The second explanation that has been put forward is that there may be a distribution of pores and N becomes a weighted average of the various sizes. For example, if $N_{BET}=2.5$, this could mean that half of the pores accommodate two layers, and half accommodate three.

Notice that regardless of the value for N_{BET}, the value for n_m, which according to BET is interpreted to be the monolayer coverage, is extractable. This is a physical quantity that theories that preceded the BET were unable to extract.

This approach to determining porosity has been attempted for several cases, but has been found not to be successful in most publications. Part of the reason for this is that the determination depends upon pressures outside the BET "range" and would be considered invalid. One rarely sees this used in the recent literature.

The DR and DA equations

The Dubinin-Radushkevich [3] (DR) and the Dubinin-Astakhov [4] (DA) equations may be expressed as:

$$\ln\left(\frac{n_{ads}}{n_p}\right) = B\left(\frac{T}{\beta}\right)^k \ln^k\left(\frac{P_{vap}}{P}\right) \qquad (6.7)$$

where n_p is the number of moles that fill the pore volume. The constant β is interpreted to be an energy term. The interpretation of the parameters in equation other than n_p is of little practical importance. Eq. (6.7) may be derived using the following assumptions:

- The quantity $\partial(RT \ln(P/P_{vap}))/\partial T = 0$ at constant n_{ads}. Dubinin called this the "thermodynamic criterion."
- The energy of adsorption follows a distribution function, specifically the Weibull distribution curve, $\mathbf{D_w}$, which was originally used to describe particle size distributions [5]. This distribution is given by:

$$\mathbf{D_w}(x, k, \beta) = \frac{k}{\beta} \left(\frac{x}{\beta}\right)^{k-1} \exp\left(-\frac{x}{\beta}\right)^k \mathbf{U}(x) \qquad (6.8)$$

Thus, the parameters of Eq. (6.7) are related to this distribution function.

The parameter k may be any value, with $k=2$ being the special case of the DR equation. In general, these equations are referred to as Dubinin-Polyani (DP) equations.

One of the advantages of Eq. (6.7) is that one can plot $\ln(n_{ads})$ as a function of $\ln^k(P_{vap}/P)$ and adjust k to obtain a straight line in the plot. With today's computers, adjusting k to obtain the best straight line is a trivial task. The intercept on the ordinate, $\ln n_{ads}$, axis yields the value for $\ln n_p$. For a wide range of micropore sizes and energies, one is able to find a fairly long range in the transformed isotherm where a straight line fit is appropriate [6]. If the external surface area is negligible compare to pore volume, such an analysis is not necessary since it is simple to extrapolate the untransformed isotherm to $P_{vap}/P = 1$. The DR-DA extrapolation, however, works well even in the presence of significant external surface area. Fig. 6.9 shows an example of a DA fit to some real data. This particular data is N_2 adsorption on 5A zeolite by Danner and Wenzel [7] (DW) (chosen at random from many sets of data) and is quite typical. Often there is a slight upswing or curvature in the plot near the n_{ads} axis, which indicates significant external surface area. In this case, the external surface area was too small to cause this problem. Table 6.1 provides a summary of the DA analysis of the data by Danner and Wenzel. One thing to notice is that in order to obtain a straight-line fit, k has a considerable range.

The analysis using χ theory is presented after the explanation of how this analysis is performed using a generalized standard curve method.

Standard curve analysis using distributions

In Chapter 3, micropore analysis using a standard curve was presented. It was assumed that the system of pores was very simple in that analysis. The simplification was that there is one energy of adsorption and one pore size. This is very unlikely to be the case, so in this section additional parameters will be introduced into the standard curve analysis. These additional parameters are in turn also introduced

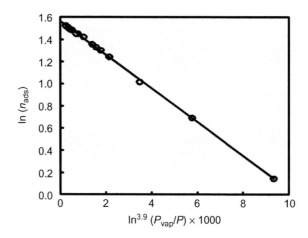

FIG. 6.9

An example of a DA plot illustrating the straight line fit. The data is N_2 adsorption in 5A zeolite by Danner and Wenzel.

Table 6.1 DA analysis: N_2 on 5A and 10X

Adsorbent	Adsorbate	V_p /mL g^{-1}[a]	k
10X	O_2	0.46	4.2
	N_2	0.54	2.4
	CO	0.62	2.3
5A	O_2	0.40	3.8
	N_2	0.42	4.3
	CO	0.48	2.1

[a] Full liquid $\bar{V}$ assume to make this calculation.
Data by Danner and Wenzel.

into the χ plot, and take into account the distributions that are lacking in the section on ultramicropores in this chapter.

In principle, any standard curve may be used in this analysis provided the standard curve is descriptive of a homogeneous, non-porous material of identical surface composition. This is quite an order and there are only a few materials for which one could with some confidence say that the standard curve used might be appropriate. Such materials include silica, alumina, beryllium, thoria, and perhaps other pure ceramics. The χ theory formulation, however, does not need a separate standard. This is its main advantage over a calibrated standard curve. In the following analysis, the χ curve will be used due to its simplicity.

The following analysis need not be interpreted in terms of physical dimensions. Thus it yields an analytical form that one could use more easily with more traditional pore size analysis systems as well as χ theory or DFT. Included in the traditional digital methods are the pore length method originated by Wheeler [8] and developed by Shull [9], the BJH method (Barrett Joyner and Halenda [10]) and the Cranston and Inkley method [11]. The present approach, however, is easier to visualize. It may also be possible that once the parameters for a particular isotherm are obtained, one could attach different meanings to them. Indeed, the χ plot representation has been presented [12] as a method to construct empirically an analytical expression for the standard curves.

For the curve fitting, it will be assumed that there is a distribution of energies, E_as, and a distribution of pore sizes. Furthermore, some of the surface area is not inside the pores and is referred to as external. The pore radius is reflected in a cutoff in the standard curve, or in terms of χ there is a mean value $\langle \chi_p \rangle$ for which the standard curve in the pores is terminated. The PMF (probability mass function) distribution will be used with standard deviations for energy and pore size. Any reasonable distribution could be used and the parameters expanded—for example, to include skewness, etc.—but usually the experimental data would not justify this. However, as demonstrated previously in the section on microporosity, there may also be more than one classification of pores present in a sample. These two caveats will be ignored. Thus, there are six parameters plus a seventh after the pore distribution becomes negligible:

1. $\langle \chi_c \rangle$ = the mean value of the start of the standard curve $\Rightarrow \langle E_a \rangle$.
2. $\sigma_c(E_a)$ = the standard deviation of χ_c in a distribution function. If one had extensive low-pressure data, it would be possible to formulate any energy distribution based on the second derivative of the χ plot.
3. $\langle \chi_p \rangle$ = the mean value of the shutdown of adsorption due to the restriction of the pores
4. σ_2 = the standard deviation of $\chi_p \Rightarrow D(r_p)$ which is assumed here to be the PMF.
5. n_m = monolayer equivalent moles $\Rightarrow A_s$ = total surface area (including pore surface area).
6. n_p = moles inside the pores $\Rightarrow V_p$ = the volume inside the pores.
7. Derived constant: n_{ex} = monolayer equivalent moles for the external surface = final slope (if possible) $\Rightarrow A_{ex}$ = external surface area.
8. Definition: $\Delta \chi_p = \chi_p - \chi_c$ (a combination of two parameters)

At this point, no geometry will be assumed. With the assumption of the conversion to geometry, other quantities, such as pore radius, may be calculated.

The distribution for the energies implies:

$$\frac{\partial^2 n_{s+}}{\partial \chi^2} = \frac{n_m}{\sigma_p \sqrt{2\pi}} \exp\left(-\frac{(\chi - \langle \chi_c \rangle)^2}{2\sigma_c^2}\right) \tag{6.9}$$

where the symbol n_{s+} indicates the amount adsorbed on all the surfaces and would continue to adsorb with increasing pressure if there were no porosity restriction. Likewise the distribution in the pore size implies:

$$\frac{\partial^2 n_{s-}}{\partial \chi^2} = \frac{n_p}{\sigma_2 \sqrt{2\pi}} \exp\left(-\frac{(\chi - \langle \chi_p \rangle)^2}{2\sigma_2^2}\right) \tag{6.10}$$

where the symbol n_{s-} indicates the amount of material that is not adsorbed due to the pore restriction. σ_2 reflects the cumulative distribution for both the energy and the pore sizes, and is related to these through the well-known statistical relationship for non-correlated distributions:

$$\sigma_2^2 = \sigma_c^2 + \sigma_p^2 \tag{6.11}$$

The problem with Eq. (6.11) is that a correlation could exist between the energies of adsorption and the pore sizes. For example, χ theory predicts that the smaller a cylindrical pore, the higher is its energy of adsorption. Thus, Eq. (6.11) may not be followed in some unlikely circumstances.

Obviously to get the entire isotherm, Eq. (6.10) must be subtracted from Eq. (6.9) and the results doubly integrated from $\chi = -\infty$ (which is $P/P_{vap} = 0$) to whatever χ is of interest. (A similar equation is given in Chapter 5 in the discussion of the Freundlich and Dubinin-Polanyi isotherms. There the match to the second derivatives was used as a more sensitive test.) This yields a rather messy but quite useable and easily calculated equation:

$$n_{ads} = n_m \mathbf{Z}(\chi, \langle \chi_c \rangle, \sigma_c) - n_p \mathbf{Z}(\chi, \langle \chi_p \rangle, \sigma_2) \tag{6.12}$$

where:

$$\mathbf{Z}(x, y, s) = \frac{s}{\sqrt{2\pi}} \exp\left(\frac{-(x-y)^2}{2s^2}\right) + \frac{x-y}{2}\left(1 + \mathrm{erf}\left(\frac{x-y}{s\sqrt{2}}\right)\right) \tag{6.13}$$

Normally $n_m = A_s/A$ and $n_p = V_p/V$. The six parameters are then: n_m, n_p, $\langle \chi_c \rangle$, $\langle \chi_p \rangle$, σ_c, and σ_2. With six parameters one should be able to fit almost any isotherm that resembles a type I isotherm. Indeed, in many cases these are too many parameters so the following could be attempted to yield five parameters. For the σs:

1) If the very low pressure data is unavailable, set σ_c to zero (or for practical purposed to use the same program, set it to a very low number such as 1×10^{-5}).
2) With $\sigma_c = 0$, then $\sigma_2 = \sigma_p$.
3) If σ_2 drifts in the calculation to a smaller number than σ_c, then either try setting $\sigma_2 = \sigma_c$ or recognize that $\sigma_2 < \sigma_c$ is possible and Eq. (6.11) is not valid in this case.

In many cases, σ_c is not determinable. This is unfortunate since the value of $\overline{E}_a$ is an important parameter for the analysis. Since there are always uncertainties (sigmas) in the calculations, obtaining $\overline{E}_a$ by an extensive extrapolation is tricky.

The simplest method to obtain the parameters for Eq. (6.12) is to run a minimum search routine. This is easily accomplished with a simple spreadsheet. Some reasonable starting parameters would be 0.01 for both σs, -2.8 for χ_c, and -1.5 for χ_p, n_m, and n_p could initially be set equal to each. Alternatively, one could use the simple method in Chapter 3 with the results being the starting values and 0.01 for both σs. It is also advised to have a graphical representation of the data and the fit in order to have a visual guide for the initial estimates. If the starting parameters are very far from correct, the calculation can drift off to a false minimum. In such a case, the graphical representation will make it obvious. The criterion for minimization should be the minimization of the sum of squares of the difference between the calculated n_{ads} values and the experimental values. Uncertainty in the pressure values should be relatively small.

For illustration, the data by Danner and Wenzel [7] for adsorption of CO, N_2, and O_2 on 10X and 5A zeolite at 144.3 K are plotted in Figs. 6.10 and 6.11 with the calculation from Eq. (6.12) shown as solid lines. Table 6.2 provides the results' uninterpreted parameters. Several of these data sets illustrate the provisos listed above and the number of parameters probably should be five instead of six since $\sigma_2 < \sigma_c$. It is unlikely that this is correct but there is not enough low-pressure data to confirm this unlikelihood as real. The data for these particular isotherms do not extend low enough or well enough to separate σ_2 into σ_p and σ_c.

The same precaution applies to the value of n_{ex} listed as quantity 7 above. The value of n_{ex} in Table 6.2 was determined by the final slope. This should agree with the difference between n_m and n_p. This is clearly not the case for several of the experiments. The reason is that the PMD has not decayed away enough to make the ending slope the same as the difference between n_m and n_p.

In Table 6.3 are the interpretations of the parameters of Table 6.2. Thus, the designation $t_p \Delta \chi_p$ is at cutoff for the pores obtained from $\Delta \chi_p$. However, since it would be for a flat surface, then it is simulating the radius—that is, the distance from the wall to the center—of the slit pores. There should, therefore, be a factor to two to yield the answer in the radius of pores. (The classical t designation for this is explained in the next section.) The other method to calculate r_p is the value of $V_p A_p$. The parameter g given in the last column is the ratio of t_p, described below, to $V_p A_p$. The value of g indicated there is a problem with the calculation of the pore radius by one or both methods. The problem is most likely in the values of χ_c or χ_p since there is

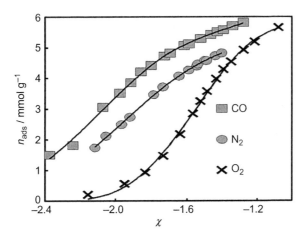

FIG. 6.10

Adsorption of CO, N_2, and O_2 on 10X zeolite by Danner and Wenzel.

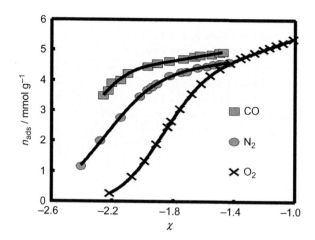

FIG. 6.11

Adsorption of CO, N_2, and O_2 on 5A zeolite by Danner and Wenzel.

not much data in those regions that would yield reliable values. This would be another area of potential further research—that is, what relationship does $t_{\chi,p}$ have to r_p, if any?

Again, reporting these value in Table 6.3 might yield some insight but it is still recommended to present *the uninterpreted parameters* in any report or publication, or, even better, the original data. The reason is that the interpretations may change with future developments, regardless of the theory used.

χ theory interpretation of the distribution fit
Surface areas and pore volume calculations

For microporosity, χ theory indicated that Eqs. (6.12) and (6.13) yield the parameters needed. The parameter n_m is indicative of the total surface area inside and outside the pores, whereas n_p is indicative of the pore volume. For microporosity, the difference between n_m and n_p yields the external monolayer equivalence, n_{ex}, which is the surface area including the pore opening area. Thus, the total surface area is given by:

$$A_s = n_m \overline{A} \tag{6.14}$$

where $\overline{A}$ is the molar area that is selected from various possibilities. In this book, this is the molar area consistent with the IUPAC convention. Likewise, the molar volume is assumed to be that of the liquid. The problem with this assumption is the question of what temperature for the liquid to select, since the density of the liquid varies with temperature. Between the normal boiling point of a liquid and its critical point, a factor of two or three is likely. The liquid density at the boiling point is selected here, with the acknowledgment that this assumption is also an open question. The pore volume in the simple analysis was obtained by extrapolation the high linear portion

Table 6.2 Fit to Eq. (6.12) for the data by Danner and Wenzel—uninterpreted

	n_m /mmol g^{-1}	n_p /mmol g^{-1}	n_{ex} /mmol g^{-1}	$\langle\chi_c\rangle$	σ_c	$\langle\chi_p\rangle$	σ_2	$\Delta\langle\chi_p\rangle$	%A_p /%	σ(fit) /mmol g^{-1}
10X O$_2$	12.50	8.77	3.61	−1.745	0.287	−1.386	0.098	0.359	71.1	0.041
10X N$_2$	6.93	5.27	0.02	−2.323	0.330	−1.563	0.330	0.760	99.8	0.026
10X CO	6.36	3.07	2.21	−2.525	0.318	−1.786	0.134	0.738	65.2	0.065
5A O$_2$	8.74	6.83	1.91	−2.116	0.177	−1.635	0.138	0.481	78.2	0.047
5A N$_2$	6.22	5.29	0.974	−2.596	0.023	−1.930	0.154	0.666	85.0	0.017
5A CO	6.35	5.58	0.78	−2.760	0.338	−2.072	0.131	0.687	87.8	0.035

Table 6.3 Fit to Eq. (6.12) for the data by Danner and Wenzel—interpreted

	$\bar{E}_a$ /kJ mol^{-1}	V_p /mL g^{-1}	A_s /m^2 g^{-1}	A_p /m^2 g^{-1}	A_{ex} /m^2 g^{-1}	t_p /nm	$V_p A_p$ /nm	x2 ⇒ r_p /nm	g
10X O$_2$	6.87	0.248	1067	759	308	0.119	0.327	0.682	0.36
10X N$_2$	12.24	0.183	678	676	2	0.270	0.270	0.541	1.00
10X CO	14.99	0.147	631	412	219	0.264	0.358	0.715	0.74
5A O$_2$	9.96	0.193	746	583	163	0.159	0.332	0.663	0.48
5A N$_2$	16.05	0.183	608	517	91	0.236	0.355	0.710	0.66
5A CO	18.95	0.198	631	554	77	0.246	0.358	0.715	0.69

of the isotherm back to $\chi = \chi_c$. This should give a good approximation. In the analysis with Eqs. (6.12) and (6.13) one may use the parameter n_p:

$$V_p = n_p \overline{V} \tag{6.15}$$

The external surface area, A_{ex}, is therefore:

$$A_{ex} = (n_m - n_p)\overline{A} \tag{6.16}$$

The external surface area includes at least two types of surfaces: the surface of the adsorbent that is not in the pores, which will be referred to as the "wall" area, A_w, and the surface area of the filled pore openings, A_o. Since n_p includes the opening of the pores but not beyond the outer surface, then:

$$A_{ex} = A_w + A_o \tag{6.17}$$

The slope of the χ plot after the pores are filled should not include the openings. This may be difficult to determine experimentally. This can cause problems in the analysis of porosity unless the external surface area is very small compared to the surface area within the pores. If one subtracts the wall surface area from the total surface area, one might be able obtain the surface area inside the pores, A_p. It should, therefore, be safe to use a possibility of a spread of A_p:

$$A_s > A_p > A_s - A_{ex} \tag{6.18}$$

Calculation of pore size assuming a geometry

There are two ways of calculating the pore size. For the first, one needs to assume a pore geometry. For cylindrical pores, the pore radius is given by the simply derived geometrical relationship:

$$r_p = \frac{2V_p}{A_p} \tag{6.19}$$

If slit-like pores were assumed, then:

$$r_p = \frac{V_p}{A_p} \tag{6.20}$$

where r_p is the distance from one slit wall to half the distance to the other wall. These are the primary types of pores. Other types would require other relationships.

Calculating r_p from $\Delta\chi_p$

An alternate derivation for Eq. (6.20) is to calculate r_p from the value of $\Delta\chi_p$ ($\Delta\chi_p = \langle\chi_p\rangle - \langle\chi_c\rangle$). The amount adsorbed up to any $\Delta\chi$ is given by:

$$n_{ads} = \frac{A_s \Delta\chi}{\overline{A}} \tag{6.21}$$

The classical thickness is obtained from the molar volume and area covered:

$$t = \frac{\overline{V} n_{ads}}{A_s} \tag{6.22}$$

regardless of the geometry. Substituting, one obtains:

$$t_{\chi,p} = \frac{\Delta\chi_p \overline{V}}{\overline{A}} \quad (6.23)$$

where the χ has been inserted to indicate that this is from χ theory and not classical. This works for a flat surface. For a restricted geometry, the $t_{\chi,p}$ remains the same, but the amount adsorbed will vary. In a cylindrical pore with a radius of r_p, the amount adsorbed from the pore wall to a distance t from the wall is:

$$n_{ads} = \frac{A_s}{\overline{V}}\left(t - \frac{t^2}{2r_p}\right) \quad (6.24)$$

or using Eq. (6.21), t is related to $\Delta\chi$ by:

$$\Delta\chi = \frac{\overline{A}}{\overline{V}}\left(t - \frac{t^2}{2r_p}\right) \quad (6.25)$$

Obviously for cylindrical micropores at $\Delta\chi_p$, $t = r_p$ and $\Delta\chi_p$ is given by:

$$r_p = \frac{2\overline{V}\Delta\chi_p}{\overline{A}} \quad (6.26)$$

Thus, for cylindrical pores the pore radius from classical considerations is twice that of the χ thickness. This constant then expresses the conversion to radius of the volume-to-surface ratio (inverse of SA:V with units.) This geometric factor will be given the symbol g. For a flat slit $g=1$, for a perfect cylinder $g=0.5$, and for a sphere $g=0.33$. The more uneven the inside of the pore is, the higher this number.

An example of the interpretation is the information provided in Table 6.3. There is a range for specified by σ_2 for $\Delta\chi_p$. The final column of this table lists the g found for each. There is little that is systematic about the numbers. Since these zeolites have cylindrical pores, $g \approx 0.5$ or maybe a little more due to roughness or interconnectedness. However, one of these values was clearly lower than 0.5. The data by Goldmann and Polayni in Table 6.4 in the next section yielded consistently a value $g \approx 1$, indicating slit pores.

Another potential problem with the analysis is that the value of $\overline{E}_a$ may need correction for surface curvature. Although this does not affect the original fit, it should shift the $\overline{E}_a$ to higher energies, and thus lower χ_c, yielding a larger value for $r_{\chi,p}$. (The formula has both χ in a ln term and outside the ln term requiring successive approximations.)

Example with significant distribution

For another example, in Table 6.4 is a reinterpretation of the classic data by Goldmann and Polanyi (GP) [13] for various adsorbates on activated charcoal. This data was analyzed as having slit pores for the carbon samples, and significant σ values are assumed. Every sample revealed a σ_2 and all but n-pentane reveal a finite σ_c. This is a little easier, since there is no need to correct the E_a for surface curvature. In Table 6.3, the range for r_p from Eq. (6.20) is presented. None of the samples tested well for mesoporosity. (This could be true since it might be that, for at least some slit types, the condition for collapse does not exist. That is, these slits are more like an open flat surface without a concave surface to initiate collapse.)

Except for one of the data sets marked as having too few data points (6), there seems to be very good agreement at least within the adsorbate set and fair agreement across adsorbate sets. There are several case where the calculated σ_c is greater than σ_2. This indicates that there is some correlation between σ_c and σ_p or that the data is not good enough to yield a reliable σ_c. This latter possibility is likely since the data for the low-pressure ranges are lacking. The contrast between the sets where one can obtain σ_c and where one cannot is obvious from the examples in Figs. 6.12–6.15.

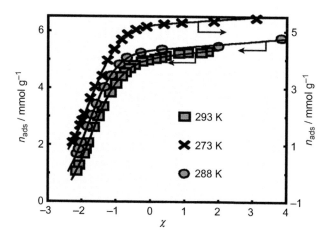

FIG. 6.12

Adsorption of ethyl ether on carbon by GP.

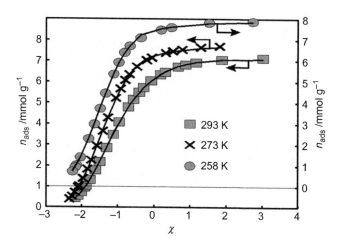

FIG. 6.13

Adsorption of ethylene chloride on carbon by GP.

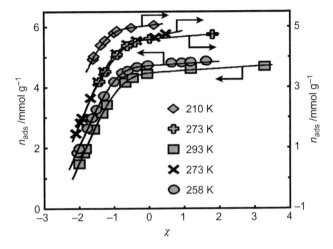

FIG. 6.14

Adsorption of n-pentane on by Goldmann and Polayni.

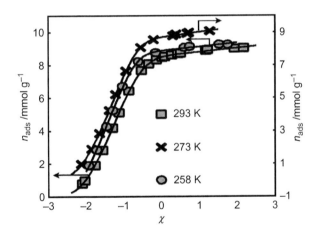

FIG. 6.15

Adsorption of CS_2 on carbon carbon by Goldmann and Polayni.

In Fig. 6.12 there seems to be enough data at the low-pressure end to determine the value of σ_c, In contrast, the low-pressure data in Fig. 6.14 (n-pentane) are missing.

The data in Table 6.4 are uninterpreted except for d_p ($= 2r_p$) which is calculated from both V_p/A_p and $\Delta\chi_p(\overline{V}/\overline{A})$. These two methods of calculation should yield the same answer for slits. However, V_p/A_p depends upon determination of the slope after the PMD of $\Delta\chi_p$ decays to nearly zero. Therefore, it could be used if the high-pressure data is cut short.

In spite of the problem with the estimate of σ_c, there is fair agreement with any particular data set for r_{slit} (to distinguish it from the statistical quantity r^2) except,

Table 6.4 Data from Goldmann and Polanyi

Adsorbate		T(K)	$\langle \chi_c \rangle$	σ_c	$\langle \chi_p \rangle$	σ_2	n_m/mmol g^{-1}	n_p/mmol g^{-1}	n_{ex}/mmol g^{-1}	r^2	σ (fit)	r_{slit}[a]	GP table #
Chloroethane	●	257.85	−2.299	0.325	−0.666	0.415	4.56	7.23	0.13	0.9999	0.032	0.741	2
Chloroethane	×	273.15	−2.233	0.336	−0.611	0.432	4.44	6.89	0.20	0.9999	0.026	0.736	3
Chloroethane	■	293.15	−2.261	0.339	−0.910	0.298	5.29	6.47	0.50	0.9999	0.019	0.613	4
Ethyl ether	●	257.85	−2.381	0.576	−1.311	0.592	4.84	0.506	0.11	0.9998	0.022	0.548	5
Ethyl ether	×	273.15	−2.310	0.536	−1.306	0.665	5.07	0.499	0.09	0.9998	0.019	0.514	6
Ethyl ether	■	293.15	−2.294	0.446	−1.334	0.728	5.06	0.472	0.15	0.9998	0.019	0.491	7
n-Pentane	●	257.85	−2.813	-0-	−0.966	0.402	2.52	4.54	0.06	0.9988	0.026	0.976	8
n-Pentane	×	273.15	−2.635	-0-	−1.059	0.416	2.76	4.01	0.09	0.9997	0.018	0.833	9
n-Pentane[a]	+	278.39	−3.013	-0-	−0.765	0.362	2.01	4.32	0.09	0.9997	0.010	1.189	10
n-Pentane	■	293.15	−2.572	-0-	−0.885	0.469	2.62	4.31	0.07	0.9998	0.024	0.892	11
n-Pentane	▼	209.5	−3.255	-0-	−1.241	0.329	2.42	4.60	0.13	0.99991	0.013	1.065	12
CS$_2$	●	257.85	−2.192	0.378	−0.678	0.309	5.75	8.34	0.24	0.9997	0.051	0.646	13
CS$_2$	×	273.15	−2.108	0.428	−0.624	0.309	5.72	8.03	0.31	0.9999	0.032	0.633	14
CS$_2$	■	293.65	−2.035	0.418	−0.521	0.379	5.57	8.06	0.24	0.9999	0.036	0.645	15

Note: -0- implies insufficient data to calculate this quantity.
[a]Interpreted value—all other physical quantities are uninterpreted.

again, for n-pentane. Thus, it appears that a good estimate of σ_c is not necessary in order to obtain the other physical quantities to a sufficient degree of accuracy. Notice that every one of the samples registers a low external monolayer. This simply means that the external surface area is insignificant compared to the overall surface area. The graphs seem to indicate a continued increase at high pressures, but this is due to the broadness of the pore distribution affecting high χ values.

The reason that the n-pentane seems to be erratic may have something to do with the long chain and unsymmetrical nature of the molecular stereochemical conformation. In all of the calculations, an ideal symmetric molecule is assumed and, although configuration-wise n-pentane has some symmetry, the non-rigid nature of the molecule breaks the symmetry.

The method so far outlined for treating microporosity has yielded very good fits for the isotherm. This might be the most important outcome even if one does not subscribe to the theory presented to justify the fit. With a few parameters, one is able to do an excellent fit with r^2 factors usually greater than 0.999 and standard deviation in the range of 1% or less. Normally it is difficult to obtain the data itself at this level. Thus, the advantage of this method is that it is at least a very good representation of the data. Alternative theoretical developments need to take this into consideration and be able, in some way, to mimic these equations.

Continuing with more examples, another data series where the value for σ_c was, in most cases, not possible to obtain is from the data by Wisniewcki and Wojsz [14] (WW). However, continuing to allow the calculation of σ_c yields a value. The values are highly variable. The analysis this data is given in Table 6.5. There was apparently no mesoporosity, so the fit is for microporosity only. Even though the possibility of mesoporosity was left in the computation "on."

The fitting minima were also very shallow and the average 2σ (of the fit) is about $0.6 \, \text{mmol g}^{-1}$. This is not sufficient precision, especially for the low pressure data which is most important for determining σ_c. In this case, the most reliable values seem to be n_p and $\langle \chi_p \rangle$. Setting $\sigma_c = 0$ made very little difference for all but the HY. In the case of HY the two calculations for r_p, with either $\sigma_c = 0$ or floating, are not in agreement. This does not bode well for this particular analysis. More precise, accurate, and higher-resolution data would hopefully improve this situation.

Figs. 6.16 and 6.17 show the χ plots of water adsorbed on zeolites NH$_4$Y, HY, NaY, CaY, and MgY. The symbols given in the first column of Table 6.5 are the key to which zeolite is plotted. Notice that the sharper the turn for the corner in the figures, the lower the number for σ_2 in the table, as one would expect.

The negative curvatures for the cutoff in adsorption due to the microporosity are quite clear. However, the positive curvatures for the energy distribution at the beginning of the plots are missing due to lack of low-pressure data. (In this case the χ value of -2 is a pressure of about 0.02 torr, which could have been their limit of detection.) In such a situation it is probably best to set the value of σ_c equal to nearly zero. For example $\sigma_c = 1 \times 10^{-6}$, since the general program does not allow exactly 0.

These examples are presented here to illustrate some problems that one might encounter both in the fitting of the standard curve in general and in the interpretation

Table 6.5 Microporosity analysis: water on Y-zeolites by Wisniewcki and Wojsz

Zeolite symbol	n_m	n_p	n_{ex}	n_{meso}	$\langle\chi_c\rangle$	σ_c	$\langle\chi_p\rangle$	σ_2	$\Delta\langle\chi_p\rangle$	$\%A_p$	σ (fit)
	/mmol g^{-1}									/%	/mmol
NH$_4$Y	15.48	12.90	2.81	0	−2.251	0.302	−1.165	0.261	1.086	81.9	0.286
CaY ×	10.37	15.59	0.77	0	−2.527	0.001	−0.902	0.206	1.624	92.6	0.245
HY ■	14.89	9, 83	0.99	0	−1.861	0.824	−1.121	0.569	0.740	93.3	0.240
MgY +	14.13	17.45	0.75	0	−2.489	0.484	−1.185	0.296	1.304	94.7	0.305
NAY ▼	46.26	16.80	1.04	0.004	−2.094	0.172	−1.723	0.372	0.372	97.8	0.363

Note: -0- implies insufficient data to calculate this quantity.

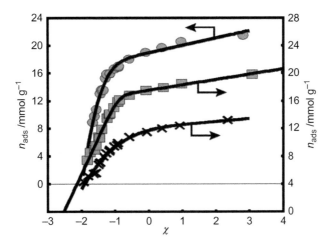

FIG. 6.16

χ plot of water adsorption on Y-zeolites at 298 K according to WW.

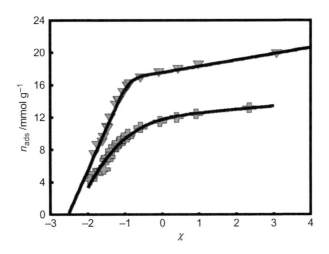

FIG. 6.17

χ plot of water adsorption on Y-zeolites at 298 K according to WW.

by χ theory. One cannot blindly analyze the data but rather must pay particular attention to the low-pressure data.

In all three cases, the use of the mesoporosity calculation is deemed unsuitable. The calculation for mesoporosity is performed along with microporosity in the next section. By turning on the mesoporosity calculation, one can determine if it is present, although in the χ-plot it is usually obvious. For example, when the Goldmann-Polanyi data were checked for mesoporosity they consistently failed by a large margin. For the Wisniewcki and Wojsz data, when the mesoporosity calculation was turned on, the amount of capillary filling was either extremely small or zero and made an insignificant difference in the answer.

On the other hand, using the data by Qiao et al. in the next section, the mesoporosity was very significant and the microporosity was very small. This next section presents a method based on χ theory to determine the simultaneous analysis of microporosity and mesoporosity. The method has both strengths and weaknesses and one can easily fly off into a far from correct false minimum, so they should not be used blindly. Future developments might provide programs that will be more robust and require little oversight.

Before presenting the uninterpreted method using an equation similar to Eq. (6.12), some background for possible reasons and interpretation for pore prefilling, or mesoporosity, are presented.

Analysis of mesoporosity—Classical

The signals in the isotherm that indicate mesoporosity are the presence of type IV and V isotherm or feature 3 in the standard plot designation. Feature 3 is a positive curvature in the standard or χ plot at a pressure well above the threshold. This would be a practical definition for purposes of analysis. The IUPAC definition is pore size greater than 2 nm but less than 10 nm. This high value might be extended in the future due to advances in experimental control and more precise measurements for the high-pressure region.

Some wording needs to be modified here to match the Russian school of vocabulary. This differs slightly from the IUPAC since the IUPAC does not have the full set. Indeed, the IUPAC does not even make the following distinctions, but rather uses adsorbent metrics. These definitions are not new but an emphasis may be needed here to make clear what is being addressed. The definition of a "film" includes the recognition that it has two interfaces of molecular dimensions and, most likely, some adsorbate material between the interfaces. The film tension is designated by the letter γ. The interfaces are controlled by interface tension designated by the letter σ. Associated with the material between the interface is the tensor referred to as the "disjoining pressure" given by the symbol Π[3]. The disjoining pressure should be at a minimum for a bilayer.

It seems to be universally accepted that the explanations for the mesopore filling phenomenon depends upon the formation of a normal film with two distinct interfaces, at least for the system boundary separating the gas and liquid phases. It is reasonably assumed that the adsorbate-adsorbent boundary is already present. It is also often assumed that the adsorbate-gas boundary is always present, which may not be the case.

---

[3]The disjoining pressure should be at a minimum for a bilayer. For free-standing surfactant stabilized films this is the limit for the so-called "black films."

Nearly all the theoretical explanations of mesoporosity starts with the Kelvin-Cohan [15] (KC) formulation. Foster [16] proposed the Kelvin equation for the effect of vapor pressure on capillary rise but did not anticipate its use for very small capillaries where the adsorbate "thickness" is a significant geometrical perturbation. Cohan formulation subtracts the adsorbate "film thickness" from the radius of the pore to yield the modified Kelvin-Cohen equation:

$$-RT \ln \left(\frac{P}{P_{\text{vap}}} \right) = \frac{m_c \gamma_{gl} \overline{V}}{(r_p - t)} \qquad (6.27)$$

where γ_{gl} is the gas-liquid surface tension and m_c is a constant (KC notation) which depends upon the geometry of the interface. The most common values of m_c are:

- $m_c = 1$ for a cylindrical interface, herein referred to as the two-dimensional case or 2D case; and
- $m_c = 2$ for a spherical interface, herein referred to as the three-dimensional case or 3D case.

Whether one is referring to a 2D or 3D case, it is not necessarily the same as the pore geometry and is embedded in whatever theory is being used. This could be a confusing point. So, in this book a 2D or 3D interface refers to the *gas-liquid interface*. There could be intermediate cases between strictly a cylindrical interface and a spherical interface so that $1 < m_c < 2$. In principle it could be less than 1 since obviously for flat surfaces $m_c = 0$ and $P = P_{\text{vap}}$. "P_{vap}" will always be used for the flat surface vapor pressure. (The symbolism "P_o" has been used in the literature for the vapor pressure over a pure liquid that possibly has a curved interface. Therefore it will be avoided here.)

Comments about the standard plot of determining mesoporosity

The most common method for determining the mesoporosity from a standard plot was presented in Chapter 3 in some detail. This need not be repeated here. This simple method might be good enough, but if one requires more detail it will not suffice. However, "more detail required" may very likely mean that one needs high-definition data that goes to lower pressures using an ultra-high vacuum system.[4]

Before the gas solid interface is formed, the adsorbate follows the multilayer calculation provided by χ theory. This is sometimes also (confusingly) referred to as a film, even though only one interface has been formed.[5] In this section, when the word

---

[4] See the publication by Silvestre-Albero et al. [1] for an excellent explanation for the system and the requirements.
[5] Attempts to introduce language to keep these straight has been met with vigorous opposition by reviewers. The terms "subliquid" and "subfilm" had been proposed with no success. The IUPAC may someday come up with something.

"film" is used it indicates a fully formed film with two distinct interfaces. On a flat surface, the difference between the adsorbate and a film fades as the number of layers increase. By 10 layers, there is a trivial difference. For curved surfaces, however, the outlook is different.

There are several explanation for what is referred to as "pore filling," "prefilling" or "collapse." One proposal has already been given with respect to the Disjoining Pressure Theory in Chapter 1. The Kelin-Cohan equation has also been given in that chapter, but since it is the most widely quoted theory, the implications are given here.

Although these theories may provide some insight and possibly be useful for predictions, they are not needed for analysis. The methods of standard curve or χ theory should be sufficient for this purpose.

Kelvin-Cohan equation

One of the postulates is that the hysteresis is the difference in the geometry of film formation from the adsorbate and the adsorbate formation from the film. If the metastable 2D situation should arise upon adsorption and the 3D arise upon desorption, this would lead to a hysteresis. Some points to notice about the Kelvin-Cohan equation when the desorption branch is subtracted from the adsorption branch are as follows:

1) If there is a switch in the film geometry from adsorption to desorption, there is a difference in m_c, Δm_c. Therefore, there is a shift in the pressure of onset of the film collapse upon adsorption and the film breakage upon desorption as:

$$\frac{P_a}{P_d} = \exp(-\Delta m_c)\exp\left(\frac{\gamma_{gl}\bar{V}}{r_p - t}\right) \propto \exp(-\Delta b) \quad \Delta b \approx \frac{\Delta m_c \gamma_{gl}\bar{V}}{r_p - t} \quad \text{note}: \lim_{r_p \to \infty}\left(\frac{P_a}{P_d}\right) = 1 \quad (6.28)$$

where P_a and P_d indicate the hysteresis break point for adsorption branch versus desorption branch respectively, The proportionality assumes that t is consistent.

2) Since the film must have enough material to at least form two complete interfaces, then t is at least the thickness of two monolayers. Thus, mesoporosity should not be observed by this phenomenon below $\Delta \chi = 2$.
3) t, in fact, varies with P. This generates a messy equation:

$$\frac{\exp(-\Delta\chi_a)}{\exp(-\Delta\chi_d)} = \frac{m_{c,a}(r_p - \bar{L}\Delta\chi_d)}{m_{c,d}(r_p - \bar{L}\Delta\chi_a)}, \quad \bar{L} = \frac{\bar{V}}{\bar{A}} \quad (6.29)$$

which probably needs to be solved with successive or series approximations.

The relationship between the two pressures may also be determined using Eq. (3.32) from Chapter 3 since r_p is the only parameter in common between adsorption and desorption.

One of the problems is the question, "Where is P_a and P_d?" Is it at the low-pressure end or the high-pressure end, or is it at the midpoint of the upswing? Up to this point, it is assumed that the increase in the χ plot is vertical, thus the question would be irrelevant. However, complicating this is the now known phenomenon of the pore energy distribution, which is tied in with E_a distribution, and size distributions, which, of course, affects the χ plot. Even E_a can change from adsorption to desorption for a variety of reasons, therefore shifting P_a and P_d which in itself can cause a hysteresis effect. This latter fact is usually not taken into account.[6] The pore and energy distributions are not just a problem with the standard curves; the DFTs also predict a vertical increase. This situation for χ theory will shortly be corrected below.

The constant m_c is the factor that is different for a 2D or 3D interface. It may also be that the pores are more complex than simple cylinders or slits. It only implies that, given other alternatives, the lowest Gibbs' free energy situation is for the sudden appearance of a 2D or 3D interface. Eq. (6.27) and the dependence of the free energy for χ theory:

$$-RT \ln\left(\frac{P}{P_{vap}}\right) = \mathbf{E}_a := \overline{E}_a \exp(-\Delta\chi) \tag{6.30}$$

should at some point cross. Crossing after $\Delta\chi = 2$ would cause a shift from Eq. (6.30) to Eq. (6.27). However, physically it would seem that if they crossed before $\Delta\chi = 2$, the adsorbate would continue to follow Eq. (6.30) since there is not enough adsorbate to form a film. Evidence from Tables 6.2, 6.4, and 6.5 indicate that this is indeed the case up to this point.

The most widely used theory for calculating the mesoporosity other than the Kelvin-Cohan method is the Broekhoff-deBoer method. This is presented next.

The Broekhoff-deBoer (BdB) theory

The Broekhoff-deBoer theory (BdB) [17] relates to the capillary filling of cylindrical pores. It makes the following assumptions:

1. The adsorbed layer is continuous with the density of the liquid phase, thus with a sharp outer boundary[7] with a γ_{gl} regardless of the value of n_{ads}.
2. The chemical potential of the adsorbed layer is what determines the thickness of the film using the same functional dependence as with a flat adsorbed layer.
3. The Kelvin-Cohen equation determines the chemical potential for a curved adsorbate layer. For cylindrical pores, this is the 2D use of the equation.
4. The surface tension, γ_{gl}, of the gas-liquid (distinction between adsorbate and film not made) is a constant.

---

[6] In most cases, investigators are probably unaware of these problems. This is especially true of the variation of χ_c since very few are aware of its existence. There are hints of such considerations, but no outright conclusion. Again Ref. [1] provides the seed for considering all these possibilities.
[7] Note: this assumption conflicts with χ.

Given these assumptions and some rather fundamental thermodynamic relationships, some equations are derived for a generalized isotherm. The isotherm function is written in terms of the gas pressure, P, and the vapor pressure over a flat surface, P_{vap}, as:

$$-RT \ln\left(\frac{P}{P_{vap}}\right) = \mathbf{F}(t) \qquad (6.31)$$

where $\mathbf{F}(t)$ is an arbitrary monotonic function. The layer thickness, t, may be found either theoretically or experimentally. One need not know the theory behind $\mathbf{F}(t)$ so long as one can write a reasonably good analytical form for it[8]. Alternatively, one may write this in terms of the chemical potentials of the liquid, μ_{liq} (again over a flat surface) and the adsorbed layer μ_{ads}:

$$\mu_{liq} - \mu_{ads} = \mathbf{F}(t) \qquad (6.32)$$

Using the following thermodynamic relationship, one can obtain the equilibrium condition.

$$dG|_{P,T} = \mu_c dn_{ads} - \mu_g dn_{ads} + \gamma_{gl} dA_{gl} \qquad (6.33)$$

where μ_c is the chemical potential of the condensed phase flat or otherwise, μ_g is the chemical potential of the adsorbent, and A_{gl} is the area of the adsorbate layer-gas interface. The equilibrium condition may be obtained knowing dA_{gl}/dn_{ads}. Since cylindrical pores are being addressed here, the inside area of the adsorbed layer in the pore of radius r_p is given by:

$$A_{gl} = 2\pi(r_p - t)l \qquad (6.34)$$

where l is the total length of all the pores. The number of moles adsorbed is given by:

$$n_{ads} = \frac{2\pi}{\overline{V}}\left(r_p^2 - [r_p - t]^2\right)l \qquad (6.35)$$

if it were a film. Differentiating both of these equations and combining, one obtains:

$$\frac{dA_{gl}}{dn_{ads}} = \frac{-\gamma_{gl}\overline{V}}{r_p - t} \qquad (6.36)$$

This is the 2D assumption since what is being considered here is the adsorbed film with a cylindrical shape. Since for equilibrium $dG|_{P,T}/dn_{ads} = 0$, then:

$$\mu_{cyl} - \mu_g = \frac{\gamma_{gl}\overline{V}}{r_p - t} \qquad (6.37)$$

---

[8] This means that assumption 1 was not necessary.

where the subscript "cyl" stands for cylinder. Utilizing assumption 2 or $\mu_{ads}=\mu_{cyl}$ and adding Eq. (6.32) to Eq. (6.37):

$$\mu_{liq} - \mu_g = \mathbf{F}(t) + \frac{\gamma_{gl} V_m}{r_p - t} \qquad (6.38)$$

or the modified isotherm is:

$$-RT \ln\left(\frac{P}{P_{vap}}\right) = \mathbf{F}(t) + \frac{\gamma_{gl} V_m}{r_p - t} \qquad (6.39)$$

One way of viewing this equation is to think of the chemical potential inside the adsorbed film as the sum of the chemical potential of the gas adsorbent plus the change in the chemical potential due to the hydrostatic pressure produced by the gas-liquid interface. This is very similar to the effect of osmotic pressure with the gas-liquid interface acting as a semipermeable membrane. Fig. 6.18 illustrates this schematically. The last term on the right sides of Eqs. (6.37)–(6.39) then is the hydrostatic correction term.

For thermodynamic stability the condition (a minimum and not a maximum in the Gibbs' free energy change):

$$\frac{d^2 G|_{P,T}}{dn_{ads}^2} \geq 0 \qquad (6.40)$$

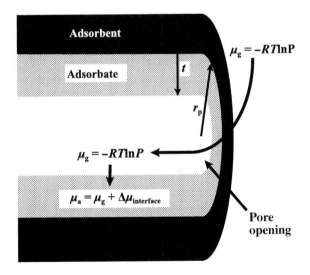

FIG. 6.18

The Broekhoff-deBoer model for adsorption in a cylindrical pore. The adsorbate-gas interface creates a hydrostatic pressure which changes the chemical potential of the adsorptive.

must be met. In this case, the condition is stable provided that:

$$\frac{d\mathbf{F}(t)}{dt} + \frac{\gamma_{gl} V_m}{(r_p - t)^2} \leq 0 \tag{6.41}$$

Thus there could be some value of t for which the right side of this equation becomes zero. This is referred to as the critical thickness, t_{cr}, and a corresponding pressure, P_{cr}, for which the layer becomes unstable. Above P_{cr} the condition of the one-dimensional, i.e., a cylindrical coordinate, adsorbed layer may no longer exist and the capillary filling begins to convert to the two-dimensional (i.e., spherical coordinate) condition. Notice that the second term of equation must always be positive, therefore the slope of $\mathbf{F}(t)$ must be negative, that is, $\mathbf{F}$ is a decreasing function with P. Upon examination of the definition in Eq. (6.31), this must be the case. The question then is whether the value for $|d\mathbf{F}/dt|$ is large enough to satisfy Eq. (6.41).

If a critical radius, t_{cr}, exists, then one can determine the free energy change from this value to other values by substituting into Eq. (6.32) and integrating.

$$\int_{t_{cr}}^{t} d\mathbf{G} \bigg|_{P,T} = \int_{t_{cr}}^{t} \left[-RT \ln\left(\frac{P}{P_{vap}}\right) - \mathbf{F}(t) - \frac{\gamma \overline{V}}{r_p - t} \right] dn_{ads}, \quad t \leq r_p \tag{6.42}$$

where $\mu_{cyl} - \mu_g$ has been replaced. Converting dn_{ads} into terms of t using the pore length, l_p:

$$\Delta G = \frac{\pi l_p}{V_m} \left\{ -RT \ln\left(\frac{P}{P_{vap}}\right)(t - t_{cr})(2r_p - t - t_{cr}) - 2\int_{t_{cr}}^{t} \mathbf{F}(t)(r_p - x)dx - 2\gamma V_m(t - t_{cr}) \right\} \tag{6.43}$$

For equilibrium, this value of ΔG is set to zero to obtain the desorption pressure at which the filled capillary will spontaneously revert to an adsorbed layer. Since for the filled pore $t = r_p$, this value should be different from the spontaneous capillary filling value, t_{cr}. Thus, Eq. (6.43) becomes:

$$-RT \ln\left(\frac{P}{P_{vap}}\right) = +\frac{2\int_{t_{cr}}^{r_p} \mathbf{F}(x)(r_p - x)dx}{(r_p - t_{cr})^2} + \frac{2\gamma V_m}{(r_p - t_{cr})} \tag{6.44}$$

This should, according to BdB theory, yield the desorption branch. There is a very close resemblance between Eqs. (6.39) and (6.43) with the former containing the 2D form of the Kelvin-Cohan equation and the latter the 3D form. Notice that by L'Hospital's rule as $t_{cr} \to r_p$ the first term on the right side will approach $\mathbf{F}(t_{cr})$, thus yielding the 3D form (Fig. 6.19).

Fig. 6.20 illustrates the capillary filling and capillary emptying as envisioned by the BdB theory. The difference in the filling and emptying geometry is the postulated reason for hysteresis. The sequence from left to right is as follows:

Analysis of mesoporosity—Classical **315**

A just before core collapse;
B just after collapse;
C at fully filled;
D desorption at same geometry as B;
E just before capillary emptying; and
F just after emptying

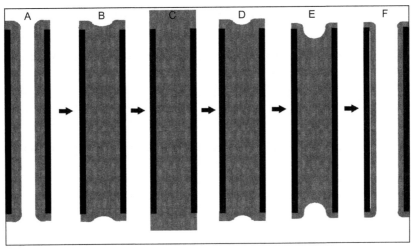

FIG. 6.19

The BdB model of pore filling and pore emptying. Notice there is no form "E" in the adsorption branch.

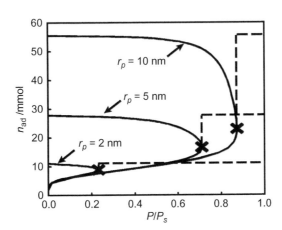

FIG. 6.20

The isotherms for adsorption on porous SiO_2 according to the Broekhoff-deBoer theory.

It would be instructive to show some plots of the isotherm predicted by Eq. (6.39) to see what this equation means. Fig. 6.20 shows some plots in terms of moles adsorbed for a 2-, a 5-, and a 10-nm pore radius. This calculation uses the α-s plot for the function $\mathbf{F}(t)$. At the points marked with $\mathbf{X}$s the critical thickness is reached and the isotherm follows the dotted lines. The point of capillary filling as predicted by Eq. (6.41) and the amount of capillary filling are indicated by the dashed lines. Fig. 6.21 shows the dependence of t and P/P_{vap} on the pore radius. A comparison of the BdB theory with the Kelvin-Cohan equation, both the 2D and 3D forms, is shown in Fig. 6.22.

Several modifications have been made to the theory including proposals for the function $\mathbf{F}(t)$ and making γ a function of t (or $r_p - t$) as well. For example, Kowalczyk et al. [18] (KJTKD) proposed a double form for $\mathbf{F}(t)$ similar to ESW (and χ):

$$\mathbf{F}(t) = \Pi_1 \exp\left(-\frac{t}{\lambda_1}\right) + \Pi_2 \exp\left(-\frac{t}{\lambda_2}\right) \tag{6.45}$$

where the subscripted quantities are basically parameters which are interpreted in terms of the disjoining pressure theory (see Chapter 4). Along with this equation, a dependence of γ_{gl} on the core radius as proposed by Miyahara et al. [19,20] (MKYO) was used. This relationship was given as:

$$\gamma_{gl}(r_c) = \gamma_\infty \left(1 + \frac{\delta}{r_c}\right) \tag{6.46}$$

where ($r_c = r_p - t$ and) the value of δ is about the same value as the van der Waal radius. (For consistency and practical purposes, r_c and r_p are positive throughout this book.) At the low end of the mesopore range this could yield about a 40% correction for nitrogen adsorption. This is about a 10% correction for a pore radius of 3 nm.

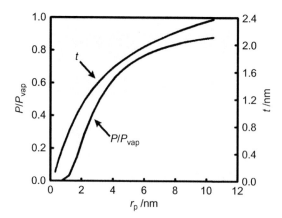

FIG. 6.21

The relationship between the pore radius in SiO_2 as the critical thickness and the critical relative pressure according to the Broekhoff-deBoer theory. The lines correspond to the $\mathbf{X}$s in Fig. 6.20.

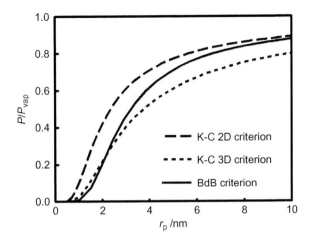

FIG. 6.22
Comparison of the Broekhoff-deBoer theory with the Kelvin-Cohan calculation for the switch to capillary filling. F(t) uses the α-s for nitrogen on SiO$_2$ as the model isotherm.

Table 6.6 Some δ values for the Tolman equation to correct γ for surface curvature

Liquid	δ (nm)
Argon	0.314
Nitrogen	0.330
Cyclohexane	0.503
Benzene	0.461
Water	0.274

A similar correction to γ has been calculated by Ahn et al. [21]. The calculations are rather involved but yield results similar to the derived equation by Tolman [22]:

$$\frac{\gamma_{gl}(r)}{\gamma_\infty} = \left[1 + \frac{2\delta}{r}\right]^{-1} \tag{6.47}$$

which is more convenient. The values for the *parameter* δ (his notation) are approximately the same as the diameter of the liquid molecule, i.e., van der Waal radius. Table 6.6 sets out some values for this parameter. Clearly the two corrections for γ_{gl} are not the same, with the latter being more extreme than the former. For liquid N$_2$ the first correction is about 16% for a 2-nm pore, whereas the second one is about 50%. Whether this correction is required or not is still a question.

The original BdB theory, due to assumption 1, cannot be applied to slit-like pores. Indeed, BdB theory predicts that slit-like mesopores should behave the same as micropores with no capillary filling upon adsorption.

What χ theory says about prefilling

Although χ theory can detect and analyze the prefilling, this does not tell us why it happens. At this point in development, there is not enough high-resolution data to test any proposal that is made. Therefore, what follows is speculation.

In the population of the layers, according to χ theory n-layer calculations, there is a distribution which is not molecularly sharp, but extends over at least three layers, and molecules in the individual layers might not "line up." Thus, it may be that the prefilling is a self-alignment of the adsorbate molecules to make it possible to create an interface. This is a reasonable, but unobserved, postulate.

However, if the above speculation is correct then the question is, "What are the energies involved that cause this?" On the one hand there is the χ theory energy, which is obviously stronger than the liquid-gas surface tension and, at least up to some pressure, does not predict a sharp boundary. However, as more and more adsorbate is added, this first layer energy drops. There will be a point at which the incremental energy of adsorption and the liquid-gas interface energy become equal. It seems obvious that it has something to do with the gas-liquid interface, possibly with the formation of the interface being the initiator. The film formation is assumed for all the theories dealing with mesoporosity and hysteresis, and so it is here also. Here, however, the twist introduced by χ theory is introduced. In other words, before collapse the adsorbate is not a film. This implies that the energies involved are associated with σ_{sg} and $\mathbf{E}_a$. These are associated with A_{sg} and A_{sa} where $A_{sg}+A_{sa}=A_s$. After the collapse the energies are associated with σ_{sg}, $\mathbf{E}_a$, and σ_{lg} with areas of A_{sg}, A_{sa}, and A_{gl}. Without a concave surface, however, the energy balances do not yield a break-point. This reverts back to something like the Kelvin-Cohan equation.

The modification of the Kelvin-Cohan equation to fit into χ theory has been discussed in Chapter 3. The results yielded Eqs. (3.30)–(3.32). Some modifications due to energy distributions and pore size distributions follow. However, at this point, prediction of the isotherm based on the geometry of the pores is uncertain.

Given the uncertainty about the translation of geometry to adsorption behavior, it should not be surprising that the hysteresis problem is unresolved. The Broekhoff-deBoer explanation seems reasonable, but qualitatively does not work for all cases. This leads to the tentative conclusion that the phenomenon of hysteresis could be due to several situations. One of these is the possibility of the E_a (the energy of the first adsorbed molecule) changing from adsorption to desorption to a higher exothermic energy. This would explain some type H3 and H4 loops, since a shift in the χ plot would show up most prominently in the classical isotherm at intermediate pressures. Why E_a would change is a challenging question. However there is a correlation between the change in ($\Delta\chi_{c,a}-\Delta\chi_{c,d}$) and ($\Delta\chi_{p,a}-\Delta\chi_{p,d}$) with $r^2=0.83$ for data by Qiao et al. [23].

There are other explanations. One unlikely explanation is that contaminates are being purged during the adsorption process. This would mean that the vacuum systems are not clean enough. This seems highly unlikely, given the reported reproducibility of the phenomenon. Swelling and shrinking have also been proposed, assuming that some reason exists that the adsorbent has a delayed reaction to shrinking. For this to happen the compression stress energy upon adsorption needs to be less than the tensile stress energy upon desorption to retain the same sequence in the hysteresis. There are probably more explanations available.

Is it microporous or mesoporous and does it matter?

This is an obvious question. What if one were to treat a microporous sample as a mesoporous sample or vice versa? Furthermore, how can one really tell if the sample is microporous or mesoporous? What precisely is the boundary between the two?

Combined mesopore/micropore equation χ

To answer these questions, a few calculations are in order. The following simulations are based upon the ideas presented previous for the analysis of microporosity and mesoporosity. These two methods can be combined into one formulation with a special interpretation for mesoporosity. Using the χ donation, the following has been postulated. For a single energy of adsorption and a single pore size and no distributions:

$$n_{ads} = n_m \left(\Delta\chi U(\Delta\chi) - q\Delta\chi_p U(\Delta\chi_p) \right) + n_p U(\Delta\chi_p) \tag{6.48}$$

where $\Delta\chi$ has the usual meaning, $\chi - \chi_c$, $\Delta\chi_p$ is the difference between the zero adsorption point and the pore-filling, i.e., $\Delta\chi_p = \chi_p - \chi_c$. q is the fractional amount that is in the pores, so that $(1-q)$ is the amount external. In this equation it is assumed that at $\Delta\chi_p$ is the position of either the micropores or the mesopores. An obvious expansion on this is to modify the equation to have both, in which case the first set would be for micropores, $\Delta\chi_{micro}$, and the second for mesopores, $\Delta\chi_{meso}$. Converting Eq. (6.48) to dimensions:

$$n_{ads} = \frac{A_s}{A} \left(\Delta\chi U(\Delta\chi) - q\Delta\chi_p U(\Delta\chi_p) \right) + \frac{V_p}{\overline{V}(\Delta\chi_p)} U(\Delta\chi_p) \tag{6.49}$$

The molar volume of the adsorbate, $\overline{V}$, could be a function of $\Delta\chi$ and as such as shown as $\overline{V}(\Delta\chi)$. This possibility will be ignored since it is usually assumed that the prefilling is fully dense liquid.

If instead of a single energy and a single pore size there are distribution of energies, $\mathbf{D}_1$, and a distribution of pore sizes, $\mathbf{D}_2$, then the delta functions that created the step function are replaced with the appropriate integral expressions:

$$n_{ads} = \int_{-\infty}^{\chi} dx \int_{-\infty}^{\chi} G\mathbf{D}_1(y, \chi_c, \sigma_c, \ldots) - H\mathbf{D}_2(y, \langle \chi_{p,2} \rangle, \sigma_2, \ldots) dy$$
$$+ J \int_{-\infty}^{\chi} \mathbf{D}_3(y, \langle \chi_{p,3} \rangle, \sigma_3, \ldots) dy \tag{6.50}$$

If there is only one pore type, then the sumscripts 1 and 2 are the same. For simplicity, given this assumption and for interpretation to geometry:

$$n_m = A_s/\overline{A} \text{ and } n_p = V_p/\overline{V}(\Delta\chi) \qquad (6.51)$$

where the parameters $G = n_m$, $H(=qG)$, and $J(V_p/\overline{V})$ have been introduced to remove the equation from interpretation. The distributions could be any arbitrary distribution. An obvious requirement for the **D**s is that the values approach zero when the value of y approaches either either $+\infty$ or $-\infty$. The question is whether anything more complicated than a PMF is justified by the precision of the data. This might be determined by the second technique below. The number of parameters in the distribution, other than the position on the standard (or χ) axis, mean value and the standard deviation, is also arbitrary. Again, beyond these three distribution parameters, the data usually does not justify more. Altogether then, Eq. (6.50) has seven parameters. If the forms of both **D**s are PMFs, then Eq. (6.50) becomes upon integration:

$$n_{ad} = G\mathbf{Z}(\chi, \langle\chi_c\rangle, \sigma_c) - H\mathbf{Z}(\chi, \langle\chi_p\rangle, \sigma_2) + \frac{J}{2}\left(1 + \text{erf}\left(\frac{\chi - \langle\chi_p\rangle}{s\sqrt{2}}\right)\right) \qquad (6.52)$$

which is simply Eq. (6.12) with an added term. The function **Z** is the same as before, that is, Eq. (6.13). Eq. (6.52) could be used as a non-interpreted fit to the isotherm. Again, as with Eq. (6.12), the simplest method of determining the parameters is a minimum search routine. This routine is the one that was used previously but to make it Eq. (6.12) J was set to 0 and the increment in J was set to 0.

Using an arbitrary function to determine porosity

The above equation should easily provide the answer for both mesoporosity and microporosity, but it has the disadvantage of a minimum search routine. On a personal computer this can be tedious. So here a method is presented to provide the answers without Eq. (6.52).

If one were to have a function that fitted the isotherm very well then one can see from Eq. (6.52) that the first and second derivatives of the data are needed. It is definitely not recommended to attempt to obtain these derivatives by differences using the data but rather to fit the entire data, either in whole or piecewise, to an arbitrary function for which the derivatives can be easily obtained. The obvious choice is a non-weighted polynomial fit. In Appendixes IIA and IIB are two routines that might work. The first one is called OPALSPA and is written in Fortran. This was used to aid the Relativistic Augmented Plain Wave calculation, and is relatively free of problems. The problem is that the standard deviations of the individual parameters and fit are not included in the routine. The deviation of the fit is a relatively easy add-on. The other is a routine that works in a spreadsheet. It is very conventional but sometimes has problems with singularities. If such a problem arises the recommended quick fix is add an off-set to the data, obtain the fit, and subtract the off-set to get the fit needed. How many parameters to use is a judgement call, so it is always a good idea to look at the fit and the standard deviations that are obtained.

Experience with this polynomial method technique is that the independent variable should be χ and the dependent variable n_{ads}, i.e., a χ plot, usually works the best. However, a n_{ads} versus $\ln(P/P_{vap})$ may work out better. The differentiation with respect to χ in the latter case is a little more difficult, but this may actually be the plot that one needs. For example, ultramicroporous materials would best fit this and the interpretation would take an entirely different direction.

Another observation is that with mesoporosity, a polynomial fit is often unstable due to the sudden increase for the pore-filling. Thus, the suggestion of a piecewise fit. Remember that it does not matter what function is used so long as the first and second derivatives can be easily and reliably obtained.

Interpretation of mesopore equation using standard curve

The interpretation of Eq. (6.52) presented here is classical in its approach and should work for any standard curve. The χ interpretation of the standard curve is used here since it has some distinct advantages.

Firstly, a practical matter. χ theory can provide some guidance for the initial estimates of the parameters. The estimates are the same as for Eq. (6.12) but the same advice concerning a graphical guidance applies. This latter method is the best way to get an initial estimate of J.

The interpretation of the parameters is basically the same for the mesopore analysis as it is for the micropore analysis with additional relationship with respect to the presence of the parameter J. Thus, Eq. (6.20) is a test for the validity of the calculation of the pore radius. Eqs. (6.14) and (6.16) yield the total surface area and the final external surface area (wall plus pore openings), respectively. Eq. (6.15) is modified by the addition of the parameter J:

$$V_p = V_m \left(H \Delta \chi_p + J \right) \tag{6.53}$$

In addition, r_p may be calculated from $\langle \chi_p \rangle$ and $\Delta \chi_p$ using the Kelvin-Cohan Eq. (6.27) and Eq. (6.23). In this case $t \neq r_p$ and is specified by Eq. (6.25). Converting to χ notation:

$$r_p = \frac{2\gamma_{gl} V_m}{RT} e^{\Delta \chi_p} + t \tag{6.54}$$

(One could solve for t in Eq. (6.25) and substitute into this equation and solve for r_p or, as a practical matter, simply leave it as is and make a circular calculation to solve for r_p.)

The above analysis, which includes the last term of Eq. (6.52), will be referred to as the "mesopore analysis." An analysis without this last term, which is identical to the analysis for microporous materials previously described, will be referred to as "micropore analysis." Essentially, the non-interpreted micropore analysis uses Eq. (6.52) without the last term and sets the pore radius, in place of Eq. (6.27) equal to t obtained from Eq. (6.23). (Simply doing this does *not* yield the same value for t as obtained from the mesopore analysis due to the interactions between the parameters in the fitting routine.)

The boundary between mesopores and micropores

Using the above equations, one could model, using the χ n-layer method, what the isotherm should look like as a function of pore size. The most sensitive representation is the standard plot representation, whereas with the use of P/P_{vap}, some details in the low-pressure portion become obscured. Fig. 6.23 shows the modeled standard plots for modeled adsorption of N_2 at 77 K on porous silica for the pore radii of 0.5, 1.0, and 1.5 nm (pore sizes 1, 2, and 3 nm). (The other input quantities are: $\chi_c = -2.8$, $\sigma_c = 0.20$, $\sigma_2 = 0.25$ thus $\sigma_p = 0.15$, percent in pores $= 95\%$, $\gamma = 8.8$ mJ m^{-2}. The surface area and χ_p were adjusted to yield the desired r_p with the pore volume held constant for scaling purposes.) The sample with pore radius of 1.5 nm is from the diagram, obviously mesoporous, and the sample with a pore radius of 0.5 nm would clearly be declared microporous. For the 1.0 nm pore radius, the answer is not so obvious, even over the full range.

A common range for measurement is indicated by the dotted box in the figure which would make the 1.0 nm sample appear very much as if no mesoporosity were present. Possibly in these cases, a plot of the "log law" would give more insight.

From this, one might well conclude that there is a continuous transition from "mesopore" to "micropore"—the quotes indicating that this is a rather artificial definition based upon judgement. The next question is, "Does it matter in the answer?"

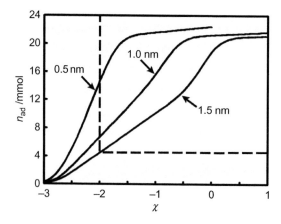

FIG. 6.23

Generated standard plot using the modeling that includes mesopores to illustration the transition from "micropores" to "mesopores."

Should one use a micropore or a mesopore analysis?

There are some complications in answering this question. Firstly, there is the question of the precision of the data. If Eq. (6.52) is to be fitted to the data, the simplest method is to run a minimum search routine comparing the equation to the data. Unfortunately, such a method can get stuck in the many false minima that even very precise data provides. Graphically adjusting the input parameters to get an approximate fit before running such a routine helps. One could also start by doing the simple fits described in Chapter 3. Secondly, if the low-pressure data are not present, leaving σ_c as a parameter is very likely to create an error. Therefore, in such cases χ_c should be set to a very low value (zero is not tolerated in the PMF) and not allowed to change. Thirdly, data that lack the low pressure points will make it difficult to separate the various parameters and large uncertainty arises in the final answer.

Nevertheless, it is instructive to attempt an analysis on some modeled data to see what happens.

In the following, the data generated for Fig. 6.23 is analyzed in four different ways:

1. All the data are used and the capillary filling part of the equation is used.
2. All the data are used but the capillary filling is ignored and r_p determined as for the micropore case.
3. The higher-pressure data (above $\chi = -2$) are used and the capillary filling part retained.
4. The higher-pressure data (above $\chi = -2$) are used and the micropore analysis is performed.

The results of this exercise for the 1.0 and 1.5 nm were very far from correct, as expected. Table 6.7 contains the results for 0.5 nm modeling. Several attempts were made with differing starting approximations which led to a large spread in the calculations for the microporous analysis assumption. Unfortunately, in the microporous analysis the fit looked graphically very good for all the fits obtained, so there does not appear to be a way to discern from the numbers which is correct. Keep in mind that this occurs using "perfect data." For experimental data, the problem must surely be worse. The mesoporous analyses work very well with the values rebounding nicely to about the same value.

The answer seems to be that it does indeed make a difference whether the mesoporous portion of the analysis is used. For "microporous" samples, it should be noted that the collapsing core is small compared to the amount already adsorbed in the pore, thus an error in the value of χ_{gl} does not lead to a large error in the answer.

Table 6.7 r_p analysis of modeled data by the two techniques and by data

"micro" all data used	"meso" all data used	"micro"—only $\chi \geq -2$ used	"meso"—only $\chi \geq -2$ used
0.39 nm → 0.48 nm	0.50 nm	0.39 nm	0.49 nm

Table 6.8 The dependence of γ_{gl} on r_p. r_p given as a percent of the original

Pore size	Percent change in γ_{gl}	
	150%	75%
0.50 nm	111%	95%
1.00 nm	111%	93%
1.50 nm	125%	88%

Of course for mesopores, χ_{gl} needs to be approximately correct. The larger the mesopores the more critical the value of γ becomes, but the possible dependence of χ_{gl} on t would not be a problem. Table 6.8 provides an analysis of the effect of changing χ_{gl} on the answer for the pore radius provided by the simulations.

The trend makes sense since the proportion of the amount in the adsorbed layer before capillary filling versus the amount of core that is filled is relatively greater for the smaller pores. This is consistent with the conclusions made above concerning the qualitative appearance of the isotherm.

Real data examples

So far, questions have been answered using simulated data, which is fine if comparisons are made. The question remains, "How well does the method work on real data?" Not much work has been performed to answer this question. Some analysis of data by Qiao, Bhatia, and Zhao [23] (QBZ) for adsorption of N_2 on MCM-41 porous materials has been successfully performed [24]. MCM-41 material has been described extensively in the literature since its discovery and development (see: Beck et al. [25], and Zhao et al. [26].) It is a regular uniform mesoporous material for which the pore size may be varied depending upon preparation. The advantage of the specific data used is that X-ray analysis of the material was performed that yielded the packing distances between pores. With an assumption about the wall thickness between the pores, the pore radius is easily calculated.

Table 6.9 gives a summary of the mesopore analysis using the above method. From the mesopore analysis and the X-ray data, the wall thickness is calculated. With the exception of the desorption data for the last data set, which is designated C-22, the wall thickness is calculated to be between 0.60 and 0.87 nm, which is fairly reasonable according to the criterion of Eq. (6.20). (The middle column is for adsorption and the last column is for desorption.) C-22 also shows hysteresis that is not explained by an energy shift proposal described below.

Figs. 6.24 and 6.25 illustrate how well the theory agrees with the data. (In Fig. 6.24 the data are stacked by adding 10 and 20 to the numbers for n_{ads} for the top two isotherms in order to provide space in the figures. Likewise in Fig. 6.25 the offset is for χ of -1 and -2 for the top two isotherms.)

Table 6.9 Mesopore analysis of the data by Qiao et al.

Sample designator	X-ray d_{100}/nm	$\Delta\chi_p$ adsorption	$d_{\chi,p}$ adsorption /nm	Wall thickness /nm	$\Delta\chi_p$ desorption	d_p desorption /nm	Wall thickness /nm
C-10	2.87	1.99	3.08	0.87	2.01	3.24	0.81
C-12	3.25	2.31	3.58	0.78	2.32	3.59	0.73
C-14	3.56	2.47	3.80	0.77	2.57	3.97	0.75
C-16	3.87	2.67	4.12	0.71	2.74	4.23	0.70
C-18	4.24	2.86	4.42	0.60	2.93	4.52	0.68
C-22	4.88	3.08	4.75	0.69	2.95	4.55	0.97

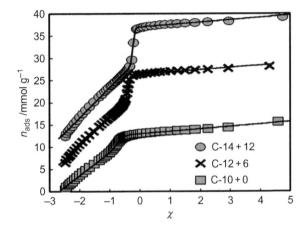

FIG. 6.24

Data and fit for N_2 adsorption on MCM-41 by QBZ (off-set by n_{ads}).

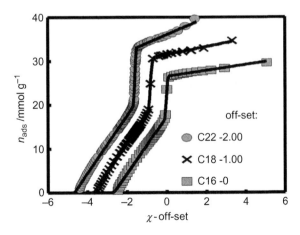

FIG. 6.25

Data and fit for N_2 adsorption on MCM-41 by QBZ (off-set by χs).

In Fig. 6.26 the individual adsorption parts of the C-10 are shown. The mesoporosity distribution is from the second derivative. It should be noticed that the mesoporosity might have been missed without this treatment. For the other MCM-41 by QBZ, there is obvious mesopore prefilling.

Fig. 6.27 illustrates two points:

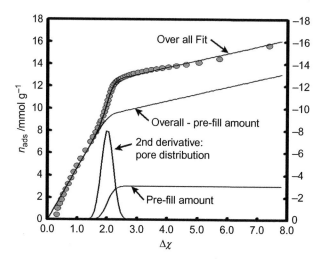

FIG. 6.26

Details of the adsorption on C-10 by QBZ.

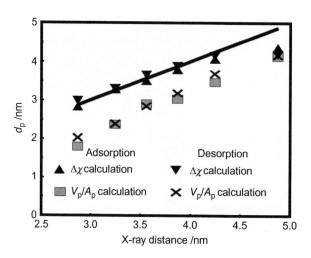

FIG. 6.27

Methods to calculate the pore "radius" versus to X-ray measurements.

1) The pore size as given by $\Delta\chi_p$ is well coordinate with the X-ray measurement. The X-ray 100 values are drawn in as a straight line for comparison. The size from the χ_p starts at about $\Delta\chi > \sim 2.5$ much as predicted.
2) The calculate pore radii from V_p/A_p is highly correlated but low compared to the X-ray.
3) All the pore radii from $\Delta\chi$ agree with the X-ray except C-22, which creates a poor correlation for this set of data.

(In Fig. 6.27 the "$\Delta\chi$ calculation" uses Eq. (6.26) and the "V_pA_p calculation" is from $r_p = 2V_p/A_p$.)

Another example of MCM-41 is the adsorption of Ar by Krug and Jaroniac [27], which is shown in Fig. 6.28. An attempt was performed using a polynomial fit to fit this data and was not successful. A polynomial up to the eighth order was used. It seems apparent that large jumps, characterized by mesoporosity are too severe a challenge. Perhaps a piecewise fit would work. However, using Eqs. (6.48)–(6.52) yielded good results.

Uninterpreted:	Interpreted:
$\chi_c = -1.987 \Rightarrow$	$E_a = 5.28$ kJ mol^{-1}
n_{ads} prepore $= 6.41$ mmol g$^{-1} \Rightarrow$	A_p total $= 550$ m^2 g^{-1} (54% of BET)
$n_p = 33.1$ mmol g$^{-1} \Rightarrow$	$V_p = 0.947$ cm^3 g^{-1}
$\Delta\chi_p = 2.51 \Rightarrow$	t_p(collapse) $= 0.910$ nm
	$2V_p/A_s \Rightarrow r_p = 3.74$ nm (as cylinders)

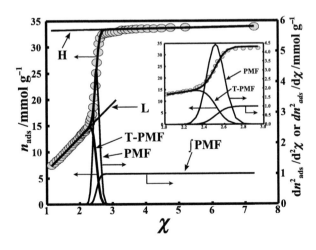

FIG. 6.28

χ plot and pore distribution of Ar adsorbed on MCM-41 and the mesopore fit.

Data by Kmg and Jaroniac.

The various parts of the fit are labeled as:

- H—line for the high slope;
- L—line for the low slope;
- PMF—pore distribution;
- T-PMF—adsorbate material remaining after pore material is subtracted; and
- ∫PMF—cumulative distribution of pore sizes.

The 2σ for the fit is 2.5% which is higher than with Qiao et al. However, it is obvious from the fit that the distribution is not quite a PMF. The PMF might be fine for practical purposes, but there needs to be further work on this topic.

The analysis revealed the following parameters:
Some speculation about the numbers is as follows:

1) The film formation started when there was enough to form a film, $2 < \Delta\chi < 3$.
2) In this case the collapse occurred when the potential thickness of the film to be formed was 25% of what was need to fill the pores entirely.
3) The pore diameter by X-ray is approximately 4–5 nm and the full pore radius, r_p, in this case is ~3.74 nm or roughly one monolayer less, as is the case with Qiao et al. However, taking into account (2), this number would be 4.7 nm.

None of these observations provides a reliable method to predict the commencement of the collapse to the dense film. The Broekhoff-deBoer theory still seems to be the best alternative. The only qualification is the requirement to have enough adsorbed to form at least a bilayer. Thus, it would seem at this point that the film cannot form before $\Delta\chi = 2.0$.

What does χ theory say about hysteresis?

As mentioned above, hysteresis is undoubtedly a real phenomenon. It has been widely reported and reproducibly observed. The Broekhoff-deBoer theory and the theories that propose a switch from 2D to 3D meniscus are capable of explaining it, although whether they calculate it properly is open to question. It is unclear whether this is an experimental problem, which is a matter of kinetics, or not. χ theory does not explain hysteresis except for the following caveat. This caveat should be taken into account for any calculation that may be attempted.

Referring to the data by Qiao et al., in the untransformed isotherm there appears to be hysteresis for nearly all the samples. However, in the analysis it should be noticed that the r_p for adsorption is nearly the same as that for desorption, indicating little or no hysteresis. Thus, in the plots of n_{ads} versus $\Delta\chi$, independently obtained for adsorption and desorption, instead of χ the adsorption data and the desorption data almost coincides (or indicate a small "negative" hysteresis). This is true for all the samples except the largest pore size sample, C-22 (which interestingly enough has a pore size just exceeding the value specified by the BdB theory where one should observe hysteresis). Even for sample C-22 in the plot n_{ads} versus $\Delta\chi$ instead of χ, the hysteresis for is considerable less—about half. The absence of hysteresis on a $\Delta\chi$ plot for all samples except C-22 and the decrease in the hysteresis for C-22 would

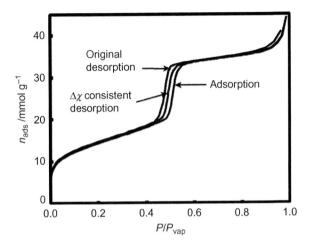

FIG. 6.29

The hysteresis loop for the data by Qiao et al. showing the original data and the postulated energy correction for the adsorption data.

indicate, in this case, that the value of E_a increases from the adsorption branch to the desorption branch. There are several explanations for this phenomenon, one of which is that the adsorption process eliminates pre-adsorbed gases which would artificially lower the adsorption energy. Adsorbed gases such as CO or H_2 are very difficult to avoid even in very good ultrahigh vacuums.

To illustrate the decreased amount of hysteresis, the data of sample C-22 for desorption is modified by shifting the P/P_{vap} value an energy amount required by the difference in χ_c of the adsorption versus the desorption. Therefore an untransformed plot of "energy corrected desorption" may be obtained to compare to the adsorption branch. The plot thus obtained along with the original adsorption and desorption data is illustrated in Fig. 6.29. Although this explains some of the hysteresis, it does not explain all of it. The use of the nearly half-power relationship mentioned with respect to Eq. (6.29), or using 1 in place of 2 in Eq. (6.27), overestimates the hysteresis by a considerable amount. It also does not explain the nearly total absence of hysteresis for the other samples. The Broekhoff-deBoer theory also considerably overestimates the magnitude of the hysteresis.

χ theory n-layer simulation

The n-layer calculation was described in Chapter 4 for a flat surface. With this information, χ theory can be used to simulate the adsorption layer-by-layer. In the section "Ultramicropore analysis" earlier in this chapter, this was taken advantage of to make calculations needed for pores that restrict the adsorption to one monolayer thickness. This yielded the often-observed "log law" from which one can extract the required parameters. There were some other calculations showing what the isotherm on the "log law" plot would look like. Figs. 6.4–6.7 demonstrated what the χ plot looks like

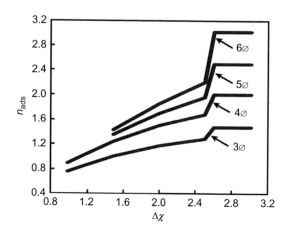

FIG. 6.30

n-layer χ simulation which show the isotherm "jump" as a function of diameter.

up to a slit separation of 10 ($r=5\phi$) layers. This is well into the mesopore region as one would expect, prefilling at about a $\Delta\chi$ of about 2.5–3. Inspection of the $\Delta\chi$ plot indicates for slits that they are filled to only 68% filled at that point. Contrast this with three layers ($r=1.5\phi$) where the pores are more than 95% filled at this point. Thus, with pores at $r=1.5\phi$, if prefilling were to start, it would be barely noticeable.

Fig. 6.30 shows some simulations using the n-layer calculation from χ theory to illustrate how the isotherm for adsorbents that have various pore sizes. It will be assumed that the prefilling starts at 2.5 and cylindrical pores are involved. The external area is assumed negligible.

It is evident that the larger the pores, the greater the jump, but, of course, the relative amount present before prefilling is also greater. The relative amounts, however, are increasing. The ratio of prefilled amount, that is, before the jump, to the final amount goes from 0.19 for 3ϕ to 0.80 for 6ϕ.

Conclusions and summary

All theories of porosity require:

- a reliable measurement of the surface area;
- a reliable standard curve against which to compare the porous materials;
- a method to determine if ultramicropores are present;
- a mechanism for the cutoff of adsorption for micropores;
- a mechanism for the enhanced adsorption for mesopores; and
- an explanation for the switch-over from microporous behavior to mesoporous behavior.

These are the classical concerns that calculations normally address and are emphasized. χ/ESW theory looks are other physical quantities as more fundamental, such as monolayer equivalence, energy of the first adsorbed molecule, where the deviations from linearity in terms of $\Delta\chi$, etc.

The classical viewpoint is usually based on the BET equation, since it is the only classical theory that reputedly yields these quantities. This is because the first requirement is normally fulfilled using the BET equation for the low-pressure adsorption. The rest of these quantities depends either directly or indirectly on obtaining this critical quantity. An exception to this is the DP theories which yield a value for microporosity. The usual procedure is to determine the standard curve and the BET surface area associated with it, and to use this standard curve to analyze the porous material. There are several problems with this:

- The linear range, deemed to yield the "correct" surface area, varies widely from adsorbate-adsorbent pairs. For some, a linear range is not possible to find. The range of 0.05–0.35 P/P_{vap} is appropriate for SiO_2 materials
- For microporous materials, the actual surface area may be much larger than the surface area determined from a BET-based standard plot. This is due to the difference between the accessible surface, i.e., that not covered by filled pores, and the real surface area under the adsorbate.
- For non-porous materials, the actual surface area is much less than what BET predicts. Often the BET answer has an error of 300% or more. Basically, it is no better than randomly placing a finger on the isotherm and reading the answer.
- There is much controversy about the validity of the BET equation as it relates to adsorption within the liquid "film" temperature range. A large number of references [28] have pointed out the weakness of the theory.
- Even if the BET yielded the correct surface area for a standard, it is very difficult to create standards that mimic exactly the surface properties of the porous material.

The alternative to the BET approach is to use the χ/ESW theory approach, with χ derived from quantum mechanics, backed up by ESW theory, derived from thermodynamics. The ESW approach is identical if one takes into account the factors presented in Chapter 3. χ theory is basically a sample-determined standard curve approach and as such could be reinterpreted in terms of any theory, for example, with the Broekhoff-deBoer theory or the Kelvin-Cohan theory. With the χ theory approach, one does not need a separate standard curve; it is internal with the theory. The principal problem with χ theory is that it has not been sufficiently tested and several aspects are still in question.

For the cutoff in adsorption for microporous materials, all theories assume the same postulate, that is, the adsorption stops when the pores are fully filled. An exception to this is the problem associated with the change in density of the adsorbate with the amount adsorbed. This problem was first discovered by Dubinin et al. [29], and seems to be predicted by χ theory [6]. If this were the case, then the values for the microporosity listed in tables of micropore radii would be low.

The enhanced filling associated with mesoporosity is dependent in all cases upon the Kelvin equation in some way. Some of the theories, such as the original Cohan formulation or the Broekhoff-deBoer theory, assume a fully formed liquid film with a sharp liquid-gas interface. Other theories, such as χ theory or the Tóth theories of adsorption, do not make this assumption. The conversion is a transition from an adsorbed state, possibly an ill-formed film, to the liquid phase due to formation of the liquid-gas interface. This can happen only in pores with some concave curvature due to energy balances. Thus, there are three approaches, and maybe more, for the capillary filling:

1. a transition from a 2D to 3D meniscus;
2. a transition from an adsorbate with no interface to either a 2D or 3D meniscus; and
3. a transition from the adsorbate with no interface to a 2D meniscus such that upon desorption the interface first encountered is a 3D interface.

For the first case, the hysteresis is explained in cylindrical pores as a 2D to 3D transition upon adsorption and a 3D to 2D transition upon desorption which yields a hysteresis. For the first case, this precludes the possibility of hysteresis by the most common explanations. For the second case, the 2D case is associated with slit pores and the 3D case is associated with cylindrical pores and the explanation for hysteresis is unknown. For the third case, cylindrical pores would be characterized with a hysteresis. Indeed adsorbent with pores that have concave surface, not necessarily circular cross-sections but elliptical etc., would according to the major theories evidence hysteresis to some extent.

Several explanations exist for the transition from the micropore to mesopore behavior in the literature. Since it appears to matter which assumption is made, it is important to have some method to decide between the two possibilities. Some postulated guidance would be as follows:

1. If the measurements are made above the critical point for the adsorptive, then no liquid-gas interface should form. This is obviously the case since $\gamma_{gl}=0$ and the micropore calculation is appropriate.
2. If no discernable positive curvature is present in the adsorption range above 0.5 monolayer, then probably only the micropore calculation is appropriate
3. If in the mesopore analysis J of Eq. (6.52) is zero (or nearly so), then mesopores are probably not present.
4. If in the mesopore analysis and χ interpretation the criterion of Eq. (6.20) is not met but is for micropore analysis, then the micropore calculation is probably appropriate.

References

[1] J. Silvestre-Albero, A.M. Silvestre-Albero, P.L. Llewellyn, F. Rodríguez-Reinoso, J. Phys. Chem. C 117 (2013)16885.
[2] S. Brunauer, L.S. Deming, W.E. Deming, E. Teller, J. Am. Chem. Soc. 60 (1938) 309.
[3] M.M. Dubinin, E.D. Zaverina, L.V. Radushkevich, Zh. Fiz. Khim. 12 (1947) 1351.
[4] M.M. Dubinin, V.A. Astakhov, Izv. Akad. Nauk SSSR, Ser. Khim. 1971 (1) (1971) 5–17.
[5] P. Rosin, E. Rammler, The laws governing the fineness of powdered coal, J. Inst. Fuel 7 (1933) 29–36.
[6] J.B. Condon, Microporous Mesoporous Mater. 38 (2000) 359.
[7] R.P. Danner, L.A. Wenzel, AIChE J 32 (1986) 1263.
[8] A. Wheeler, Catalysis, vol. 2, Reinhold, New York, 1955, p. 118.
[9] C.G. Shull, J. Am. Chem. Soc. 79 (1948) 1410.
[10] E.P. Barrett, L.G. Joyner, P.H. Halenda, J. Am. Chem. Soc. 73 (1951) 373.
[11] R.W. Cranston, F.A. Inkley, Advances in Catalysis, 9 Academic Press, New York, p. 143 1957.
[12] J.B. Condon, Langmuir 17 (2001) 3423.
[13] F. Goldmann, M. Polanyi, Phys. Chem. 132 (1928) 321.
[14] K.E. Wisneiwski, R. Wojsz, Zeolites 12 (1992) 37.
[15] L.H. Cohan, J. Am. Chem. Soc. 60 (1938) 433.
[16] A.G. Foster, Trans. Faraday Soc. 28 (1932) 645.
[17] J.C.P. Broekhoff, J.H. deBoer, J. Catal. 9 (1967) 8.
[18] P. Kowalczyk, M. Jaroniec, A.P. Terzyk, K. Kaneka, D.D. Do, Langmuir 21 (2005) 1827.
[19] M. Miyahara, H. Kanda, T. Yoshioka, M. Okazaki, Langmuir 16 (2000) 4293.
[20] H. Kanda, M. Miyahara, T. Yoshioka, M. Okazaki, Langmuir 16 (2000) 6622.
[21] W.S. Ahn, M.S. Jhon, H. Pak, S. Chang, J. Coll. Interface Sci. 33 (1972) 605.
[22] R.C. Tolman, J. Chem. Phys. 17 (1949) 333.
[23] S.Z. Qiao, S.K. Bhatia, X.S. Zhao, Microporous Mesoporous Mater. 65 (2003) 287.
[24] J.B. Condon, Microporous Mesoporous Mater. 84 (2005) 105.
[25] J.S. Beck, J.C. Vartuli, W.J. Roth, M.E. Leonowicz, C.T. Krege, K.D. Schmitt, C.T.-W. chu, D.H. Olson, E.W. Sheppard, S.B. McCullen, J.B. Higgins, J.L. Schlenker, J. Am. Chem. Soc. 114 (1992)10834.
[26] D. Zhao, Q. Huo, J. Feng, B.F. Chmelka, G.D. Stucky, J. Am. Chem. Soc. 120 (1998) 6024.
[27] M. Krug, M. Jaroniec, Microporous Mesoporous Mater. 44–45 (2001) 723–732.
[28] G. Halsey, J. Chem. Phys. 16 (1948) 931.
[29] M.M. Dubinin, E.G. Zhukovskaya, K.O. Murdmaa, Ivza. Acad. Nuak USSR, Ser. Khim. 1966 (1966) 620.

CHAPTER 7

Density functional theory (DFT)

Introduction

Density functional theory (DFT) as applied to adsorption is a classical statistical mechanic technique. For a discussion of DFT and classical statistical mechanics, with specific applications to surface problems, the textbook by Davis [1] is highly recommended. Here the more commonly used symbol for number density of $\rho(r)$ is used. Davis uses $n(r)$, so one will have to make an adjustment for this text. The calculations at the moment may be useful for modeling but are questionable for analysis with unknown surfaces. The reason for this is that the specific forces, or input parameters, required for calculations are dependent upon the atoms assumed to be present on the surface. For unknown surfaces, a reversion to the use of the BET equation is often employed.

DFT, and for that matter the Monti Carlos techniques, are methods for calculating the modeling of adsorption given certain assumptions. These assumptions usually include site-wise attractions between the surface atoms and the adsorbate molecules, and attractions between the adsorbate molecules. Interaction potentials and surface spacings are assumed. The configuration of the adsorbate molecules is adjusted to yield a minimum in the overall free energy of the system. In DFT this adjustment in the configuration is performed by adjusting the number density as a function of distance from the surface, primarily.

It is difficult to find in the literature a complete explanation of how DFT works, so in the following an attempt is made to explain the technique. There are several parts that need to come together in DFT calculations, so it may seem that the sections reviewed here are not related until they are finally compiled.

One of the assumptions of the various DFT theories is that there exists an "external potential" that has a distance decay constant of about one monolayer. This is the same assumption that the disjoining potential theory uses. It is also the assumption that Brunaur [2] scathingly criticized in reference to the deBoar-Zwikker theory. It is an assumption that seems to work but has no basis in fact, neither for the potential nor for the distance decay constant. However, the thermodynamics of the ESW does not really depend upon this assumption, but rather upon the decay of the excess surface work, regardless of the cause. The reason that this assumption works was demonstrated in the derivation of the quantum mechanical χ theory which does not assume an "external potential" or any other unexplained potential, but rather simple intermolecular forces.

The early DFT calculations assumed that the attractions between the adsorbate and adsorbent molecules were essentially local. This did not work very well. Another approach was taken, which is referred to as non-local DFT or NLDFT. This assumes that the adsorbate-adsorbent interaction operates over a range greater than one layer. This worked better, but as can be seen from the literature it did not fit the isotherm very well due to apparent layer-by-layer results. This means that for the standard curves, the fit oscillates in n_{ads} about the experimental isotherm. A further refinement is the Quenched Solid DFT (QSDFT) method. In this method, the assumption that Fuller made in his ASP theory is used to justify the potentials that are so formed to fit the isotherm with highly mobile adsorbate. This is still using a classical mechanical calculation and is difficult to justify without the QM background supplied by either χ theory of ESW theory. Thus, it reverts to the Tarazona potential hypothesis explained later in this chapter. Nevertheless, this calculation yields very good fits for the isotherm. This is the case since, as seen in the derivation of the ESW from χ, it does not matter if one uses the virtual distance-from-the-surface potential in place of the perturbed potential of the QM model. The calculations, however, are still very difficult to do and best performed with many series of prepared kemels, essentially models, to match the isotherm. This approach still seems to leave the question of what the standard material for a similar porous material should be.

A possible large advantage of the QSDFT is that the surface tensions of the three interfaces, γ_{gs}, γ_{1s}, and γ_{1g} should be amenable to calculation, thus providing the answer for many forms of the hysteresis. Obtaining these values experimentally is not for some adsorbates or adsorbents easy to obtain. Without these constants, this means that up until now, the χ or ESW calculations are not able to predict the pore filling, but only fit it to yield the pore volume and cross section.

This chapter is not intended to be a comprehensive review of the DFT methods. Indeed, an entire textbook could be written on the subject. Two reviews are recommended specifically for adsorption and porosity calculations. There are reviews by Ravikovitch et al. [3], and by Landers et al. [4] (LGN). Attempts have been made to get around the problem of needing an atlas of kemel and being able to de-convolute point-wise an otherwise ill formed problem; an integral that is central to DFT has been presented by Olivier and Neimark. It remains to be seen this is practical since it appears that it is very sensitive to the precision and accuracy of the data points.

The latest versions of DFT, the no-local DFT (NLDFT), and the (quenched solid DFT) seem to be quite promising for calculating pore filling. However, one of the problems clear from the review by Landers, Gennady, and Neimark is, so far, the inability to match the nonporous isotherm. This is illustrated by Fig. 7.1. Matching to the standard curve is also somewhat off, as seen from Fig. 7.2. There is reason to hope that these theories will improve with further research.

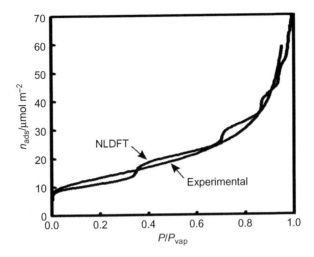

FIG. 7.1

NLDFT attempt to simulate the non-porous portion of the isotherm. Calculation by LGN.

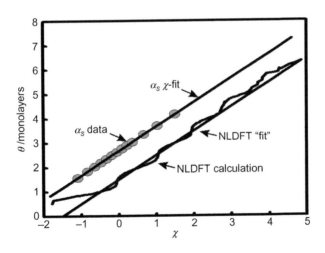

FIG. 7.2

Demonstration of the mismatch between a standard curve and the NLDFT. NLDFT by LGN.

What is a functional?

Firstly the question is; what is a functional? One may think of a functional as a function of a function. Thus, one writes $F(y(\mathbf{r}))$, where y is a function of the vector $\mathbf{r}$ and F depends upon y. One might ask what is the difference then between a function and a functional since the mapping of **r** to the final values of F is like a function. Indeed, if the function y remained constant, then F would simply be a function, but this is not necessarily the case with functionals. Take an example from quantum mechanics. The expectation value for a particular physical quantity is given by a functional that contains the wave functions in the functional. For example using 1 dimension for purposes of illustration, the energy is given by:

$$E_n = \frac{\int \psi_n^*(x)\hat{H}\psi_n(x)dx}{\int \psi_n^*(X)\psi_n(X)dx} \tag{7.1}$$

The subscripts n are more than just numbers, they change the function ψ and not simply its argument x. For another example, examine the functional:

$$F = \int_0^1 (y(x) + x)dx \tag{7.2}$$

Clearly, F has a definite value and can be determined provided y is provided. One would not however expect to get the same for different functions of y. Take the series $y = x^n$ as an example. The values obtained for F for some of the functions y are given in Table 7.1.

The function inside the functional need not be an analytic function. For example, what approximately is the average age for all Iowans given the number of people who have had their 1st, 2nd, 3rd, etc. birthday? This could be approximated by:

$$F = \sum_i i n_i / \sum_i n_i \tag{7.3}$$

where i is the last birthday passed and n_i are the number people that have obtained that ith birthday status.

Obviously, this functional will vary with time as the distribution in n changes. Thus a functional may be dependent upon an arbitrary function, even a digitally specified function, as is normally the case for the density distribution for adsorbate molecules. Notice that for the birthday functional the size of the population is not relevant in the answer, which has led to the idea of statistical sampling, i.e., given a sufficiently large random sample one can get a good estimate of F without sampling the entire population.

Table 7.1 The values of the functional, F, from Eq. (7.2) given the function y

$y = x$	$F = 1$
$y = x^2$	$F = 5/6$
$y = x^3$	$F = 3/4$
$y = x^4$	$F = 7/8$
$y = x^5$	$F = 2/3$

The challenge in modeling adsorption is first to construct a functional of number density of the adsorbed molecule that is capable of calculating the free energy of the system. The density, which is a function position, is then adjusted to minimize this free energy.

The functional derivative

One of the steps in DFT will be to find the most probable arrangement, i.e., the most probable physical distribution of the adsorbate molecules. Assuming one is able to write the free energy as a function of the distribution, then minimizing this energy by rearranging the distribution will solve the entire adsorption problem. Between the distribution of where the molecules are, referred to as the density distribution, and the free energy is the construction of a model to relate the two. For whatever model is proposed, the free energy will be a functional of the density and the minimization will require a type of derivative. This derivative is referred to as a functional derivative.

How does one minimize, or for that matter find other extrema, of a functional? Referring to the functional as $F(y(x))$ the question is how are total changes of y for the entire range of x (from a to b) going to change F. Thus the entire range of x is to be considered and if one were to select probe values of x for this, one would add these up, so:

$$F(y(x) + \varepsilon(x)) = \sum_{i \ni a < x < b} \frac{\partial F(y(x_i))}{\partial y_i} \varepsilon(x_i) \tag{7.4}$$

with ε being a small incremental change in the function y.

To form this into a continuous expression in place of a sum we can make the increments between the x_j be of constant size and change the sum to an integral. Note, the following may not be rigorous but should give a "feel" for what the functional derivative is. Define another functional $G(y, x)$ such that:

$$F = \int_a^b G(y, x) dx \tag{7.5}$$

Then:

$$F(y(x) + \varepsilon(x)) = \int_a^b G(y(x) + \varepsilon(x), x) dx \tag{7.6}$$

Expanding G:

$$F(y(x) + \varepsilon(x)) = \int_a^b \left[G(y(x), x) + \frac{\delta G}{\delta y} \delta y + \cdots \right] dx \tag{7.7}$$

where the δy is related to ε.

Evaluating the first term of the integral, which is simply $F(y(x))$, and subtracting this from the left-hand side, thus making this dF:

$$dF = \int_a^b \frac{\delta G}{\delta y} \delta y \, dx \qquad (7.8)$$

Since a limit is taken to obtain $dF (\lim \varepsilon \to 0)$ the higher terms of G may be ignored. The functional $\delta G/\delta y$ is referred to as the "functional derivative" of F and is simply given symbolism as $\delta F/\delta y$ rather than using a new letter. One important property of a functional derivative is obtained from the mathematics involve with Euler-LaGrange relationships. If F is of the form:

$$F(y(x)) = \int G(y(x), x) \, dx \qquad (7.9)$$

that is, with G being a functional only of x and y, then the functional derivative of F is readily obtained by:

$$\frac{\delta F}{\delta y(x)} = \frac{\partial G(y(x)x)}{\partial y} \qquad (7.10)$$

By setting $\delta F/\delta y$ to zero should obtain the extrema for the variation of F as a function of y as evaluated for the entire range from a to b. As with functions, whether a particular extremum is a local minimum, maximum, or a (vertical) inflection point may be determined by the second and third derivative.

The extension of the functional to higher dimensions follows the same principles. For n classical particles, one can construct a functional describing the positions and velocities of all the particles, in which case there would be $6n$ dimensions.

Correlation functions

A well-known relationship in statistics is if two sets of observations are independent of each other, then the variances are additive. However, if they are not, then there is what is referred to as a correlation between the observations. In terms of probability, this can be expressed as follows.

Given a probability that particle #1's position is at the coordinate position $\mathbf{r}_1$, i.e., $P\{\mathbf{r}_1\}$ regardless of the position of all the other particles and likewise for particle #2 at position $\mathbf{r}_2$, i.e., $P\{\mathbf{r}_2\}$, if they are independent of each other than the combined probabilities, $P\{\mathbf{r}_1, \mathbf{r}_2\}$ is equal to:

$$P\{\mathbf{r}_1, \mathbf{r}_2\} = P\{\mathbf{r}_1\} P\{\mathbf{r}_2\} \qquad (7.11)$$

If this is not true, then there is a correlation between the two probabilities designated by $g(\mathbf{r}_1, \mathbf{r}_2)$ defined as:

$$P\{\mathbf{r}_1, \mathbf{r}_2\} = P\{\mathbf{r}_1\} P\{\mathbf{r}_2\} g(\mathbf{r}_1, \mathbf{r}_2) \qquad (7.12)$$

The function g is referred to as the (two-body) "correlation function." It is convenient to define a number density by the following. The number density, $\rho\{\mathbf{r}_1\ldots\mathbf{r}_M\}$, for M particles in a system of N total particles is:

$$\rho\{\mathbf{r}_1\ldots\mathbf{r}_M\} = \frac{N!}{(N-M)!}P\{\mathbf{r}_1\ldots\mathbf{r}_M|N\} \qquad (7.13)$$

Obviously, the number density inspecting one particle is $NP\{\mathbf{r}_1\}$. Since N is normally very large then:

$$\rho\{\mathbf{r}_1,\mathbf{r}_2\} = \rho\{\mathbf{r}_1\}\rho\{\mathbf{r}_2\}g(\mathbf{r}_1,\mathbf{r}_2) \qquad (7.14)$$

The velocity components of P have no cross-terms where the velocity of one component depends directly on another particle ($\tfrac{1}{2}m_1v_1^2$, for example, has no sub 2 etc. term and the kinetic energies are additive in the exponent of P) and, therefore, cancel.

Determination of g is very important. Given the function g and the inter-particle and external potentials for the entire system in question, one may calculate all of the thermodynamic functions as well as $\rho(\mathbf{r})$.

If one were to know g for the entire system in question (including its dependence as a function of position) and the distribution molecular velocities or kinetic energies (using the Maxwell distribution since what is referred to here is classical), then all thermodynamic functions can be determined.

A quick trip through some partition functions

In calculations of statistical mechanics only the two-body correlations are important, although there may be many particles that have an influence upon a particular particle. The reason for this is simply that the forces acting upon a body from multiple directions are additive in a vector sense. Since the force is the divergence of potential energy, the calculated potential energies generated from the particles are also additive at each point in space. Or for forces, $\mathbf{F}$, and potential energies, u:

$$\mathbf{F}_j = \sum_i \mathbf{F}_{i,j} \text{ thus}: \nabla u_j = \sum_i \nabla u_{i,j} \;\therefore\; u_j = \sum_i u_{i,j} + c \qquad (7.15)$$

where c is the inevitable arbitrary offset for potential energies.

Here the index i,j indicates the quantity expected with only particle i and j are present and the index j indicates the total force or potential on the particle labeled "j." Thus, only pair-wise interactions and correlations are of importance. The above vectors $\mathbf{F}$ all have the same coordinate system. The most convenient, of course, would be with the origin at the center of particle j.

Using very simple arguments (for most, this is simply a review for orientation purposes. For a more thorough and rigorous explanation see for example Denbigh [5]) one can relate probabilities to energy states. In the following, the probability is in a general fashion related to the energy state, that is, $P\{E_j\}=f(E_j)$ where $P\{E_j\}$ is the probability of a particle being in designated state which has the energy E_j. For two isothermal bodies of constant volume that are in contact the probabilities are multiplicative. Furthermore, the first law of thermodynamics dictates that the total energy for the overall probability, $P\{E_j+E_J\}$, is simply the sum of individual energies, E_i+E_j. Thus:

$$P\{E_i\} = C_i e^{\beta E_i}, \quad P\{E_j\} = C_j e^{\beta E_j} \tag{7.16}$$

Therefore since Ps are real they must be of the form:

$$P\{E_i+E_j\} = P\{E_i\}P\{E_j\} \tag{7.17}$$

and so:

$$P\{E_i+E_j\} = C_{ij} e^{\beta(E_i+E_j)} \tag{7.18}$$

β and the Cs are arbitrary constants. Since the sum of all probabilities is 1, i.e., $\Sigma P_i = 1$, the values for the Cs are given by:

$$C = \frac{1}{\sum_i e^{\beta E_i}} \equiv \frac{1}{Q} \tag{7.19}$$

The normalizing factor Q is referred to as the "partition function." Further arguments relating these relationships to thermodynamics by analogy reveals that $\beta = -kT$ (or RT on a per mole basis).

Partition functions in general can usually be separated into separate multiplicative parts, such as rotational, vibrational, electronic, and translational. For the following discussions, the internal portions that are vibrational and electronic are being ignored and the molecules in question are assumed to be spherically symmetrical, or nearly so, so the rotational portion is also ignored.

Examining the classical notation, the probability that N particles will have the positions $\mathbf{r}_1$ for particle #1, $\mathbf{r}_2$ for particle #2, and $\cdots \mathbf{r}_N$ for particle #N, and velocities $\mathbf{v}_1$ for particle #1, $\mathbf{v}_2$ for particle #2, and $\cdots \mathbf{v}_N$ for particle #N (or more precisely #1 between $\mathbf{r}_1$ and $d^3\mathbf{r}_1$, etc.) is designated by:

$$P\{\vec{\mathbf{r}}_1\ldots\vec{\mathbf{r}}_N, \vec{\mathbf{v}}_1\ldots\vec{\mathbf{v}}_N\} = \frac{e^{\frac{-H(\vec{\mathbf{r}}_1\ldots\vec{\mathbf{r}}_N,\vec{\mathbf{v}}_1\ldots\vec{\mathbf{v}}_N)}{kT}}}{\int\cdots\int e^{\frac{-H(\vec{\mathbf{r}}_1\ldots\vec{\mathbf{r}}_N,\vec{\mathbf{v}}_1\ldots\vec{\mathbf{v}}_N)}{kT}} \prod_{i=1}^{N} d^3\vec{\mathbf{r}}_i d^3\vec{\mathbf{v}}_i} \tag{7.20}$$

where H is the total of the classical potential and kinetic energy (Hamiltonian) and the rs and vs inside the integrals are dummy variables that are integrated over all space and velocity, versus the specific rs and vs in the numerator.

The denominator of this expression is a normalizing factor so that all the singular probabilities or combinations of probabilities that would include all the particles add up to 1, the certainty probability. For the classical system, i.e., which uses continuous variation in $\mathbf{r}$ and $\mathbf{v}$ versus states, the summation for the partition function is replaced by an integral:

$$Q = \int\cdots\int e^{-H/kT} d^3\vec{\mathbf{r}}^N \quad \left(\text{defining } d^3\vec{\mathbf{r}}^N \equiv \prod_{i}^{N} d^3\vec{\mathbf{r}}_i\right) \tag{7.21}$$

In general, the total energy may be separated into velocity dependent (kinetic) and position dependent (potential) portions yielding a product in the integrals in Eq. (7.20). Thus:

$$\int \ldots \int e^{-\frac{1}{kT}\left(1/2 \sum_{i=1}^{N} m_i v_i^2 + \sum_{i=1,j=i+1}^{N} u_{ij}\right)} d^3\vec{r}^N d^3\vec{V}^N \equiv \int \ldots \int e^{-\frac{1}{kT}\left(1/2 \sum_{i=1}^{N} m_i v_i^2\right)} d^3\vec{V}^N \int \ldots \quad (7.22)$$

The second term of this equation is defined as the "configuration partition function" for N particles, Z_N:

$$Z_N \equiv \int \ldots \int e^{-\frac{1}{kT}\left(\sum_{i=1,j=i+1}^{N} u_{ij}\right)} d^3\vec{r}^N \quad (7.23)$$

In likewise fashion, one may define a configuration probability distribution function for particles 1 through S in a total population of N particles:

$$P\{\vec{r}_1 \ldots \vec{r}_S | N\} = \frac{\int_{S+1} \ldots \int_N \exp\left(-\frac{1}{kT} \sum_{i=S+1,j=i+1}^{N} u_{ij}\right) d^3\vec{r}^{N-S}}{Z_N} \quad (7.24)$$

In Eqs. (7.22) and (7.24) the factors that are obtained from the kinetic energy cancel since

$$\int \exp\left(-\frac{m v_i^2}{2kT}\right) d^3 \mathbf{v}_i = \left(\frac{2\pi kT}{m}\right)^{3/2} \quad (7.25)$$

Of interest in adsorption are systems that are open,[1] that is, where particles are able to come and go with a certain over-pressure. To take this into account, one could imagine that energy is brought in and out of the system by piggybacking on particles (again this is not rigorous; most readers are probably already familiar with the grand canonical partition function anyway). With this in mind, one modifies Eqs. (7.20) and (7.21) to add a term $N\mu/kT$ to the exponents. A similar normalizing factor to Eq. (7.22) is obtained:

$$\Xi = \int \ldots \int e^{-1/2 \sum_i m_i v_i^2 + \sum_{i,j>i} u_{in}(\vec{r}_i - \vec{r}_j) + \sum_j u_{ex}(\vec{r}_j) + \frac{N\mu}{kT}} d^3\vec{r}^N d^3\vec{V}^N \quad (7.26)$$

which is the Grand Partition Function. In this function, the potential energy due to external force has been added in using the symbol u_{ex} to distinguish it from the potential energy due to inter-particle forces, which is designated by u_{in}. The density distribution for m particles from a population of N particles in an open system is thus given by:

$$\rho\{\vec{r}_1 \ldots \vec{r}_m\} = \sum_{N=m}^{\infty} \frac{1}{(N-m)!\Xi} \int \ldots \int e^{-1/2 \sum_i m \vec{r}_i^2 + \sum_{i,j>1} u_{in}(\vec{r}_i - \vec{r}_j)}$$
$$+ \sum_i u_{ex}(\vec{r}_i) + \frac{N\mu}{kT} d^3\vec{r}_{m+1} \ldots d^3\vec{r}_N \quad (7.27)$$

---

[1] See Denbigh case (3c). Along with (3c), maximizing case (2b) was used for χ theory. (2b) avoids the need to use functional derivatives, even for the binary case.

Direct correlation functions

The equation for a non-ideal gas could be written as the ideal gas equation plus additional terms. The chemical potential could likewise be written:

$$\mu = -kT \ln\left[Q_{int}\left(\frac{2\pi MRT}{h^2}\right)^{3/2}\right] + u_{ex}(\vec{r}) - kT \ln(\rho(\vec{r})) + C^{(1)}(\vec{r}) \quad (7.28)$$

where $C^{(1)}(\mathbf{r})$ is the correction factor to the ideal gas chemical potential and is referred to as the singlet direct correlation function.

$C^{(1)}(\mathbf{r})$ is related to Eq. (7.27), which yields the direct correlation function by functional derivatives as demonstrated below. A functional φ is defined as:

$$\varphi(\vec{r}) = u_{ex}(\vec{r}) - \mu \quad (7.29)$$

Using Eq. (7.27) one can demonstrate that:

$$\rho\{\vec{r}\} = -kT \frac{\delta \ln \Xi}{\delta \varphi(\vec{r})} \quad (7.30)$$

and:

$$-kT \frac{\delta \rho\{\vec{r}\}}{\delta \varphi(\vec{r}')} = (kT)^2 \frac{\delta^2 \ln \Xi}{\delta \varphi(\vec{r}') \delta \varphi(\vec{r})} = \rho\{\vec{r}, \vec{r}''\} - \rho\{\vec{r}\}\rho\{\vec{r}'\} + \rho\{\vec{r}\}\delta(\vec{r}' - \vec{r}) \quad (7.31)$$

where the symbolism $\delta()$ indicates the Dirac delta function.

Using the functional derivative chain rules and rearranging one ends up with the following equation:

$$\frac{\delta \varphi(\vec{r})}{\delta \rho\{\vec{r}'\}} = -kT\left[\frac{\delta(\vec{r} - \vec{r}')}{\rho\{r\}} + C^{(2)}(\vec{r}, \vec{r}')\right] \quad (7.32)$$

where $C^{(2)}$ includes several ρ terms and is referred to as the "direct correlation function."

On the other hand, by taking the functional derivative of (7.27) with respect to $\rho(\mathbf{r})$ one obtains:

$$\frac{\delta \varphi(\vec{r})}{\delta \rho\{\vec{r}'\}} = -kT\left[\frac{\delta(\vec{r} - \vec{r}')}{\rho\{\vec{r}\}} + \frac{\delta C^{(1)}(\vec{r})}{\delta \rho\{\vec{r}'\}}\right] \quad (7.33)$$

so it's obvious that:

$$C^{(2)}(\vec{r}, \vec{r}') = \frac{\delta C^{(1)}(\vec{r})}{\delta j(\vec{r}')} \quad (7.34)$$

With a few mathematical manipulations and the definition of g given in Eq. (7.14), one can obtain the equation:

$$g\left(\vec{r},\vec{r}'\right) - 1 = C^{(2)}\left(\vec{r},\vec{r}'\right) + \int C^{(2)}\left(\vec{r},\vec{r}''\right)r\left(\vec{r}''\right)\left[g\left(\vec{r}',\vec{r}''\right) - 1\right]d^3\vec{r}'' \quad (7.35)$$

This equation is known as the Omstein-Zernicke equation and if one were able to solve it, all of the thermodynamic quantities, along with the density profiles etc., would be known. To do so, $C^{(2)}$, or alternatively $C^{(1)}$, is required at least as a function of the other functions in Eq. (7.35).

The Percus-Yevick approximation uses $C^{(2)} = g[1 - \exp(u_{in}/kT)]$ to obtain the physical quantities of a homogeneous fluid. Before examining this and the Carnahan-Starling approximation for hard spheres, some manipulations for 1D rods are presented in order to get a feel for the methods.

The hard rod approximations

The reason for studying the hard rod approximations is to obtain some qualitative of the consequences of various assumptions. By simplifying to one dimension, rather than three dimensions, the mathematics is simpler, albeit still messy in some places. More than one type of particle can also be included. To make the following discussion simple, only two molecular diameters will be assumed, a_1 and a_2, for species 1 and 2.

The canonical partition function is given as:

$$Q_N = \frac{\int_{x=0}^{L}\int_{x=0}^{L} e^{-\frac{u(x_1\ldots x_2)}{kT}} dx_1\ldots dx_N}{N_1!N_2!\Lambda_1\Lambda_2} \quad (7.36)$$

where

$$\Lambda_{1or2} = \left(\frac{h^2}{2\pi m_{1or2}kT}\right)^{1/2} \text{ and } \begin{array}{l} u(x) = \infty, |x| < d \\ u(x) = \infty, |x| > d \end{array} \quad (7.37)$$

and where d is the distance $(a_1+a_2)/2$, $(a_1+a_1)/2$ of $(a_2+a_2)/2$, depending upon what is appropriate and L is the length of the 1D box which contains these particles.

The integral in the numerator of this equation is the configuration partition function, Z_N. Given the conditions with respect to u, which are the hard-rod conditions, this can be simplified to:

$$Z_N = (L - N_1 a_1 - N_2 a_2)^N \quad (7.38)$$

For the open system, one needs the grand canonical partition function or:

$$\Xi = \sum_{N_1=0}^{\infty}\sum_{N_2=0}^{\infty} e^{\frac{N_1\mu_1 + N_2\mu_2}{kT}} Q_N \quad (7.39)$$

or, using the above considerations:

$$\Xi = \sum_{N_1=0}^{\infty} \sum_{N_2=0}^{\infty} \frac{e^{-\frac{N_1\mu_1 + N_2\mu_2}{kT}}}{N_1! N_2! \Lambda_1^{N_1} \Lambda_2^{N_2}} (L - N_1 a_1 - N_2 a_2)^N U(L - N_1 a_1 - N_2 a_2) \qquad (7.40)$$

The unit step function, **U**, is inserted to account for the obvious fact that the total length of the rod cannot exceed the length of the box. For each individual N_1 and N_2 the probability, $P\{N_1, N_2\}$, is:

$$P\{N_1, N_2\} = \frac{1}{\Xi} \frac{e^{-\frac{N_1\mu_1 + N_2\mu_2}{kT}}}{N_1! N_2! \Lambda_1^{N_1} \Lambda_2^{N_2}} (L - N_1 a_1 - N_2 a_2)^N U(L - N_1 a_1 - N_2 a_2) \qquad (7.41)$$

The 1D pressure,[2] P, is obtained from Ξ by analogy to the 3D case, that is the length, L, replaces the volume, so:

$$P = kT \left(\frac{\partial \ln \Xi}{\partial L} \right)_{T,\mu_1,\mu_2} \qquad (7.42)$$

or:

$$P = \frac{NkT}{L - N_1 a_1 - N_2 a_2} \quad \text{provided} \quad L > N_1 a_1 + N_2 a_2 \qquad (7.43)$$

The extension to more than two molecular species should be clear from the above by simply adding additional terms for species 3, 4, etc. It is easy to model the above on a spreadsheet. To obtain an idea of what this would look like, consider the case of only 1 species. For the probabilities as a function of the distance, L, Eq. (7.41) becomes:

$$P\{N\} = \frac{1}{\Xi} \frac{e^{-\frac{N\mu}{kT}}}{N! \Lambda^N} (L - Na)^N U(L - Na) \qquad (7.44)$$

The number density is:

$$\rho = \sum_{N=0}^{\infty} \frac{NP\{N\}}{L} \qquad (7.45)$$

and the total 1D pressure would be:

$$\rho = \sum_{N=0}^{\infty} \frac{NkT}{L - Na} \qquad (7.46)$$

From these equations, one can easily calculate the profiles of these quantities. In Figs. 7.3–7.5 are the results of these calculations. The values for Λ and kT are arbitrary and scaled for clarity. These figures represent the various quantities as a function of layer thickness and not a distance between confining walls. The extension to multiple adsorbates is obvious from the above equations but requires some little more calculations since for a total of N particles there may be several combinations for N_1 and N_2.

---

[2]Notice the difference between $P\{\}$ probability and P pressure. Probability will always have "{ }" and pressure P will never be followed by "{ }"

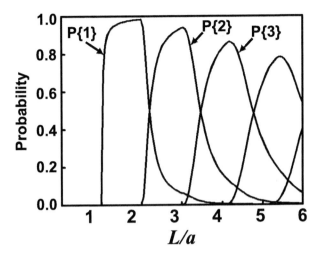

FIG. 7.3
Probabilities for the number of layers to be 1, 2, 3, etc. for the hard rod calculation as a function of layer thickness.

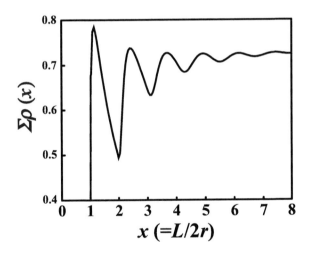

FIG. 7.4
Number density (total) for the hard rod case as a function of thickness.

Hard rods between two walls

Another relatively easy modeling is for hard rods confined between two walls. The mathematics is a little messier and will not be completely given here (again, see Davis or other statistical mechanic books). The modeling can also include an external

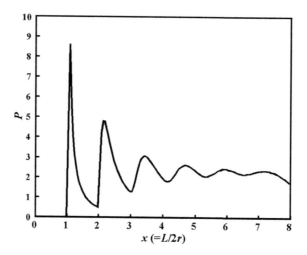

FIG. 7.5

One-dimensional pressure for the hard rods as a function of thickness.

field, which is also instructive. Using y and z for the position of the walls, Q, the canonical partition function for this case is:

$$Q_N = \frac{\int_y^z \cdots \int_y^z e^{-\sum_{i>j=1}^N \frac{u_J l(x_1 \ldots x_2)}{kT} \sum_{i=1}^N \frac{v_I(x_I)}{kT}} dx_1 \ldots dx_N}{N_1! N_2! \Lambda_1 \Lambda_2} \tag{7.47}$$

With a considerable amount of reworking, a reformulation of this is obtained in what is referred to as the p formulation. The p formulation separates the solution into two solutions, one from each wall. The solutions are:

$$p_\pm(x) = \sum_i w_i(x \pm a_i/2) e^{\mp \int_x^{x \pm a_i} p_\pm(z) dz} \tag{7.48}$$

with the function w_i for the ith size rod given by:

$$w_i(x) = \frac{e^{\frac{\mu_i + v_i(x)}{kT}}}{\Lambda_i} \tag{7.49}$$

The number density for the ith size rod is obtained from:

$$\rho_j(x) = w_i(x) e^{-\int_{-\infty}^x [p_+(z + a_i/2) - p_-(z - a_i/2)] dz} \tag{7.50}$$

The solution to these equations is rather messy because of the shifts in x that are required.

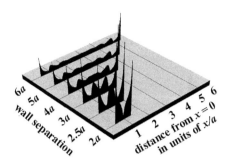

FIG. 7.6

Number density as a function of slit width for one type of rod with a length of a.

Notice that in Eq. (7.48) there is a shift from $x \pm {a_i}/{2}$ to x. Numerical techniques are obviously called for to perform this calculation. Restricting the calculation to one sized rod is relatively simple for a spreadsheet calculation. In Fig. 7.6 is a series of calculations for various slit widths (varying distance between the walls) with the chemical potential and temperature held constant and the externally imposed potential, $v(x)$, set to zero. For this calculation, one wall was held as $x = 0$ and the other wall allowed to move. Since the center of the rod cannot approach any closer to the fixed wall than the distance $a/2$, n is zero up to this point. A similar comment is in order for the wall that is allowed to move.

With the above equations it is simple to add in an external potential to see how the adsorption is affected. In Fig. 7.7 a catenary potential has been added, that is:

$$u_{\text{ex}}(x) = c\left(e^{-bx} - e^{-b(L-x)}\right) \tag{7.51}$$

As one would expect, the density is suppressed at the walls and enhanced in the middle with such a field present.

The hard rod modeling should not be taken too seriously as reflecting the situation in adsorption other than in a qualitative sense. One could start adding such features as a Leanard-Jones 6-12 potential for the walls as an external potential and add in inter-particle potentials. Such modeling does not seem to be justified for the 1D case. Firstly, molecules are not hard spheres and secondly the 1D picture is not very accurate since even the hard spheres would not line up exactly like a string of beads. It does, however, indicate that the density functional approach is at least qualitatively reasonable.

Percus-Yevick solution expansion for hard spheres

Almost all of the DFT calculations require a hard-sphere equation of state as part of the calculation. The van der Waal and other approximations have been used, but the most widely used approximation is the Carnahan-Starling (CS) approximation. The

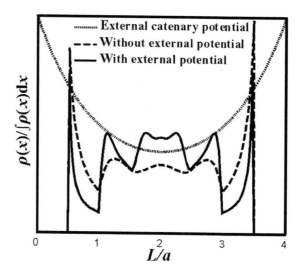

FIG. 7.7

The effect of an external field on the number distribution.

following sequence utilizes the Ornstein-Zernicke equation and makes some assumptions in order to solve the equation. The first of these assumption yields a solution by Percus and Yevick [6]:

$$\exp\left(\frac{u(\vec{r})}{kT}\right)g(\vec{r}) = 1 + \langle \rho \rangle \int \left(\exp\left(\frac{u(\vec{r}')}{kT}\right) - 1\right)g(\vec{r}')\left(1 - g(\vec{r}-\vec{r}')\right)d^3\vec{r}' \quad (7.52)$$

with u here being u_{in}.

This equation was rewritten in a form easier to solve with:

$$\tau(\vec{r}) := \exp\left(\frac{u(\vec{r})}{kT}\right)g(\vec{r}) \quad (7.53)$$

to give:

$$\tau(\vec{r}) = 1 + \langle \rho \rangle \int \tau(\vec{r}')\left(\exp\left(-\frac{u(\vec{r}')}{kT}\right) - 1\right)\left(\exp\left(-\frac{u(\vec{r}-\vec{r}')}{kT}\right)\tau(\vec{r}-\vec{r}') - 1\right)d^3\vec{r}' \quad (7.54)$$

With this equation, Percus and Yevick were able to extract various thermodynamic quantities and the virial coefficients. The virial coefficients agreed very well with the results of Monti Carlo calculations, lending credibility to the approach.

Thiele analytical approximation

Thiele [7], having noticed the precision of the Percus-Yevick equation, postulated that an exact analytical solution could be found. Starting with this equation and after considerable mathematical manipulations he arrived at the equation for pressure:

$$P = nkT \left[\frac{1 + 2z + 3z^2}{(1-z)^2} \right] \quad \text{where}: z = \frac{\pi a^3 p}{6} \tag{7.55}$$

and a is the diameter of the hard sphere.

The Carnahan-Starling approximation

An accurate reduced equation of state for the hard sphere approximation using the virial expression was determined by Ree and Hoover [8]. The first 6 virial coefficients[3] are given by:

$$\frac{PV}{nRT} = 1 + 4z + 10z^2 + 18.36z^3 + 28.2z^4 + 39.5z^5 \tag{7.56}$$

where z has the same meaning as above.

It was noticed by Carnahan and Starling [9] that the expression:

$$\frac{PV}{nRT} = 1 + 4z + 10z^2 + 18z^3 + 28z^4 + 40z^5 + \ldots \tag{7.57}$$

is very close to the virial equation shown above, and may be (Padé) approximated by:

$$\frac{PV}{nRT} = 1 + \frac{4z - 2z^2}{(1-z)^3} \tag{7.58}$$

Eq. (7.58) has been written here in the form of two terms. The first term on the right-hand side is the same as the ideal gas. One may think of the second term as a correction to the ideal gas. In Fig. 7.8 is a comparison of Eq. (7.58) with the virial equation derived by Ree and Hoover. It is apparent that this is a good approximation above a value of $V/V_m(1) \approx 2$.

Comparison to experimental data is difficult since, firstly, there is no such gas represented by hard spheres and secondly, experimental virial coefficients, even for gases such as argon, are not readily available to the fifth term. This, however, seems to be a reasonable starting point for modeling.

Notice that this does not include any attractive potential as one would add in, for example, the van der Waal equation. Some authors have added in an attractive term dependent upon the square of the molecular density to give what is referred to as Carnahan-Starling-van der Waal approximation. That is:

$$P = \rho kt \left(1 + \frac{4z - 2z^2}{(1-z)^3} - \frac{a_{vdW}\rho}{N_A} \right) \tag{7.59}$$

where a_{vdW} is the usual van der Waal constant associated with pressure.

---

[3]The units for the coefficients are not given but they are such that each term of the virial equation on the right hand side is dimensionless. Likewise, the second term of Eq. (7.58) is dimensionless overall.

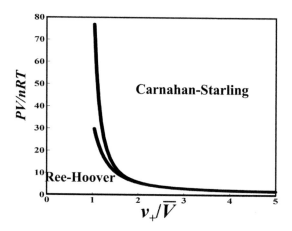

FIG. 7.8

A comparison of the Carnahan-Starling approximation with the Ree and Hoover hard sphere calculation.

Helmholtz Free Energy from the CS Approximation

There are several forms similar to Eq. (7.58) which could be used to arrive at a Helmholtz free energy. The CS form, however, is presently the most widely used. In the following the internal contributions, vibrational electronic etc., are not considered. The molar Helmholtz free energy, $\mathbf{F}$, is related to the pressure of the fluid at constant temperature whether ideal or not by:

$$d\mathbf{F}|_T = -PdV \tag{7.60}$$

where V is the molar volume.

Thus, for an ideal gas:

$$d\mathbf{F}_I|_T = -\frac{RT}{V}dV \tag{7.61}$$

Integration of Eq. (7.61) yields the Helmholtz free energy of the ideal gas with the question being what the integration constant is. This, however, is known from quantum statistical thermodynamics, i.e.:

$$\mathbf{F}_I = -RT\left[\ln\left(\frac{V}{N}\right) + \frac{3}{2}\ln\left(\frac{mkT}{2\pi\hbar^2}\right) + 1\right] \tag{7.62}$$

The second term of Eq. (7.58) may be integrated, keeping in mind that the ideal gas is applicable as the volume approaches infinity and so the total Carnahan-Starling Helmholtz free energy, A_{cs}, is:

$$\mathbf{F}_{cs} = RT\left[\ln\left(\frac{\rho(\vec{r})}{n_Q}\right) - 1 + \frac{4z^2 - 3z}{(z-1)^2}\right] \tag{7.63}$$

where n_Q is referred to as the "quantum density" (or $n_Q = 1/\Lambda^3$, where Λ is the deBrolie wave length).

Non-local density functional theory (NLDFT)

The distinction between the local density functional theory and the non-local is the assumption for the local that the fluid is structureless for purposes of calculating the long-range interactions between fluid particles. This assumption works when there are no strongly interacting boundaries but breaks down for surface adsorption. Intuitively, this seems obvious from the calculation made for the 1D hard rod case above. As seen in the figures, the walls have a considerable influence upon the number density as does the strong catenary potential. In adsorption it is common for the adsorption potential to be 5 to 10 times greater than the inter-particle potential, so the non-local assumption is called for.

In order to compensate, Nordholm et al. [10] introduced a non-local calculation based upon the van der Waal model. Percus [6] provided a general framework for the non-local density functional theory that follows these lines.

A reference Helmholtz free energy[4] and a perturbation energy are assumed to compose the overall Helmholtz free energy, so:

$$\mathbf{F} = \mathbf{F}_{ref} + \lambda \mathbf{F}_p \tag{7.64}$$

A_{ref} consists of the following parts: (1) the external field contribution, $A_{external}$, (2) the ideal gas contribution, A_I, and (3) an excess free energy functional term, A_{excess}. $A_{external}$ is given by:

$$\mathbf{F}_{external} = \int \rho(\vec{r}) u_{external}(\vec{r}) d^3\vec{r} \tag{7.65}$$

and A_I is given by Eq. (7.62) above. The free energy approach developed by Tarazona [11] and Evans [12] has been the most successful modeling approach so far. In this modeling, the excess free energy term is obtained by using a smoothed density functional. This is given by:

$$\mathbf{F}_{excess} = \int \rho(\vec{r}) \Delta\psi(\bar{\rho}(\vec{r})) d^3\vec{r} \tag{7.66}$$

where $\bar{\rho}$ is a smoothed density function.

From the derivative of pressure with respect to volume from the Carnahan-Starling equation (Eq. 7.58 above) one has for $\Delta\psi$:

$$\Delta\psi(\rho) = \frac{kT(4z - 3z^2)}{(1-z)^2} \quad \left(Z = \frac{\pi a^3 \rho}{6} \right) \tag{7.67}$$

---

[4]The standard IUPAC Helmholtz free energy symbol used here is A. Many physics paper use the symbol "F" for this. I have also expanded the subscripts to be more descriptive.

The smoothed density functional, $\bar{\rho}$, is expanded to a quadratic series in order to make the homogeneous fluid match the Percus-Yevick using the expression:

$$\bar{\rho}(\mathbf{r}) = \sum_{i=0}^{2} (\bar{\rho})^i \int W_i(\vec{\mathbf{r}} - \mathbf{r}') \mathbf{r}(\vec{\mathbf{r}}') d^3\vec{\mathbf{r}}' \qquad (7.68)$$

where the w functions are referred to as weighting functions.

The conditions for a homogeneous fluid also require:

$$\int w_0(\vec{\mathbf{r}} - \vec{\mathbf{r}}') d^3\vec{\mathbf{r}} = 1 \text{ and}$$
$$\int w_{1or2}(\vec{\mathbf{r}} - \vec{\mathbf{r}}') d^3\vec{\mathbf{r}} = 0 \qquad (7.69)$$

The functions w_0 through w_2 are evaluated as a function of $\mathbf{r}/a$. The weighting function w_0 for the homogeneous fluid is given simply as:

$$w_0 = \frac{3}{4\pi a^3} \quad |\mathbf{r}| < a \qquad (7.70)$$
$$w_0 = 0 \quad |\mathbf{r}| > a$$

which fulfills the first integral of Eq. (7.69) (reflecting that the volume of a hard sphere is simply $4\pi a^3/3$). Using only this weighting function yields a generalized van der Waal modeling. Thus, the higher powers in the smoothed density are the more subtle (but important) corrections to the vdW approach.

In order to obtain w_1 and w_2 the following strategy is used.

1. The direct correlation, defined in Eq. (7.34), is related to the excess free energy, Eq. (7.66) by:

$$C(\vec{\mathbf{r}}, \vec{\mathbf{r}}') = -\frac{1}{kT} \frac{\delta \delta \int \rho(\vec{\mathbf{r}}) \Delta \psi(\bar{\rho}(\vec{\mathbf{r}})) d^3 \vec{\mathbf{r}}}{\delta \rho(\vec{\mathbf{r}}) \delta \rho(\vec{\mathbf{r}}')} \qquad (7.71)$$

so that evaluating the functional derivatives for the homogenous fluid with a density of ρ_0:

$$-kTC(\vec{\mathbf{r}} - \vec{\mathbf{r}}') = \frac{\partial \Delta \psi(\rho_0)}{\partial \rho(\mathbf{r})} \left[\frac{\delta \bar{\rho}(\vec{\mathbf{r}})}{\delta \bar{\rho}(\vec{\mathbf{r}}')} \right]_{\rho_0} + \rho_0 \int \frac{\delta^2 \bar{\rho}(\vec{\mathbf{r}}'')}{\delta \rho(\vec{\mathbf{r}}) \delta \rho(\vec{\mathbf{r}}')} d^3 \vec{\mathbf{r}}''$$

$$- \frac{\partial^2 \Delta \psi(\rho_0) \rho_0}{\partial \rho \partial \rho} \int \left. \frac{\delta \bar{\rho}(\vec{\mathbf{r}}'')}{\delta \rho(\vec{\mathbf{r}})} \right|_{\rho_0} \left. \frac{\delta \bar{\rho}(\vec{\mathbf{r}}'')}{\delta \rho(\vec{\mathbf{r}}')} \right|_{\rho_0} d^3 \vec{\mathbf{r}} \qquad (7.72)$$

2. The function $\Delta \psi$ is obtained from the Carnahan-Starling approximation or, as recommended by Tarazana, from the original virial expansion expression for the hard sphere. The derivatives are then easily obtained.

3. The derivatives of $\bar{\rho}$ are obtained from Eq. (7.67) with w being expanded into a power series in ρ:

$$w(\vec{r}, \rho) = w_0(\vec{r}) + w_1(\vec{r})\rho + w_2(\vec{r})\rho^2 + \ldots \tag{7.73}$$

The terms beyond ρ^2 are assumed to be small.

4. This information is substituted back into $\bar{\rho}$, Eq. (7.67), and into the derivatives of $\bar{\rho}$.
5. $\rho, \Delta\psi$ and their derivatives are then substituted into the direct correlation function thus getting a power series for C.
6. The power series for C must agree with the direct correlation function results from the Percus-Yevick calculation for the homogeneous fluid over a large range. Thus, a match is made to obtain the appropriate functions for the w_is for Eq. (7.73). These functions are available in either Davis' book or in the original article by Tarazona.

Modeling with the presence of a surface

The presence of a surface is modeled with an external potential simulating the solid surface. The external field portion is typically modeled as an infinitely high potential hard wall or, with more sophistication, a Lennard-Jones potential. The former model, used by Tarazona, can be used for a slit pore with the conditions:

$$u_{ex}(x) = \infty \quad x < \frac{a}{2} \quad \text{and} \quad x > L\frac{a}{2}$$
$$u_{ex}(x) = 0 \quad \frac{a}{2} < x < L\frac{a}{2} \tag{7.74}$$

Using only the x direction the weighing function coefficients are appropriately adjusted. This condition is equivalent to making $\rho = 0$ when $x \leq 0$ or $x \geq L$. Thus, the integrals may end a 0 and L.

The results of the calculation are in excellent agreement with Monte Carlo calculations, which require considerably more computer computations. In Fig. 7.9 is the results of this calculation from one side of a hard-wall slit with a comparison with the expected results from χ theory. The χ theory calculation used a harmonic oscillator approximation to the LJ 6-12 potential:

$$E_{LJ} = \varepsilon_{LJ}\left[\mathbf{r}_{LJ}^{12}\mathbf{r}^{-12} - 2\mathbf{r}_{LJ}^{6}\mathbf{r}^{-6}\right] \tag{7.75}$$

to calculate the normal direction profile and the position is the center of the adsorbate molecule. The χ theory calculation is broader and deeper than the Monte Carlo calculation, whereas the DFT calculation is almost indistinguishable from the Monte Carlo calculation.

It is desirable to replace the hard-wall assumption, Eq. (7.74), with a wall potential. This potential could be a detailed Lennard-Jones 6-12 potential (see the first

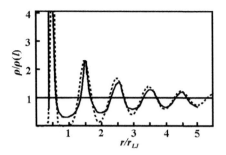

FIG. 7.9

Results of the NLDFT calculation by Tarazona (solid line) and results of harmonic oscillator approximation from χ theory (dashed).

equation in Appendix II H) or possibly an average-type potential such as the Steele [13] 10-4-3 potential. This modifies $A_{external}$ accordingly.

The part left for inclusion is A_p in Eq. (7.64). For this perturbation, the interparticle forces are normally chosen to be a Lennard-Jones 6-12 potential. For with the perturbation fully in force, that is $\lambda = 1$:

$$F_p = \frac{1}{2}\int \rho(\vec{r})\rho(\vec{r}')g(\vec{r},\vec{r}')E_{LJ}(|\vec{r}-\vec{r}'|)d\vec{r}\,d\vec{r}' \tag{7.76}$$

where $|\mathbf{r}-\mathbf{r}'|$ in the Lennard-Jones potential, E_{LJ}, is the distance between the centers of the molecules, that is r of Lennard-Jones 6–12 equation in Appendix II H.

The chemical potential of the adsorbent/adsorbate is given by the derivative of the Helmholtz free energy with respect to the number density:

$$\mu = \frac{\delta F}{\delta \rho(\vec{r})} \tag{7.77}$$

Combining all the parts of A together (substituting the L-J 6–12 Eq. (4.67) into (7.76), substituting (7.67) into (7.66), and adding these to (7.62) and (7.65)) and differentiating with respect to $\rho(\mathbf{r})$ one obtains the relationship between the density profile and μ (on a number basis):

$$\mu = kT\ln\left(\Lambda^3\rho(\vec{r})\right) + \Delta\psi\left(\rho(\vec{r})\right) + \int \rho(\mathbf{r})\frac{\delta\Delta\psi\left(\bar{\rho}(\vec{r})\right)}{\delta\bar{\rho}}\delta\bar{\rho}\delta\rho d\vec{r}\,'$$
$$+ \int \rho(\vec{r}\,')E_{LJ}(|\vec{r}-\vec{r}\,'|)d\vec{r}\,' + u_{external}(\mathbf{r}) \tag{7.78}$$

The smoothing and g are those obtained by using Tarazona's weighting functions. Since the Carnahan-Starling formulation is used for $\Delta\psi$, one obtains two roots for each value of μ, one corresponding to the adsorbate and one for the adsorbent. Needless to say, solving this equation requires successive approximations for ρ for each μ specified. The amount adsorbed is then obtained by integrating the profile from the surface to a large value.

Advances on this technique for use in analyzing adsorption isotherms and porosity measurements are being pursued with some encouraging results. For example,

Olivier [14] has made calculations for the adsorption of argon and nitrogen on carbon materials. The initial results indicated a somewhat stepped isotherm that followed the experimental isotherm reasonably. The initial assumption, however, was that the correlation function, g, was the same as the homogeneous liquid regardless of the location with respect to the surface. This assumption intuitively would seem to be an oversimplification. Although one is not normally interested in the depth profiles in adsorption experiments, the calculation requires the accurate accounting for the profile due to the integration of the profile. In order to correct the profile, Olivier introduced an additional weighting function, which depends upon distance for the surface. This weighting function compensates for the postulated variation of g with respect to distance. This weighting function varies with distance from the surface and should be characteristic of the adsorbate.

Therefore, once one has this standard weighting curve for a particular adsorbate, it should apply to all isotherms. Excellent fits to the isotherms are obtained using this technique. This technique has been incorporated in some of the instruments that measure the isotherm.

There are alarge number of calculations, modifications, and explanations for NLDFT as applied to physisorption and porosity measurements. Some are mentioned here for a starting reference. Sokolowski and Fischer [15] calculated adsorption on capillary filling for generic pores, which gave insight into the observed isotherms. More specifically, adsorption on real mesoporous materials and comparison to the experimental isotherms are available. Many MCM-41 porous materials were calculated and the pore size distributions determined using NLDFT by Ravikovitch et al. [3] with fair agreement to experiment. Ravikovitch and Neimark [16, 17] have used NLDFT to calculate the surface area and porosity of the zeolite materials, designated as SBA, which have larger pores. The explanation of hysteresis appears to be within reach [16, 18, 19] with the calculation of the metastable and equilibrium branches of adsorption in mesopores.

It is anticipated that NLDFT will be very useful in the future, especially if it were combined with QM considerations.

A note about the Monte Carlo technique

The Monte Carlo technique is a brute force method to simulate how molecules might behave. For thermodynamics, the important part that is effected by this technique is the entropy terms. Roughly speaking, a geometry is selected, and energy for "binding" is selected and then molecules are allowed to interact, in this case, with the surface. There is a certain chance that the molecule will "stick" or leave depending upon energy constraints and the density and kinetic energy of the molecules.

The technique makes many simulations in a attempt to simulate the experimentally observed behavior. It is well known by those that use this technique that at least a hundred molecules are required and many steps with each molecule to come close to yielding a realistic answer. Thus, this technique requires many computations. Given that at the beginning of the calculation there is no way, within the technique, to specify an adsorption energy or a hint of the surface area, the two parameters needed to make a calculation, it is obvious that an extreme number of calculations are required.

There does not appear that any of the presently produced instruments provides this as an option for the calculations.

χ theory versus NLDFT, what are the practical differences?

Looking at the comparison results between the χ theory and the NLDFT theory, there doesn't seem to be any practical difference between the two. They certainly seem to be very different mathematically using different base theories to make the derivations. χ theory is most reasonably based on quantum mechanics, whereas NLDFT is based on classical fluid mechanics. However, χ theory can be based on classical mechanic as well using statistical assumptions about adsorption. Furthermore, there are some derivations for DFT which need to be based on quantum mechanics, for example, the well-known derivations of the ground states in molecular structure.

One of the differences is the simplicity of the χ theory compared to NLDFT. This, however, is not the most important difference. The most important difference is the starting point. The sections "Strategy of the Non-Local Density Functional Theory" and "Strategy of the χ/ESW theory" in chapter 4 are some of the important differences between the two methods. These sections describe in general the method that is followed by each of the methods.

One of the primary disadvantages of the NLDFT method is not mentioned in these tables, which is, the technique has difficulty determining the surface area of the flat surface. The relationship between the surface area and the adsorbate as well as the energy of adsorption, liquid-solid attraction, are critical for the calculation. Thus either a very imprecise value determined by trial and error is used or a simulate material non-porous material is used. In the latter case, the BET or a standard curve based on BET, is often used, which, of course, is a critical error.

NLDFT starts with a physical model with presumed pore sizes, etc. Usually the surface area is determined by the BET theory, which is frankly wrong. It is strongly advised not to use the BET theory since it yields the wrong answer for the surface area. This is not, however, necessary, but is the case in most of the literature. Sometimes other porosity determinations, such as those by Dubinin, et al. are the starting point, which also has problems. Sometimes the surface area is calculated using the disjoining pressure theory, which should yield the correct answer. There are other instances where an attempt is made to used the NLDFT to calculate the early part of the isotherm to get the surface area. This method, which yields an oscillatory answer, has large value for the uncertainty.[5]

---

[5]There are several publications that attempt to use the NLDFT to determine the surface area from the first part of the isotherm and to extract the liquid-gas parameters. The uncertainty referred to here is due to the oscillating nature of the fit about a smooth curve and the very large 2σ that results. Even so, it is better than using the BET.

With a model created, the results are the matched to the experimental results, thus starting with one of these isotherms on a simulated sample:

- the BET area, or
- the Dubinin porosity answer
- or the disjoining pressure answer

One has a starting point. Changes are then made in:

- the forces between the adsorbate and adsorbent
- the external surface area
- the pore sizes

in order to make a match. Inversion programs have been proposed to simplify the process.

The NLDFT calculations are somewhat complicated requiring computer programs to perform, and often rely on antiquated theories, such as the BET theory. In most cases, the calculations become a "black box" arrangement where the operator must accept the results blindly. There is very little way to check the validity of the results.

The χ theory method is quite different. It starts with the experimental data; usually the isotherm is measured, although the calorimetric data could be used. The data is then analyzed with a relatively simple theory to obtain the surface area and the micropore and mesopore sizes.

χ theory calculations are very simple especially for nonporous materials. Even with porous materials, most of the time, simple linear fitting to portions of the χ plot yields the important information. For a more sophisticated treatment with high resolution data, including low pressure data,[6] energy distributions and pore size distribution are obtained with a good polynomial least square root routine. χ theory does not depend upon any other theory to start and the energy calculations from calorimetry and the isotherm treatment are consistent within the χ theory. This latter point is important since no other theory has demonstrated that the extracted energies from the isotherm and from calorimetry meet the consistency requirement.

So, what is the conclusion? Firstly, they both probably could yield the correct results. If this is the case, what should one do? Here is a suggestion. First extract the adsorption energy, surface area, pore sizes, and the distributions of energy and pore sizes from χ theory. Then input these as parameters into the NLDFT to see if there are consistent results. If not, then determine what might be incorrect. For example, in χ theory, a very low energy plane could be interpreted as mesopore and possibly the NLDFT might pick this up. At any rate, any instrument with installed programs should include both theories.

---

[6] In any case, obtaining the high vacuum if not ultra high vacuum data is highly recommended.

References

[1] H. Ted Davis, Statistical Mechanics of Phases, Interfaces, and Thin Films, VCH Publishers, Inc, New York, NY, 1995. ISBN 1-56081-513-2.
[2] S. Brunauer, The Adsorption of Gases and Vapors, vol. 1, Princeton University Press, Princeton, NJ, 1945.
[3] P.I. Ravikovitch, G.L. Haller, A.V. Neimark, Adv. Colloid Interface Sci. 76-77 (1998) 203–226.
[4] J. Landers, Y.G. Gennady, V. Neimark, Colloids Suf. A Physicochem. Eng. Asp. 437 (2013) 3–32.
[5] K. Denbigh, The Principles of Chemical Equilibrium, third ed., Cambridge University Press, UK, 1971. Chapter 11.
[6] J.K. Percus, G.J. Yevick, Phys. Rev. 110 (1958) 1.
[7] E. Thiele, J. Chem. Phys. 39 (1963) 474.
[8] F.H. Ree, W.G. Hoover, J. Chem. Phys. 40 (1964) 939.
[9] N.F. Carnahan, K.E. Starling, J. Chem. Phys. 51 (1969) 635.
[10] S. Nordholm, M. Johnson, B.C. Fraesier, Aust. J. Chem. 33 (1980) 2139.
[11] P. Tarazona, Phys. Rev. A 31 (1985) 2672.
[12] P. Tarazona, R. Evans, Mol. Phys. 52 (1984) 847.
[13] W.A. Steele, Surf. Sci. 36 (1973) 317.
[14] J.P. Olivier, J. Porous. Mater. 2 (1995) 9.
[15] S. Sokolowski, J. Fischer, J. Chem. Soc. Faraday Trans. 89 (1993) 789.
[16] P.I. Ravikovitch, A.V. Neimark, Langmuir 18 (2002) 1550.
[17] P.I. Ravikovitch, A.V. Neimark, J. Phys. Chem. B 105 (2001) 6817.
[18] A.V. Neimark, P.I. Ravikovitch, Microporous Mesoporous Mater. 44-45 (2001) 697.
[19] P.I. Ravikovitch, A. Vishayakov, A.V. Neimark, Phy. Rev. E 64 (2001) 011602.

Appendixes

Appendix I: Equipment specifications

Requests were sent out to all the manufacturers that were in the first edition of this book or that according to an internet search were new. Some other sources were consulted, but no other new manufacturers were given. Several had been bought out by larger instrument companies and apparently were not manufacturing machines for the discipline. Only POROTEC answered the request for information, and from personal knowledge it is known that Micromeritics is still manufacturing machines.

Listed below are those manufacturers that are confirmed to be manufacturing or probably are. The websites are in English, but many have other languages available.

Dr. Juergen Adolphs, CEO
POROTEC Vertrieb von wissenschaftlichen Geräten GmbH
Niederhofheimer Str. 55A
65719 Hofheim am Taunus
Deutschland
website: https://www.porotec.de/en.html

Micromeritics Instrument Corporation
4356 Communications Drive
Norcross, GA 30093-2901
Website: https://www.micromeritics.com/

Quantachrome Instruments
1900 Corporate Drive
Boynton Beach, Florida 33426
website: https://www.quantachrome.com/

Particle Analytical
Agern Allé 3
2970 Hørsholm
Denmark
website: http://particle.dk/

BeiShiDe Instrument Technology (Beijing) Co., Ltd.
607, Floor 6, Bldg. 1, Yard 1, Shangdi 10th St., Haidian Area, Beijing
China
website: http://www.best17.cn/en/index.asp

Surface Measurement Systems Ltd
5 Wharfside, Rosemont Road
Alperton, Middlesex, HA0 4PE
United Kingdom
website: https://www.surfacemeasurementsystems.com/

3P INSTRUMENTS GmbH & Co. KG
Rudolf-Diesel-Str. 12
85235 Odelzhausen
Germany
website: https://www.3p-instruments.com/

Ms. Jecci Zhang, Director/CEO/General Manager
Xian Yima Optoelec Co., Ltd.
Room 1005, Building B, Western Electronic Community, Yanta District, Xi'an
Shaanxi
China (Mainland)
website: contact is through Alibaba: https://www.alibaba.com/

Ken McKeown, President
CrystalTek Corporation
2645 Financial Ct. Suite O
San Diego, CA 92117
USA
website: http://crystaltekcorp.com

Colnatec
625 N. Gilbert Road, Suite 205
Gilbert, AZ 85234
USA
website: http://colnatec.com

Appendix II: Computation programs

The following are programs for calculating polynomial fits. It has been found that the fit using polynomials is easier and better using the χ transformed abscissa, that is, $-\ln(-\ln(P/P_{vap}))$ as the x axis and using n_{ads} for the y axis.

II-A: OPLSPA—A fast least squares routine

This first program is a Fortran program that was developed in the 1960s at Iowa State University and is listed in a book by Professor Terry Louchs [1]. The program has some advantages. Firstly, it is more robust than the straight least squares routines. Secondly, one can obtain an answer for the fit to a low power polynomial and, if it is insufficient, one can use the TUWYLO (take up where you left off) feature to get a higher-order polynomial. It has the disadvantage that neither the standard deviations of the fit nor of the parameters is in the output, but must be calculated with a different routine.

```
      SUBROUTINE OPLSPA(NDEG,NPTS,X,Y,W,Q,TUWYLO)
C
C         POLYNOMIAL FITTING PROGRAM
C
          DIMENSION K(1), X(1), W(1)
          DOUBLE PRECISION Q(1), PN(1), PN(10), SUM(4), B, C, PNX, TMP
   1      N=0
          C=0
          PN(1)=1,5
          GO TO 6
   2      C=-SUM(3)/SUM(4)
          B=-SUM(1)/SUM(3)
          N=N+1
          PN1(N)=0.
          PN1(N+1)=0.
          DO 4   J=1,N
          TMP=PN(J)
          PN(J)=8*PN(J)+C*PN1(J)
   4      PN1(J)=TMP
          DO 5   J=1,N
   5      PN(J+1)=PN(J+1)+PN1(J)
   6      DO 7   K=1,3
   7      SUM(K)=0.0
          DO 11  K=1,NPTS
          PNX=1.0
          J=N
   8      IF(J)    10,10,9
   9      PNX=PN(J)+PNX*X(K)
          J=J-1
          GO TO 8
  10      SUM(1)=SUM(1)+W(K)*X(K)*PNX*PNX
          SUM(2)=SUM(2)+W(K)*Y(K)*PNX
  11      SUM(3)=SUM(3)+W(K)*PNX*PNX
          Q(N+1)=SUM(2)/SUM(3)
          IF(N)    3,3,12
  12      DO 13  J=1,N
  13      Q(J)=Q(J)+Q(N+1)*PN(J)
          IF(N-DEG)     2,14,14
  14      RETURN
          END
```

II-B: Least squares routine in a spread sheet

In the following pages is a conventional polynomial least squares routine using a spreadsheet where the polynomial desired is specified to begin with. In this case, the polynomial selected is a eighth power polynomial. To change to a different power, the spreadsheet needs to be modified. The matrixes used are laid out, but some spreadsheets can use arrays within arrays to do the same thing.

In these table listings, one should leave the " "s off. They are needed here so the codes will show and not make a calculation.

The word (element) indicates that the space should be empty at the start and will be filled by the matrix operation.

Some notes from the spreadsheet

The following were too long to put in the spaces provided:

If the spreadsheet can have array functions within array functions, you could use the @sums shown here

(1) +I$30+B4*I$31+B4^2*I$32+B4^3*I$33+B4^4*I$34+B4^5*I$35+B4^6*I$36
+B4^7*I$37+B4^8*I$38

OR

@SUM(I$30..I$38*B4^H$30..H$38)

(2) +J$31*B4^G$31+J$32*B4^G$32+J$33*B4^G$33+J$34*B4^G$34+J$35*B4^G$35
+J$36*B4^G$36+J$37*B4^G$37+J$38*B4^G$38

OR

@SUM(J$31..J$38*B4^G$31..G$38)

(3) +K$32*B4^F$32+K$33*B4^F$33+K$34*B4^F$34+K$35*B4^F$35+K$36*B4^F$36
+K$37*B4^F$37+K$38*B4^F$38

OR

@SUM(K$32..K$38*B4^F$32..F$38)

The spreadsheet may also have the capacity to have array functions within array functions for matrixes as well, eliminating the need to do all the above @ARRAY steps. For example, one could use in place of

@ARRAY(@MMULT(E16..M24,E30..E38)) in I30 with
@ARRAY(@MMULT(@ARRAY(@MINVERSE(E4..M12)),(ARRAY(@TRANSPOSE
(E27..M27)))).

This could be further compacted, but this will become a little confusing and hard to follow. Laying out the numbers can reveal where mistakes have been made. This is especially true when looking at something like the residuals.

In this table listing, one should leave the " "s off. They are needed here so the codes will show and not make a calculation.

Appendixes 365

	A	B	C	D	E	F	...	M	N
1		From "Raw Data" adsorption			0	1	...	8	9
2		x		y				S-MATRIX.....	
3	P/P_{vap}	$-\ln-\ln(P/P_{vap})$	V /mL	n_{ads} /mmol					
4	DATA	-@LN(-@LN(A4))	DATA	"=C4/22.4"	@SUM(($B4..$B64)^E1)	@SUM(($B4..$B64)^F1)	...	@SUM(($B4..$B64)^M1)	@SUM(($B4..$B64)^F1)
5	.	-@LN(-@LN(A5))	.	"=C5/22.4"	"=F4"	"=G4"	...		
6	.	.	.	.	"=F5"	"=G5"	...		
7	.	.	.	.	"=F6"	"=G6"	...		
...					.	.	...		
10	.	.	.	.	"=F10"	"=G10"	...	"=N9"	"=O9"
11	.	.	.	.	"=F11"	"=G11"	...	"=N10"	0
12							...	0	0
63	DATA	-@LN(-@LN(A63))	DATA	"=C5/22.4"					

Table continued (row numbers repeated here):

	O	P	...	S	T	U
1	10	11	...	14	15	16
2				S-MATRIX.....		
3						
4	@SUM(($B4..$B64)^O1)	@SUM(($B4..$B64)^P1)	...	@SUM(($B4..$B64)^S1)	@SUM(($B4..$B64)^S1)	@SUM(($B4..$B64)^S1)
5	"=P4"	"=G4"	...	"=T4"	"=U4"	0
6	"=P5"	"=G5"	...	"=T4"	0	0
7	"=P6"	"=G6"	...	0	0	.
...	.	.	...	.	.	.
10	"=P9"	0	...	0	0	0
11	0	0	...	0	0	0
12	0	0	...	0	0	0

	A	B	C	D	E	F	G	H	I	J	K	L	M
15	DATA	.	DATA	.					inverse s - matrix				
16					@ARRAY(@MINVERSE(E4..M12))	(element)							(element)
.		.		.		.		.		.		.	
.		.		.		.		.		.		.	
24		.		.	(element)								(element)
25													
26									Sums yx^2				
27		.		.	@SUM(($D4..$D63)*($B4..$B63)^E1)								@SUM(($D4..$D63)*($B4..$B63)^M1)
28		.		.									
29		.		.	b matrix								

In this table the columns J and K are the 1st derivative and the 2nd derivative coefficients respectively.

	A	B	C	D	E	F	G	H	I	J	K	L	M
					b vector				a-matrix	1st	2nd	σ	% error
29								0	@ARRAY(@MMULT(E16..M24,E30..E38))	(blank)	(blank)	"+Y$3/C70"	"+L30/I30"
30	@ARRAY(@TRANSPOSE(E27..M27))												
31	.	.	.	.	(element)		0	1		"+I31*H"	(blank)	"+Y$3/C71"	"+L30/I30"
32	.	.	.	.			1	2		"+I32*H"	"+J32*G32"	"+Y$3/C72"	"+L30/I30"
33	.	.	.	.	.		2	3	.		"+J33*G33"	"+Y$3/C73"	
34	.	.	.	.	.		3	4	.			.	.
35	.	.	.	.	.		4	5	.			.	.
36	.	.	.	.	.		5	6	.			.	.
37	.	.	.	.			6	7					
38	.	.	.	.	(element)		7	8	(element)	"+I38*H"	"+J38*G38"	"+Y$3/C70"	"+L38/I38"

Notice that in the case presented as a model, there were 60 data points. If there are fewer, then one should change all the B63s and D63s to the appropriate end data line number.

	A	B	C	D	E	BF	BG	BH
62	DATA	-@LN(-@LN(A62))	DATA	"=C62/22.4"	.....			
63	DATA	-@LN(-@LN(A63))	DATA	"=C63/22.4"	.....			
64	Data end - more if B63 and D63 are changed throughout							
65	1	2	3	4	counter	58	59	60
66	X:transpose							
67	@ARRAY(@TRANSPOSE(B4..B63))	(element)	(element)	.	.....	.	(element)	(element)
68								
69								
70	0	@ARRAY(@MMULT((A$67..BH$67)^A70,(B$4..B$63)^A70))	@SQRT(B70)					
71	1	@ARRAY(@MMULT((A$67..BH$67)^A71,(B$4..B$63)^A71))	@SQRT(B71)					
72	2	.	.					
73	3	.	.					
74	4	.	.					
75	5							
76	6							
77	7							
78	8	@ARRAY(@MMULT((A$67..BH$67)^A78,(B$4..B$63)^A71))	@SQRT(B78)					

	X	Y	Z	AA	AB	AC	AD
1	n cal.	$(n_{\text{data}} - n_{\text{cal}})^2$	residuals	residuals	$\chi - \chi_c$	$dn_{\text{tab}}/d\chi$	
2		Sum LS↓		max n_{tab}	for ploting*		
3		@sqrt(@sum(Y4..Y63))					
4	See (1)	"(+D4-X4)^2"	"+D4-X4"	"+Z4/D63"	"+B4 -" χ_c	See(2)	See(3)
5	"sub. B5 for B4"	"(+D5-X5)^2"	"+D5-X5"	"+Z5/D63"	.	"sub. B5 for B4"	"sub. B5 for B4"
6	"sub. B6 for B4"	"(+D6-X6)^2"	"+D5-X5"	"+Z6/D63"	.	"sub. B6 for B4"	"sub. B6 for B4"
7	.	.	.	.	.	.	.
63	"sub. B63 for B4"	"(+D63-X63)^2"	"+D63-X63"	"+Z63/D63"	"+B63 -" χ_c	"sub. B63 for B4"	"sub. B63 for B4"

*provided all points are above extrapolated χ_c.

II-C: Algebra of binary adsorption

For cases where the pressures are specified, Eq. 4.146 can be rearranged to yield two equations with two unknowns:

$$kT\ln\left(\frac{P_1}{P_{\text{vap},1}X_1}\right) + \Delta\varepsilon X_2^2 = \left(E_{a,1} + \frac{E_{a,2}\overline{A}_1}{\overline{A}_2}\right)\exp(-\theta_1 - \theta_2) - \left(\frac{E_{a,2}\overline{A}_1}{\overline{A}_2}\right)\exp(-\theta_1) \quad (A.1)$$

The second equation also has the indexes 1 and 2 switched. For simplicity, the following substitutions are made[1]:

$$V_1 = kT\ln\left(\frac{P_1}{P_{\text{vap},1}X_1}\right) + \Delta\varepsilon X_2^2 \quad W_1 = E_{a,1} + \frac{\overline{A}_1}{\overline{A}_1}E_{a,2} \quad Z_1 = \frac{\overline{A}_1}{\overline{A}_1}E_{a,2} \quad (A.2)$$

$$x = \exp(-\theta_1) \quad y = \exp(-\theta_2)$$

so that:

$$W_1 xy - Z_1 x = V_1 \quad \text{and} \quad W_2 xy - Z_2 y = V_2 \quad (A.3)$$

From the V_1 equation, solve for y:

$$x = \frac{V_1}{W_1 y - Z_1} \quad (A.4)$$

Substitute this into the equation with V_2:

$$\frac{W_2 V_1}{W_1 y - Z_1} y - Z_2 y = V_2 \quad (A.5)$$

Rearranging, we get:

$$-W_1 Z_2 y^2 + (W_2 V_1 + Z_1 Z_2 - V_2 W_1) y + V_2 Z_1 = 0 \quad (A.6)$$

and solve the resultant quadratic equation:

$$y = \frac{(W_2 V_1 + Z_1 Z_2 - V_2 W_1) \pm \sqrt{(W_2 V_1 + Z_1 Z_2 - V_2 W_1)^2 + 4W_1 Z_2 V_2 Z_1}}{2 W_1 Z_2} \quad (A.7)$$

Notice that:

$$\exp(-\theta_1)\exp(-\theta_2) \leq \exp(-\theta_1)\exp(-\theta_1)\exp(-\theta_2) \leq \exp(-\theta_2) \quad (A.8)$$

and are only equal when both are zero. So long as something is adsorbed:

$$x \geq xy \leq y \quad (A.9)$$

II-D: Differential equations for binary adsorption

It is common for the mathematics of the carrier flow system to inspect the first derivatives for the isotherm equation for the mixed adsorbents. Usually, the Langmuir isotherm is used for this purpose. However, any reasonably accurate isotherm in

---

[1] Remember that the Es are negative. Therefore $W_n = -|W_n|$ and $Z_n = -|Z_n|$.

the n_{abs} versus P/P_{vap} which is a good representation may be used. This would include purely empirical equations. Here the mathematics for the χ representation are presented.

For the first derivative of pressure f or the single adsorbate and the binary adsorbates, the following symbols will be used for easier reading:

$$a_{12} := a_1/a_2, \, a_{21} := a_2/a_1 \\ E_1^* := E_1/RT, \, E_2^* := E_2/RT \\ P_1^* = P_1/P_{1,vap}, \, P_2^* = P_2/P_{2,vap} \tag{A.10}$$

For a single adsorbate, the original χ equation can be written as follows:

$$P_1^* = \exp[-\exp(-\theta_1 + \chi_c)] \quad (\theta_1 \geq 0 = P_1^* \geq P_{1,c}^*) \tag{A.11}$$

This equation can an easily be differentiated with the chain rule. However, for the binary it is easier to perform this in the following fashion. Notice that:

$$\frac{\partial \ln P_1^*}{\partial \theta_1} \equiv \frac{1}{P_1^*} \frac{\partial P_1^*}{\partial \theta_1} \quad \therefore \quad \frac{\partial P_1^*}{\partial \theta_1} = P_1^* \frac{\partial \ln P_1^*}{\partial \theta_1} \tag{A.12}$$

and that:

$$\frac{\partial \ln P_1^*}{\partial \theta_1} = +\exp(-\theta_1 + \chi_c) \tag{A.13}$$

Thus:

$$\frac{\partial P_1}{\partial \theta_1} = +\exp(-\theta_1 + \chi_c)\exp[\exp(-\theta_1 + \chi_c)] \tag{A.14}$$

It is common practice simply to use the separate single isotherms, slightly modified, for the other adsorbate and to divide the two equations. If this is done in the present case, one would get:

$$\frac{\partial P_1}{\partial P_2} = +\frac{\exp(-\theta_1 - \theta_2 + \chi_{1,c})\exp[\exp(-\theta_1 - \theta_2 + \chi_{1,c})]}{\exp(-\theta_1 - \theta_2 + \chi_{2,c})\exp[\exp(-\theta_1 - \theta_2 + \chi_{2,c})]} \tag{A.15}$$

However, this is too great an oversimplification. Ignoring the regular solution part of the binary derivation, one can get some insight by looking at the θ portions. The regular solution portion, or any other formulation including the ideal (entropy of solution), can be incorporated into the pressure terms.

$$\ln P_1^* = (E_1^* + a_{12}E_2^*)\exp(-\theta_1 - \theta_2) - a_{12}E_2^*\exp(-\theta_1) \\ \ln P_2^* = (E_2^* + a_{21}E_1^*)\exp(-\theta_1 - \theta_2) - a_{21}E_1^*\exp(-\theta_2) \tag{A.16}$$

following #1 since #2 is identical except for the subscripts (remember that the Es are exothermic or <0):

$$P_1^* = \exp\left[(E_1^* + aE_2^*)\exp(-\theta_1 - \theta_2) - aE_2^*\exp(-\theta_1)\right] \tag{A.17}$$

and:

$$\frac{\partial \ln P_1^*}{\partial \theta_1} = -(E_1^* + aE_2^*)\exp(-\theta_1 - \theta_2) + aE_2^*\exp(-\theta_1) \tag{A.18}$$

then multiplying:

$$\frac{\partial P_1^*}{\partial \theta_1} = -\left[(E_1^* + a_{12}E_2^*)\exp(-\theta_1 - \theta_2) - a_{12}E_2^*\exp(-\theta_1)\right] \quad \text{(A.19)}$$
$$\times \exp\left[(E_1^* + a_{12}E_2^*)\exp(-\theta_1 - \theta_2) - a_{12}E_2^*\exp(-\theta_1)\right]$$

to make it look a little more like the derivation for the single adsorbate:

$$\chi_{c,12} := \ln\left(|E_1^* + a_{12}E_2^*|\right) \leq \chi_{c,1} \quad \chi_{c,2} = \ln\left(|E_2^*|\right) \quad \text{(A.20)}$$

Therefore:

$$\frac{\partial P_1^*}{\partial \theta_1} = -\left[\exp(-\theta_1 - \theta_2 + \chi_{c,12}) - \exp(-\theta_1 + \chi_{c,2})\right] \quad \text{(A.21)}$$
$$\times \exp\left[\exp(-\theta_1 - \theta_2 + \chi_{c,12}) - \exp(-\theta_1 + \chi_{c,2})\right]$$

This equation only holds for pressures above the threshold pressures for both adsorbates. Notice that as $\theta_2 \to 0$, this equation approaches Eq. (A.14), which it obviously should.

Combining the equation for #1 and #2, one gets:

$$\frac{\partial P_1^*}{\partial P_2^*} = \frac{-\left[\exp(-\theta_1 - \theta_2 + \chi_{c,12}) - \exp(-\theta_1 + \chi_{c,2})\right]\exp\left[\exp(-\theta_1 - \theta_2 + \chi_{c,12}) - \exp(-\theta_1 + \chi_{c,2})\right]}{-\left[\exp(-\theta_1 - \theta_2 + \chi_{c,21}) - \exp(-\theta_2 + \chi_{c,1})\right]\exp\left[\exp(-\theta_1 - \theta_2 + \chi_{c,21}) - \exp(-\theta_2 + \chi_{c,1})\right]}$$
(A.22)

This ratio is very important for determinating how adsorbates behave in the carrier flow system. The disadvantage of this equation is that it looks very messy. However, the advantage is that there are no derivatives of θ involved on the left-hand side. If the individual isotherms are measured and the $\chi_{c,1}$ and $\chi_{c,2}$ are calculated, then a complete mapping, as a 2D matrix, of the ratio for the left side of Eq. (A.22) is possible with very little computer power.

II-E: Interconverting for binary solutions and gas mixtures

The math for interconverting mole fractions in the gas and liquid phases is given here for your convenience. The following symbols are used:

$X_{1 or 2}$:= mole fraction of component 1 or 2 in solution ($:=X(l)$)
$Y_{1 or 2}$:= mole fraction of component 1 or 2 in the gas mixture ($:=X(g)$)
$P_{1 or 2}^*$:= the vapor pressure of component 1 or 2 over the pure liquid
P_{total}^* := the total vapor pressure over the **pure** liquid (A.23)
$p_{1 or 2}^*$:= the partial vapor pressure of component 1 or 2 over the **solution**
$p_{1 or 2}$:= the partial pressure of component 1 or 2 at total pressures than P_{total}^*
α := the Margules polynomial — it will be understood that is attaches to Xs

The laws relating the adsorptive vapor pressure to the equilibrium vapor pressure over the bulk liquid solution is given by modifying Raoult's law:

$$p_1^* = X_1 e^{\alpha X_2^2} P_1^* \quad \text{and} \quad p_2^* = X_2 e^{\alpha X_1^2} P_2^* \quad \text{(A.24)}$$

In this equation, the regular solution approximation is given with the Margules constant shown. For the rest of the mathematics, this modification will be left off. The deviation from Raoult's law can be added in with the modification of the Xs.

The total vapor pressure for the gas phase is then given by addition:
$$P^*_{\text{total}} = p^*_1 + p^*_2 \approx X_1 P^*_1 + X_2 P^*_2 \tag{A.25}$$

The mole fractions of the gases are then given by the equation:
$$Y_1 = \frac{X_1 P^*_1}{P^*_{\text{total}}}, Y_2 = \frac{X_2 P^*_2}{P^*_{\text{total}}} \quad \text{or} \quad Y_1 = \frac{p_1}{X_1 P^*_1 + X_2 P^*_2}, Y_2 = \frac{p_2}{X_1 P^*_1 + X_2 P^*_2} \tag{A.26}$$

Noticing that $Y_1 + Y_2 = 1$, with some simple steps one can solve for the Xs in terms of Ys:

$$X_1 = \frac{Y_1 P^*_2}{P^*_1 + Y_1 P^*_2 - Y_1 P^*_1} \equiv \frac{Y_1 P^*_2}{Y_1 P^*_2 + Y_2 P^*_1}, \quad X_2 = \frac{Y_2 P^*_1}{P^*_2 + Y_2 P^*_1 - Y_2 P^*_2} \equiv \frac{Y_2 P^*_1}{Y_2 P^*_1 + Y_1 P^*_2} \tag{A.27}$$

It should be observed that Eq. (A.27) for the conversion from gas mole fractions is not as straightforward as the mole fraction conversion from the solution to the gas mole fractions. Thus, the pressure terms in the χ equations which include mole fraction of solution attached to the pressure term are slightly more complex than one might expect.

For Example, introduction of the Margules constant (that is, assuming regular solution for the adsorptives) into the calculations involving the data by Arnold improves the standard deviation by a factor of 4 (12.5 versus 3.2). However, there are some problems with the data and the information available at the time of the experiments, so the additional complications of going beyond the ideal solution do not seem justified.

Data are sometimes presented as p/p^* since the total pressure is not the vapor pressure over the solution. A similar relationship to Eq. (A.24) holds for ps:
$$p_1 = X_1 e^{\alpha X_2^2} P_1 \text{ and } p_2 = X_2 e^{\alpha X_1^2} P_2 \tag{A.28}$$

Thus, at constant X:
$$\frac{p_1}{p^*_1} = \frac{P_1}{P^*_1}, \quad \frac{p_2}{p^*_2} = \frac{P_2}{P^*_2} \tag{A.29}$$

However, as mentioned in the section on transforming χ_c, the energies of adsorption are not affected by a change in the pressure reference. Therefore, either the observed relative pressure must be modified or χ_c needs to be modified. The original $\Delta\chi$ is given by:

$$\theta_1 = -\ln\left\{-\ln\left(\frac{p_1}{P^*_1}\right) + \ln\left(-\frac{\overline{E}_{a,1}}{RT}\right)\right\} := -\ln\left\{-\ln\left(\frac{p_1}{P^*_1}\right) + \ln\left(\frac{P_{c,1}}{P^*_1}\right)\right\} \tag{A.30}$$

The modification starts by recognizing the original definition of χ_c and that its values were obtained from the pure liquid adsorption. Thus, one could take this into account by substituting in from the formula relating p^*_1 to P^*_1, Eq. (A.24):

$$\theta_{1\text{or}2} = -\ln\left\{-\ln\left(\frac{p_{1\text{or}2}}{P^*_{1\text{or}2}}\right) - \ln\left(X_{1\text{ or }2} e^{\alpha X_{2\text{ or }1}^2}\right) + \ln\left(\frac{P_{c,1\text{or}2}}{P^*_{1\text{or}2}}\right)\right\} \tag{A.31}$$

or:

$$\theta_{1\text{or}2} = -\ln\left\{-\ln\left(\frac{p_{1\text{or}2}}{P^*_{1\text{or}2}}\right) + \ln\left(\frac{P_{c,1\text{or}2}}{P^*_{1\text{or}2} X_{1\text{or}2} e^{\alpha X_{2\text{or}1}^2}}\right)\right\}$$

plus a similar equation for component 2. If the mole fraction is unknown at all the pressures, then one could use Eq. (A.27) in place of X_{1or2}.

Another function that has been used is:

$$F = \ln\left(\frac{p_1 P_2^*}{p_2 P_1^*}\right) \equiv \ln\left(\frac{X_1 P_1 P_2^*}{X_2 P_2 P_1^*}\right) \quad \text{(A.32)}$$

but for a fixed mixed gas the ratios $P_1/P_2 = P_1^*/P_2^*$, thus:

$$F = \ln\left(\frac{X_1}{X_2}\right) \quad \text{(A.33)}$$

II-F: Surface liquid versus surface gas density

A rough calculation comparing the surface liquid phase to the surface gas phase utilizes the Clausius-Clapeyron concept of translating energy into PV work. A typical adsorption energy would be about $8\,\text{kJ}\,\text{mol}^{-1}$. Thus, the compressive work on a gas approaching the surface would be, at the most, this value. In reality, this energy is available only very near the surface, so this calculation is very conservative.

Assuming nitrogen adsorption, the temperature of adsorption is about 78 K. Converting the $\text{kJ}\,\text{mol}^{-1}$ to atmospheres-liters mol^{-1}, one uses a factor of $9.87\,\text{L}\,\text{atm} = 1\,\text{kJ}$.

$$E = 8\,\text{kJ} \times 9.87\,\text{L}\,\text{atm}\,\text{kJ}^{-1} = 7.9\,\text{L}\,\text{atm}$$

For adsorption experiments, there is almost always measurable adsorption at 1 torr pressure or $\sim 1.3 \times 10^{-3}$ atm. For nitrogen, the gas molar density is given by the ideal gas law:

$$\frac{n}{V} = \frac{P}{RT} \quad \text{(A.34)}$$

Thus the molar density is:

$$\frac{n}{V} \approx \frac{1.3 \times 10^{-3}\,\text{atm}}{0.08206\,\text{L}\,\text{atm}\,\text{mol}^{-1}\,\text{K}^{-1} \times 78\,\text{K}} = 2.1 \times 10^{-4}\,\text{mol}\,\text{L}^{-1} \quad \text{(A.35)}$$

Using the molar mass of nitrogen of $28\,\text{g}\,\text{mol}^{-1}$ and converting liters to cm^{-3}, one obtains the density of gaseous N_2 at 78 K using a compression energy of 8 kJ as:

$$\rho(g, \text{compressed}) = 5.9 \times 10^{-6}\,\text{g}\,\text{cm}^{-3} \quad \text{(A.36)}$$

This should then be compared to the liquid nitrogen density of $0.808\,\text{g}\,\text{cm}^{-1}$. The ratio of these is then:

$$\frac{\rho(g, \text{compressed})}{\rho(l)} = \frac{5.9 \times 10^{-6}\,\text{g}\,\text{cm}^{-3}}{0.808\,\text{g}\,\text{cm}^{-3}} \approx 7.3 \times 10^{-6} \quad \text{(A.37)}$$

Using the 2/3 power to give the 2D densities, one can see that the compressed gas is less dense by a factor of 3.7×10^{-4} compared to the surface liquid. In other words, the compressed surface gas takes up $\sim 2.7 \times 10^3$ as much space as the surface liquid.

Another way of looking at this is if the instrumentation can just barely detect adsorption at 1 torr, then the surface gas is definitely not detectable, if it exists.

II-G: The error function, etc.

The error function (**erf**) is encountered in several places. Here is some relevant information that is handy to have on hand:

The definition:

$$\mathbf{erf}(x) := \frac{1}{\pi} \int_{-x}^{+x} e^{-t^2} dt = \frac{2}{\pi} \int_{0}^{+x} e^{-t^2} dt \qquad (A.38)$$

For:

$$\mathbf{f}(x) = \left(\frac{c}{\pi}\right)^{1/2} e^{-cx^2} \qquad (A.39)$$

then:

$$\left(\frac{c}{\pi}\right)^{1/2} \int_{a}^{b} e^{-cx^2} dx = \frac{1}{2}\left(\mathbf{erf}(b\sqrt{c}) - \mathbf{erf}(a\sqrt{c})\right) \qquad (A.40)$$

For the derivative:

$$\frac{d}{dx}\mathbf{erf}(x) = \frac{2}{\sqrt{\pi}} e^{x^2} \qquad (A.41)$$

and for the integral:

$$\int \mathbf{erf}(x) dx = x\,\mathbf{erf}(x) + \frac{1}{\pi} e^{-x^2} \qquad (A.42)$$

For the normal distribution:

$$\mathbf{f}(x|\mu, \sigma^2) = \frac{1}{\sqrt{2\pi\sigma^2}} \exp\left(\frac{(x-\mu)^2}{2\sigma^2}\right) \qquad (A.43)$$

For the cumulative normal distribution:

$$\Phi(x|\mu, \sigma^2) = \int_{-\infty}^{x} \frac{1}{\sqrt{2\pi\sigma^2}} \exp\left(\frac{(t-\mu)^2}{2\sigma^2}\right) dt \qquad (A.44)$$

$$\Phi(x|\mu, \sigma^2) = \frac{1}{2}\left(1 + \mathbf{erf}\left(\frac{x-\mu}{\sigma\sqrt{2}}\right)\right) \qquad (A.45)$$

II-H: Lennard-Jones vs Harmonic Oscillator

To make the Harmonic Oscillator (HO) and Lennard Jones (LJ) 6–12 potentials match to yield at least the ground state, the following is performed. The minimums in the two curves are matched, i.e., the minimum of the curves must match, thus the first derivatives with respect to r are made equal. The curvatures at the minimum must match, thus the second derivatives at the minimum are made equal.

One form of the LJ 6–12 potential is:

$$E_{LJ} = \varepsilon_{LJ}\left[r_{LJ}^{12}r^{-12} - 2r_{LJ}^{6}r^{-6}\right] \quad (A.46)$$

and the HO equation is:

$$E_{HO} = \tfrac{1}{2}k(r_{min} - r)^2 - E_{min} \quad (A.47)$$

The first and second derivative of the LJ 6–12 are:

$$1\text{st}: \frac{dE_{LJ}}{dr} = \varepsilon_{LJ}\left[-12r_{LJ}^{12}r^{-13} + 12r_{LJ}^{6}r^{-7}\right] \quad (A.48)$$

and:

$$2\text{nd}: \frac{d^2E_{LJ}}{dr^2} = \varepsilon_{LJ}\left[156r_{LJ}^{12}r^{-14} - 84r_{LJ}^{6}r^{-8}\right] \quad (A.49)$$

For the HO, the first and second derivatives are:

$$1\text{st}: \frac{dE_{HO}}{dr} = -k(r_{min} - r) \quad (A.50)$$

and:

$$2\text{nd}: \frac{d^2E_{HO}}{dr^2} = k \quad (A.51)$$

To make the first derivation of the LJ 6–12 equal to 0, one needs to set $r = r_{LJ}$. Likewise, to make the first derivative of the HO, one needs to set $r = r_{min}$. Thus,

$$r_{LJ} = r \Rightarrow \frac{dE_{LJ}}{dr} = 0 \text{ and } r_{min} = r \Rightarrow \frac{dE_{HO}}{dr} = 0 \Rightarrow r_{min} = r_{LJ} \quad (A.52)$$

Setting r in the second derivatives equal:

$$\varepsilon_{LJ}\left[156r_{LJ}^{12}r^{-14} - 84r_{LJ}^{6}r^{-8}\right] = +k \quad (A.53)$$

and substituting to r_{min} for r:

$$\varepsilon_{LJ}\left[156r_{min}^{-2} - 84r_{min}^{-2}\right] = k \Rightarrow k = \varepsilon_{LJ}\frac{72}{r_{min}^2} \quad (A.54)$$

The remaining question is what is E_{min}? To find this, substitute back into the original equations:

$$\varepsilon_{LJ}\left[r_{LJ}^{12}r_{LJ}^{-12} - 2r_{LJ}^{6}r_{LJ}^{-6}\right] = \tfrac{1}{2}k(r_{min} - r_{min})^2 - E_{min} \quad (A.55)$$

Therefore,

$$E_{min} = \varepsilon_{LJ} \quad (A.56)$$

Note that both are exothermic and therefore negative by the convention used here (the normal thermodynamic convention).

II-I: Normal distribution and quantum oscillator relationship

Given the above relationship between LJ 6–12 and the HO, one can more easily derive the approximate ground state wave function for an adsorbate molecule.

The quantum mechanical (QM) equation for the harmonic oscillator is:

$$\hat{H} = \frac{\hat{p}^2}{2m} + \frac{1}{2}m\omega^2(r - r_{min})^2 \tag{A.57}$$

The second term corresponds to the potential given in Eq. (A.47). Thus, to be used later:

$$m\omega^2 \equiv k_{HO} \Rightarrow m\omega = \sqrt{mk_{HO}} \tag{A.58}$$

From this equation, the eigenvalues are:

$$E_n = \hbar\omega\left(n + \frac{1}{2}\right) \Rightarrow E_0 = {}^1\!/_2\hbar\omega \Rightarrow \omega = \frac{2E_0}{\hbar} \tag{A.59}$$

and the wave functions are:

$$\psi_n = \frac{1}{\sqrt{2^n n!}}\left(\frac{m\omega}{\pi\hbar}\right)^{\!1/4} \exp\left(-\frac{m\omega x^2}{2\hbar}\right) H_n\!\left(\sqrt{\frac{m\omega}{\hbar}}x\right) \quad n = 0, 1, 2\ldots \tag{A.60}$$

with:

$$x \equiv (r - r_{min}) \tag{A.61}$$

and:

$$H_n\!\left(\sqrt{\frac{m\omega}{\hbar}}x\right) \tag{A.62}$$

are the Hermite polynomials (the physics form). We will assume that the temperature of the sample is low enough that the ground state is overwhelmingly predominant. So:

$$n = 0 \Rightarrow \psi_0 = \left(\frac{m\omega}{\pi\hbar}\right)^{\!1/4} \exp\left(-\frac{m\omega x^2}{2\hbar}\right) \tag{A.63}$$

The distribution in QM terms is:

$$\mathbf{P}(x) \equiv \psi_0^* \psi_0 = \left(\frac{m\omega}{\pi\hbar}\right)^{\!1/2} \exp\left(-\frac{m\omega x^2}{\hbar}\right) \tag{A.64}$$

It would be very convenient to convert this into a normal distribution mass function. This function is the one provided by spreadsheets. This is accomplished by the following substitution:

$$\sigma = \sqrt{\frac{\hbar}{2m\omega}} \Rightarrow \frac{1}{2\sigma^2} = \frac{m\omega}{\hbar} \tag{A.65}$$

which yields:

$$\left(\frac{m\omega}{\pi\hbar}\right)^{\!1/2} = \frac{1}{\sigma\sqrt{2\pi}} \tag{A.66}$$

as the normalizing factor. Thus this equation becomes as follows:

$$P(r - r_{min}) = \frac{1}{\sigma\sqrt{2\pi}} \exp\left(-\frac{(r - r_{min})^2}{2\sigma^2}\right) \qquad (A.67)$$

which is the normal distribution. For the integrals, the cumulative normal distribution may be used, which is very handy for calculating the isotherm.

II-J: Algebra of the inflection point method

Using the quantum mechanical calculation, a simple formula may be obtained for the monolayer equivalence for a nonporous material. This utilizes the inflection point in the isotherm, which is always present for a nonporous material. The technique also reveals whether a material is porous or nonporous.

1) The first question is: where is the inflection point?

Using the χ equation, this is simple to calculate by taking the first and second derivatives. Here the designations $P^* := P/P_{vap}$ and $n := n_{ads}$ or as a function $\mathbf{n}$ are used. The starting equation is then more conveniently written as:

$$\theta = -\ln(-\ln(P^*)) - \chi_c \qquad (A.68)$$

and θ is written depending upon the function $\mathbf{n}(P^*)$:

$$\theta = \frac{\mathbf{n}(P^*)}{n_m} \qquad (A.69)$$

Differentiating once, one obtains from Eq. (A.68):

$$\frac{\partial \mathbf{n}}{\partial P^*} = \frac{-n_m}{P^* \ln(P^*)} \qquad (A.70)$$

and differentiating twice:

$$\frac{\partial^2 \mathbf{n}}{\partial P^*} = n_m \left(\frac{1}{[\ln(P^*)]^2} + \frac{1}{\ln(P^*)}\right) \qquad (A.71)$$

Setting this second derivative to zero to find the inflection point pressure designated, P_I^* yields:

$$\frac{\partial^2 \theta}{\partial P^*} = 0 \Rightarrow \ln(P_I^*) = -1 \qquad (A.72)$$

so that for the inflection point:

$$P_I^* = \exp(-1) \equiv e^{-1} \cong 0.3678... \Rightarrow \ln(P_I^*) = -1 \text{ and } P_I^* \ln(P_I^*) \equiv -e^{-1} \qquad (A.73)$$

2) The next question is: why would one want to know the inflection point?

The inflection point information can yield the monolayer equivalence and the value for χ_c or the energy of adsorption of the first adsorbate molecule. This is found by the following:

Substituting into Eq. (A.70) the slope at the inflection point is:

$$\frac{\partial \mathbf{n}(P_I^*)}{\partial P^*} = n_m e \qquad (A.74)$$

To find the tangent line to the inflection point, the intercept b in the equation is calculated at the inflection point:

$$\mathbf{n}(P^*) = n_m e P^* + b \quad \text{specifically} \Rightarrow b = \mathbf{n}(P_I^*) - n_m e P_I^* \qquad (A.75)$$

so:

$$\mathbf{n}(P^*) = n_m e P^* + \mathbf{n}(P_I^*) - n_m e P_I^* \qquad (A.76)$$

but also from Eq. (A.68):

$$\mathbf{n}(P_I^*) = n_m(-\ln[-\ln(P_I^*)] - \chi_c) \qquad (A.77)$$

and substituting and evaluating $-\ln(P_I^*)$:

$$\ln(P_I^*) = n_m(-\ln[1] - \chi_c) \equiv -n_m \chi_c \qquad (A.78)$$

thus:

$$b = n_m(-\chi_c) - n_m e P_I^* \equiv n_m(-\chi_c - 1) \qquad (A.79)$$

and the final answer for the tangent is the line:

$$n = n_m(P^* e - \Delta\chi_c - 1) \qquad (A.80)$$

and since $P_I^* = 1/e$:

$$\frac{n_I}{n_m} = -\Delta\chi_c \equiv \ln\left(-\frac{E-\varepsilon}{RT}\right) \qquad (A.81)$$

Two potential questions arise for this method:

1. Notice that the tangent line requires a value for $\Delta\chi_c$. If this cannot be determined, the value of n at P_I^* could be used to obtain the tangent line. However, in most cases one is not interested in this tangent line anyway. With good data, one could therefore determine the tangent line in this latter fashion and calculate $\Delta\chi_c$ from Eq. (A.79) or (A.80). So:
2. The second problem is if one were to attempt to determine where the inflection point is from experimental data, this would be very difficult. In the range from about P^* of 0.2 to 0.6, there is nearly a straight line—actually within 2%. Therefore, by inspection it is nearly impossible to find the inflection point. However, the point of inflection, P_I^*, at $P^* = e^{-1} (=0.3678...)$ and the monolayer equivalence can then be calculated from Eq. (A.74).

Summary:
In the isotherm using the amount adsorbed in moles, n_{ads}, versus the relative pressure, P/P_{vap}:

1. the inflection point for the curve, P_I^*, is at $P/P_{vap} = 0.3768...(e^{-1})$;
2. the slope of the isotherm at this point, P_I^*, is equal to en_m, (2.718... times the monolayer equivalence); and

3. if the $\Delta\chi_c$ or E_a (recall the E_a definition—it is not a function of n_{ads}) is required, one can obtain this from the above equations or by reworking using the following:

$$\Delta\chi_c = P_I^* e - 1 - \frac{n_I}{n_m} \qquad (A.82)$$

where n_I is the number of moles adsorbed at P_I^*.

II-K: Precision and accuracy of the BET equation

In fig. 5.62 and Tables 5.11 and 5.12, a comparison was made for the fits of the BET and the χ theory. The BET was calculated as recommended over the "BET range"—that is, the range over which there is a linear portion. There are further questions, such as:

- "How good is the error precision inside the BET range, assuming the BET answer is correct?"
- "What is the precision error in terms of monolayer equivalence for the full range, assuming the BET answer is correct?"
- "How well does it calculate the monolayer equivalence within the BET range?"

The first two questions can be answered by comparison to standard curves. The variety of standard curve is due to a variation in the energy of adsorption of the first molecule according to χ theory. Since most of the standard isotherms follow the χ theory very well, it is appropriate to compare the BET against the χ theory to obtain the answer. (If the reader is not convinced of this after reading the section on standard curves, then the rest of this argument is hopeless.) The first question is asking, assuming that the BET is accurate with respect to the monolayer equivalence: how well does it fit within the BET range? The overall precisions and F test values were already given in Tables 5.11 and 5.12, but what are the specifics with respect to P/P_{vap}? Fig. A.1 shows a graphic representation the error in terms of monolayer equivalence for the BET with respect to the BET fit.

Looking at the full isotherm, it is well known that the BET is a bad fit, but how bad is it? Fig. A.2 is an indication of this misfit. The data was discontinued past $\theta = 3.0$ since the values became very large. The last question is assuming that the quantum mechanical derivation is correct: "What is the error in the actual answer obtained from the BET?" This calculation is restricted to the linear range in which there is the best precision. Fig. A.3 shows the results of the comparison. The lowest value for the ratio of $\theta_{BET} : \theta_{actual}$ is 2.18 (which is a 118% error).

If the previous multiple criticisms of the BET theory were not enough, surely this should suffice. Since many other theories, especially theories with respect to porosity and energies of adsorption, depend upon the BET surface area measurement, this should be seriously considered.[2]

---

[2] To date, this author has only seem one publication that uses the disjoining pressure (i.e., χ theory) as a potential basis for DFT calculations, although it seems those authors abandoned the idea. Without the proper E_a, and subsequent "layering," it would seem that all the DFT calculations are incorrect by a considerable amount.

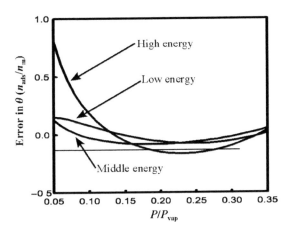

Fig. A.1

The error in precision of the BET equation inside the BET range (0.5–3.5).

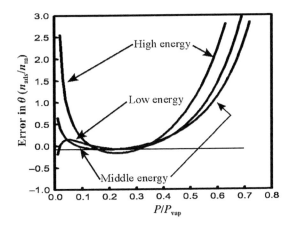

Fig. A.2

Errors that the BET equation make with respect to the monolayer equivalent determined by BET.

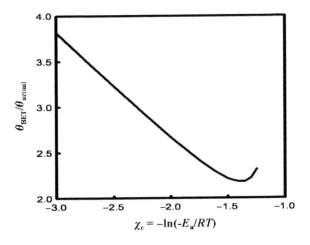

Fig. A.3

Absolute error of the BET method in terms of the ratio $\theta_{BET}/\theta_{actual}$.

Appendix III: Gregg and Jacobs' table

The Gregg and Jacobs' Table gives total heats of adsorption obtained from two methods using the BET theory in comparison to the experimental data from calorimetry. The following are the eight issues that were presented by S. J. Gregg and J. Jacobs [2] in their 1948 Faraday Society article and the table of values for the calorimetry.

Summary of article points

1. "The heat of adsorption E_1 calculated via the parameter $c[C_{BET}]$ of the theory is shown to be compatible with thermodynamic requirements as expressed in the Clausius-Clapeyron equation; but the latter are equally compatible with a value $E_1 + RT \ln C$ (C = any constant).
2. E_1 shows a moderate qualitative correspondence with the calorimetric or isosteric value, but close agreement is lacking.
3. The relationship which the theory implies between I_L and I_x, the integration constants of the vapor equation, and the adsorption isostere respectively are not found to hold in practice.
4. Arguments are adduced against the assumption made by the theory that the evaporation-condensation condition of the adsorbed multilayer differ inappreciably from those of the bulk liquid.
5. The implicit assumption that the adsorption isotherms above and below the critical temperature of the adsorbate are markedly different is shown to be improbable.
6. The values of the monolayer capacity derived from the theory and from the "phase-change" method of Gregg and Maggs are compared. Only for isotherms of Type I and II of the BDDT classification is agreement found.
7. According to the phase-change method, the adsorption isotherm invariably cannot be represented by a single equation as assumed by the BET-BDDT theory; where phase changes occur, the isotherm is split up into separate parts, each following its own equation.
8. The BET-BDDT theory does not account for hysteresis, but the phase-change conception includes it as a natural consequence."

Table symbols are:

$x =$ specific moles of adsorbate;
$x_0 =$ specific moles of a BET monolayer coverage;
$P = P/P_{vap}$;
$c =$ BET C constant (C_{BET});
$E_1 - \lambda =$ BET energy for C constant, i.e., $C = \exp(E_1 - \lambda)$;
$r =$ correction for only a fraction going to the first layer; and
$E' - \lambda = r(E_1 - \lambda)$.

References given are as follows:

11 Gregg, S. J. and K. W. S. Sing, Adsorption, first edition. Academic Press.
12 Titoff, Z. Physik. Chem. 165 (1910) 641.

14 Gregg, S. J., J. Chem. Soc. 74 (1927) 1494.
22 Gregg, S. J., J. Chem. Soc. (1943) 351.
23 Goldmann and Polanyi, Z. Physik. Chem. 131 (1929) 333.
24 Pearce and Reed, J. Physic. Chem. 35 (1931) 908.
25 Lambert and Clark, Pro. Roy. Soc. A, 122 (1929) 497.
26 Patrick and Long, J. Phys. Chem. 29 (1925) 339.
27 Felsing and Ashby, J. Amer. Chem. Soc. 56 (1934) 2226.
28 Patrick and Greider, J. Phys. Chem. 29 (1925) 1038.
29 Evans, J. Chem. Soc., (1931) 1556.

The average error for the BET differential heat of adsorption is 84%, but even worse, the r^2 is 0.14 which means that there is essentially no correlation of the BET to the energy of adsorption. If the origin is required, as it should be, $r^2 = -0.62$, which indicates a faulty correlation. See the table by Gregg and Jacobs reproduce, with permission from the Royal Society, in Fig. A.4 below.

	T /°C	x_0 /*	c	E-λ /**	x' /*	P at x'	r	λ /*	ΔH at x'/**	Q'/** BET	Q'/** Experi	Δ /**	Ref.
Carbon													
C_2H_6	0	2.9	620	14.5	1.8	0.012	1	9.41	34.3†	14.5	24.9	-10.3	14
CO_2	0	3.5	110	10.6	2.3	0.018	1	10.1	29.2†	10.6	19.1	-8.53	14
C_2H_2	0	3.3	180	11.7	2.5	0.02	1	16.3	28.9†	11.7	12.6	-0.84	14
C_2H_4	0	3.7	179	13.4	2.25	0.016	1	14.0	26.8†	13.4	12.7	+0.62	14
CO_3	0	3.7	130	10.9	2.42	0.016	1	10.1	26.8†	10.9	16.6	-5.78	13
SO_2	0	3.2	1000	20.9	2.85	0.012	1	24.5	33.5†	20.9	10.5	+9.41	22
CCl_4	25	2.5	250	14.8	1.42	0.06	1	32.3	5.86†	14.4	28.1	-13.7	24
CS_2	0	9.4	1820	7.61	7.1	0.105	1	–	9.20‡	7.61	9.20	-1.59	23
C_2H_5Cl	0	7.9	1760	7.36	6.7	0.20	1	–	11.7‡	7.36	11.7	-4.35	23
$(C_2H_5)_2O$	0	5.4	2280	9.54	4.6	0.089	1	–	14.2‡	9.54	14.2	-4.69	23
Silica gel													
C_6H_{10}	30	2.1	980	4.10	1.22	0.168	0.62	21.3	25.1‡	2.55	3.77	-1.21	26
CH_3NH_3	0	5.7	3100	13.0	5.5	0.039	0.63	26.6	6.69†	8.16	40.4	-32.2	27
SO_2	0	4.1	2100	8.79	3.5	0.076	0.66	24.5	34.3†	5.36	9.83	-4.06	28
H_2O	60	7.8	1450	6.07	6.4	0.20	0.48	46.0	46.0‡	2.93	3.77	-0.84	9
Chabasite													
NH_3	0	1.1	3000	12.6	1.05	0.56	1	21.5	12.5†	12.6	13.6	-1.09	29
Activated													
H_2O	25	2.5	2030	8.49	2.1	0.10	0.69	44.1	44.8†	5.86	0.63	+5.23	11
CCl_4	25	0.63	2370	9.92	0.40	0.23	0.16	32.2	46.9†	1.58	14.5	-12.9	11
Ferric oxide													
C_6H_6	40	1.28	1850	7.74	1.15	0.166	0.46	32.5	38.3†	3.56	6.74	-3.18	25

KEY: * = mmol g^{-1} ** = kJ mol^{-1} † calorimetric ‡ isosteric Between BET and experimeint the $r^2 = 0.14$

Fig. A.4

Detailed data table from the article by Gregg and Jacob.

The R-squared value between the BET and experimental enthalpy is 0.15, which is not good enough for physical science.

Appendix IV: Conventions used in this book
IV-A: Definition list

Term	Definition
Absolute adsorption	Adsorption that counts the liquid and gas phase in the pores—by Myers and Monson system definition
Absolute method	The Harkins and Jura method base on measure the heat of immersion
Accuracy	How close the answers from analysis is to the true value
Adiabatic calorimetry	Calorimetry that is perform in an arrangement where there is no net flow of heat from or to the system from the surroundings
Adsorbate	The molecules adsorbed on the surface of the solid material
Adsorbent	The solid material upon which the adsorbate is adsorbed
Adsorption	Addition of adsorbate to the adsorbent by increasing the adsorptive pressure
Adsorption energy	Energy released from the adsorbate-adsorbent system
Adsorption isotherm	A plot of amount adsorbed (today usually $mmol\,g^{-1}$ or $\mu mol\,g^{-1}$) versus P/P_{vap}
Adsorptive	The gas in equilibrium with the adsorbate
ASP	Auto-shielding physisorption—the classical mechanic analogue to χ theory
B point analysis	The "determination" of the surface area by the selection of the "knee" in the isotherm
BDDT	Brunauer, Deming, Deming and Teller: a theory for analyzing the isotherm
Bed porosity	The "porosity" that arises from the spaces between particles—usually macroporous
BET	Brunauer, Emmett and Teller: theory of analyzing the isotherm
BET constant	A fitting constant for the transformed BET equation which is supposed to be related to the energy of adsorption
BET range	Arbitrary range for fitting the transformed BET equation normally 0.05–0.35, P/P_{vap}, but varies depending on curve shape
Binary adsorption	Simultaneous adsorption of two adsorbates
Buoyancy correction	Gravimetric method: the correction made for the buoyancy of both the sample, the weights and other parts attached to the balance beam
Calorimetry	Method: measurement of the heat release or take-up during a reaction or phase change
Capillary filled	The postulation that surface tension drives the filling of pores in addition to adsorbate-adsorbent attraction

Continued

Appendixes

Term	Definition
Carnahan-Starling approximation	An approximation to the first six virial coefficients into a form that can be simplified by the Padé approximation
Carrier	In carrier flow method: the cover gas used to carry the active gases in a flow system
Carrier gas flow system	A system in which a column is used with a carrier gas which contains the active adsorptive
Chemical bonding	Attractions between atom and molecules that depend upon electron transfer or sharing
Chemisorption	Enhancement of the amount of gas molecules on the surface of a solid caused by covalent or ionic bonding
Correlation functions	Two functions are correlated if they are functions of the same variable and the division of one by the other yields a constant value
Dead space	Volumetric method: The space that is shared by the cold zone and zone before the inlet valve that needs to be calibrated as a volume
Desorption	Removal of adsorbate from the adsorbent by decreasing the adsorptive pressure or increasing the temperature
DFT	Density Functional Theory
Differential scanning calorimeter	A calorimeter that compares the heat flow between a standard and a well calibrated standard sample
Disjoining pressure	Disjoining Pressure Theory, a thermodynamics-based theory with one assumption; it yields the same answer as χ theory
Disjoining pressure theory	A thermodynamics-derived adsorption theory with only one assumption (also called ESW)
DR	Dubinin-Radushkevich
DRK	Dubinin-Radushkevich-Kaganer
Epitaxy	Adsorption on specific molecular sites of the adsorbent
Equation of state	An extra equation that is not specified by the four laws of thermodynamics—an example is ideal gas law
ESW	Excess Surface Work theory, the same as Disjoining Pressure Theory
Excess adsorption	A concept of what is adsorbed associated with Gibbs' excess—by Myers and Monson system definition
External area	External surface area
External surface area	The surface area of a porous sample that is not inside the pores
Fowler-Guggenheim isotherm	Eq: $bP = [\theta/(1-\theta)]\exp(-c\theta)$
Freundlich isotherm	Eq: $\theta = K_F P^{r_F^{-1}}$
Functional	A function that contains a function—see the beginning of Chapter 7

Continued

Term	Definition
Gas burette	Using a standardized volume and pressure to measure how much adsorptive is used
Gas-liquid thermometer	Using the vapor pressure over a liquid as an indication of temperature
Gibbs' adsorption equation	Eq: $d\gamma = \Gamma_1\mu_1 + \Gamma_2\mu_2$
Gibbs' Phase Rule	The thermodynamic rule that specifies the number of degrees of freedom between phases and components—see Section on phase rule
Gibbs-Duhem	At constant P and T Sum over n $(N_n d_n) = 0$, for surfaces it becomes: Sum over n $(A_n d\mu_{n,ads}) = 0$
GP	Goldmann and Polanyi
Gravimetric	Involving the use of a microbalance to measure amount of adsorptive adsorbed
Half life	The time for a reaction or system to come halfway to final state terms of amount.
Heat of adsorption	The amount of heat produced during adsorption
Henry's law	n_{ads} goes to 0 if and only if P goes to 0. Formerly an incorrect "golden standard" for an isotherm equation
Heterogeneous surfaces	Adsorbent surfaces which have a variety of adsorption energies
High vacuum	Pressures in the range 100 mPa to 100 nPa (10^{-6} to 10^{-9} bar)
Hill-deBoer isotherm	$bP = \frac{\theta}{1-\theta}\exp\left(\frac{\theta}{1-\theta}\right)\exp(-c\theta)$
Homogeneous surfaces	Adsorbent surfaces which have only one adsorption energies throughout
HV	High vacuum
Hysteresis	The phenomenon of the desorption isotherm over a limited range being greater than the adsorption isotherm
Hysteresis loop	The loop in the isotherm created by the adsorption-desorption following different paths
Hysteresis type	One of the four "looks" that hysteresis loops form
Inflection point analysis	χ analysis: Using inflection point in the isotherm at (0.3678...) P/P_{vap} to Calculate the monolayer equivalence for nonporous adsorbents
In-register	Adsorption for which the adsorbate molecules line up with the adsorbent molecular positions
Intermolecular forces	Chemical forces due to: London force and [dipole moments and/or "hydrogen bonds"]
Isosteric	In the case of adsorption, this means at constant moles adsorbate
Isotherm	The measurement of the amount adsorbed versus adsorptive pressure at constant temperature

Continued

Term	Definition
Isotherm types	Classification of isotherm base of the "looks" and characteristics of the isotherm
IUPAC	International Union of Pure and Applied Chemistry subsection of ISO
Kelvin-Cohan	The equation for the adsorbed surface film taking into account the pore curvature
KJ	Krug and Jaroniec
Knudsen number	The number that delineates whether the gas flow is laminar or turbulent
Langmuir isotherm	Eq: $\theta = \frac{KP}{1+KP}$
Layer	A concept that adsorption proceeds by forming first a monolayer, then the second layer, etc. The meaning is modified in χ theory
Layer in χ theory	A molecule is in the first layer if it is in direct contact with the adsorbate. A molecule in the nth layer is in contact with the $(n-1)$th layer
Log law	Eq: $n_{ads} = A + B \ln(P)$
Macropores	Pores with diameters greater than 50 nm (IUPAC); pores that are too large to distinguish in the isotherm (P)
Mean free path	The average distance a molecule travels before it encounters another molecule
Mesopores	(IUPAC def.) Pores with diameters between 2 nm and 50 nm* (χ def.) Pores analyzable by χ plot which have prefilling
Microbalance	A highly sensitive balance with a relative sensitive at least 10^{-6} but recommended at 10^{-9}; UHV capability recommended
Micropores	(IUPAC def.) Pores with a diameter of less than 2 nm* (χ def.) Pores analyzable by χ plot but have no prefilling
Molar area	The area that a single molecule will cover on an adsorbent. Calculated in a variety of ways
Molecular flow	Flow of molecules which appear to not be interacting
Monolayer	A theoretical uniform liquid film of adsorbate one molecular layer thick
Monolayer equivalent	The amount of adsorbate that has the same number of molecules as the theoretical monolayer; the symbol for this is n_m
Monte Carlo	A classical mechanical method to find the most probable configuration for molecules given energy constraints
N-layer formulation	Analyzing the isotherm "layer by layer"—a concept used in BDDT, DFT, and χ theories
NLDFT	Non-Local Density Functional Theory

Continued

Term	Definition
χ	$-\ln(-\ln(P/P_{vap}))$
χ feature 1	Positive curvature at the lowest pressures in the χ plot
χ feature 2	Negative curvatures in the χ plot
χ feature 3	Large positive curvature followed by negative curvature to yield a slope of the chi plot that is less than at lower pressures in the χ plot
χ feature 4	Hysteresis associated with item 3 in the χ plot
χ feature 5	A break in the straight line at moderately low pressures in the χ plot
χ method	Using calculation base on the abscissa transform to χ
χ plot	A plot of amount adsorbed (usually mmol g^{-1}) versus $-\ln(-\ln(P/P_{vap}))$
χ theory	A quantum mechanical theory of adsorption
Open system	In thermodynamics, a system that allows the input and/or output of heat, work, and matter
Packing factor	A factor that takes into account the "empty" spaces between adsorbate molecules which touch
Physical adsorption	Enhancement of the amount of gas molecules on the surface of a solid caused by intermolecular forces
Physisorption	The same as physical adsorption
PMF	Probability Mass Function
Pore filling	The premature filling of pores creating a deviation from the isotherm expected from the nonporous or microporous isotherm form
Pore surface area	surface area inside the pores
Pore volume	The volume of the pores
Porosity	The presence of pores in the adsorbent
Precision	The amount of uncertainty of reproducibility in a measurement
Prefilling	The same as pore filling
QSDFT	A modification of the NLDFT smoothing out the potential of the adsorbent
Quantum mechanical classification	Isotherm types using qualifications derived from quantum mechanics by χ equations
Reversible pore filling	A phenomenon where the adsorption and desorption isotherms are identical
rpf	Reversible pore filling
Scanning calorimetry	Calorimetry that compares a standard against the material needing measurements
Single point method	The BET-based method using a single point to analyze surface area; this is not very reliable except possibly when one is making comparisons between sampling under identical condition

Continued

Term	Definition
Specific	In this context, this means "per gram of sample." Quantities throughout the book are assume to be "specific"—always implied unless otherwise stated or irrelevant
Standard curve	The same as a standard isotherm
Standard isotherm	An isotherm that is selected to compare another isotherm against. For porous materials, a comparison is made between the nonporous material to the porous, but fails to account for energy differences
Standard plot	Refers to one of these: α-s plot, the t-thickness plot, the χ plot and others that may be specific to an adsorbate-adsorbent pair. A generalize standard plot function will be designated as $\mathbf{F}(P/P_{vap})$ in this book
Surface area	A concept that one may measure given the proper measuring device; fractal consideration makes this concept problematic
System	In thermodynamics, a 3D space designated by boundaries that correspond to some natural boundaries between phases
T-equation	An isotherm equation introduced by Tóth
threshold pressure	In χ/ESW theory and for the log law the pressure at which no absorption to the liquid-like state occurs
Type H1 isotherm	Nearly vertical and parallel adsorption and desorption branches
Type H2 isotherm	Sloping adsorption branch and nearly vertical desorption branch
Type H3 isotherm	Sloping adsorption and desorption branches covering a large range of P/P_{vap} with underlying type II isotherm
Type H4 isotherm	Underlying type I isotherm with large range for the hysteresis loop
Type I isotherm	This is characteristic of either a chemisorption isotherm (in which case the final upswing at high pressures may not be present) or physisorption on a material that has extremely fine pores (micropores)
Type II isotherm	This is characteristic of a material that is not porous, or possibly macroporous, and has a high energy of adsorption
Type III isotherm	This is characteristic of a material that is not porous, or possibly macroporous, and has a low energy of adsorption
Type IV isotherm	This is characteristic of a material that contains mesoporosity and has a high energy of adsorption. These often contain hysteresis attributed to the mesoporosity
Type V isotherm	This is characteristic of a material that contains mesoporosity and has a low energy of adsorption. These often contain hysteresis attributed to the mesoporosity

Continued

Term	Definition
Type VI isotherm	This type of isotherm is attributed to several possibilities the most likely being, if the temperature is below the adsorptive triple point, that the adsorbate is more like a solid forming a structured layer—i.e., epitaxial growth. Other possible explanations include multiple pore sizes. If the steps are at the low-pressure portion of the isotherm, then the steps may be due to two or more distinct energies of adsorption. If the steps are at the high-pressure part of the isotherm, then the steps might be due to sharp steps on the adsorbate surface
U-cup	An arrangement on a hangdown tube to assure that the temperature gradient is reproducible
UHV	Ultra High Vacuum
Ultrahigh vacuum	Pressures below 100 nanopascals
Ultramicropores	Pores in which only one "layer" of molecules can adsorb
Viscous Flow	The flow of a fluid that in the cross sectional profile is either constant or varies in a continuous monotonic manner from some high value to some low value(s)
Volmer isotherm	$bP = \frac{\theta}{1-\theta}\exp\left(\frac{\theta}{1-\theta}\right)$
Volumetric	A method of measurement of the isotherm using a gas burette

IV-B: Symbol list

In the following list are the symbols used in this book. Most of the symbols follow the third edition of the IUPAC "Quantities, Units and Symbols in Physical Chemistry" (the Green Book) [4]. In general super scripted symbols are being avoided to avoid confusion with exponents that can arise. Some of the quantity names are unfamiliar to most readers. For example, "superficial work" is the same as "surface tension," "surface energy," "surface free energy," and "specific {?} surface work."

A convention of note used here is to use the chemists' convention for molar mass—that is, the units are $g\,mol^{-1}$ and not $kg\,mol^{-1}$. The value then is M_u and the units are $g\,mol^{-1}$. The old thermodynamic convention for molar quantities will be used for this modification. There the molar mass with units of $g\,mol^{-1}$ is given the formula symbol "M."

Greek, Roman, Italic and bold

Subscripts

(IUPAC convention third edition Green Book: subscripts that are descriptions are roman letters, whereas subscripts which are indexes or numbers are italic.)

1D	1-dimensional
2D	2-dimensional
3D	3-dimensional
a	of adsorption for the first adsorbate molecule
ads	of adsorption
BET	of the BET plot, equation, etc.
c	critical
d	for the dead space
F	Freundlich
FHH	Frenkel-Halsey-Hill
i, j, k, l	for the ith, jth, kth, lth species
la	liquid to adsorbate
LJ	Lennard-Jones
m	of a monolayer (not used for molar quantities—see embellishments below)
N	for N particles
p	of or in the pore
P	Polanyi
s	of the sample
st	isosteric
vap	of vapor over bulk liquid at the same temperature as the sample
−	open space between molecules is excluded
+	open space between molecules is included
*****	indicates at the pressure picked for the experiment Symbol List for Chapters 1–6
$\overline{V}$	molar volume—used for adsorbate or liquid adsorptive
$\widetilde{p}$	fugacity
$\overline{M}$	molar mass (kg mol^{-1} IUPAC 2007)
$\overline{L}$	$\overline{V}/\overline{A}$
$\overline{H}$	molar enthalpy
$\overline{A}$	molar area
(g)	of the gas
(l)	of the liquid
(s)	of the solid
$<E_a>$	average E_a over a distribution
$<l>$	average length of pores

$\overline{G}$	molar Gibbs' free energy
α	the perturbation parameter $\alpha = E_a - \varepsilon$
α	surface coverage in the localized layer
α	constant that relates relative size of the molecule to perturbed energy $= E_0 - \varepsilon$
α	a constant relating A_w to A_{ex}, $0 < \alpha < 1$
α_{xy}	film tension (optional between phase x and phase y)
β	a constant in the Dubinin-Polanyi equations an energy term
β	$-kT^{-1}$
β_m	exponential in the calculation of θ_m
γ	film tension
Γ	adsorbate surface excess ($\approx n_{ads}$ here) used in Gibbs' adsorption equation
γ_{gl}	the surface tension between the gas and liquid phase
Γ_m	minimum value for Γ in the ESW theory
γ_{xy}	tension for the phase x-phase y interface
γ_∞	γ for fully form film value
$\Delta_l^g \overline{H}$	molar heat of vaporization
δ	Tolman corrected r_{VdW}
δ	constant in MKYO theory $\approx r_{VdW}$
$\delta()$	delta function (either type)
$\Delta\varepsilon$	$\varepsilon_{12} + \varepsilon_{21} - \varepsilon_{11} - \varepsilon_{22}$
$\Delta\theta_c$	off-set in a binary adsorption in terms of coverage
$\Delta\mu$	$\mu - \mu^*$
$\Delta\chi$	$\Delta\chi = \chi - \chi_c$
$\Delta\chi_p$	$\Delta\chi_p = \chi_p - \chi_c$
Δb	a grouping of term of which Δm_c is a part
ΔE	a group of energy terms ($=$zero if the $\overline{V_1} = \overline{V_2}$) in binary adsorption
ΔH	enthalpy change
$\Delta H()$	enthalpy function of whatever in ()
Δm_c	a change in m_c from the adsorption to the desorption branches.
$\Delta n_{ads, n}$	nth increment of amount of gas admission
Δn_c	off-set in a binary adsorption in terms of amount
$\Delta_{vap}\overline{H}$	molar enthalpy of vaporization ($=\varepsilon$/molecule)
ε_0	base value of the polarization
$\varepsilon_{i,j}$	energy of interaction between molecules i and j
η_i	a probe value for fit to pore fill by successive approximations
θ	surface coverage $= n_{ads}/n_m$
Θ	surface of the adsorbent which is *not* covered by any adsorbate
θ_1	coverage of adsorbate molecules in direct contact with the adsorbent
$\theta_{m=x}$	surface coverage of "layer" x (also $>$ and $\geq$)
λ	mean free path of gas molecules
Λ	uncertainty for the effective radius between adsorption "layers"

λ	ESW decay constant
λ_1, λ_2	two decay constants for the KJTKD theory
μ	chemical potential of a component
μ^*	chemical potential of a pure component or of a vapor over the pure liquid
μ_{ads}	the chemical potential of the adsorbate
μ_c	the chemical potential of the condensed phase flat
μ_g	the chemical potential of the adsorptive
μ_l	chemical potential of pure liquid phase at the saturation pressure
$\mu^{\ominus}$	chemical potential of a vapor at 1 bar pressure (absolute standard state)
Ξ	grand canonical
Ξ^*	the maximum term in the grand canonical ensemble partition function
Π	the disjoining pressure tensor
π	spreading pressure
Π_0	preexponential in the Π function.
Π_1, Π_2	preexponentials for the KJTKD mesopore theory
$\rho(l)$	liquid density (without (l) understood to be for liquid)
ρ_b	density of adsorptive in kg m^{-2}
ρ_1	density of the carrier gas in kg m^{-2}
ρ_s	solid density
σ	standard deviation of an overall fit
σ	interface tension
σ_2	$\sigma_2^2 = \sigma_c^2 + \sigma_p^2$
σ_c	standard deviation of χ_c
σ_p	standard deviation of χ_p
τ	time increment
τ	used for variable of integration
Φ	surface excess work function
χ	$-\ln(-\ln(P/P_{vap}))$
χ_a	χ value at commencement of prefilling on adsorption branch
χ_c	χ when $n_{ads} \to 0^+$
$\chi_{c, n}$	χ_c for the adsorbate n
χ_d	χ value at commencement of pore emptying on desorption branch
χ_p	χ at which micropores begin
χ_x	χ under conditions "x" versus under conditions "y," i.e., χ_y
ψ_n^0	the unperturbed wave function for the state n
ψ	wave function
ω	oscillator frequency
A	cross-sectional area of carrier gas tube
a	area of adsorbate molecule
$\overline{E_a}$	molar amount of E_a

A	the prelog constant in the DR or DRK equation
a_+	molecular area including spaces between molecules
a_-	molecular area excluding spaces between molecules
A_{ex}	external area $=A_w+A_o$
A_{gl}	the area of the adsorbate liquid-gas interface
A_n	surface area of the nth energy plain
A_o	total area of the pore openings
A_p	area of pore surfaces
A_s	surface area of adsorbent
$A_{s,i}$	the surface areas for ith segment or plane in χ theory
A_w	total area of the walls between pore openings
B	preexponential constant in the Polanyi formulation
b	sometimes used as an arbitrary constant except for:
b	y (ordinate) intercept of a linear fit
b	buoyancy proportionality constant
B	a constant in the Dubinin-Polanyi equations
B_1	average film permeability
B_{BET}	BET preexponential constant
c	used as an arbitrary constant
C	(Gibbs' phase rule) components
C_{BET}	BET C constant
C_{dBZ}	a constant in the deBoer-Zwikker equation $=a/v_0$
$C_{K,i}$	the ith KFG coefficient
$C_{p,\,ads}$	heat capacity at constant pressure of adsorbate
$C_{p,\,1}$	the heat capacity at constant pressure for the liquid phase of adsorptive
d	distance
D	tube diameter
$D(\chi,\chi_x,\sigma_x)$	distribution over χ with mean of χ_x and standard deviation $\sigma_x \cdot x = c \cup p$
$D(n_{ads})$	function: differential of n_{ads} wrt P/P_{vap}
D_1	a distribution function which gets doubly integrated
D_2	a distribution function which gets both doubly and singly integrated
d_{ads}	average center-to-center between "layers"
d_p	pore diameter
D_W	the Weibull distribution
E_1	$(E_a-\varepsilon)$ for adsorbate #1, E_2 for #2
E_a	energy of adsorption for the first adsorbate molecule
$E_a(\theta)$	energy of adsorption as a function of θ sometime as $\mathbf{E}_a(\theta)$
$E_{a,\,i}$	energy of adsorption for the first adsorbate molecule in the $i^{th}\chi$ plot segment
$E_{a,\,n}$	E_a for the adsorbate n
E_{BET}	BET energy from the C constant

E_{LJ}	Lennard-Jones potential
E_m	energy of adsorption for the mth adsorbate
E_q	energy of the bottom of the potential well
f	designation for a function being used in any section—meaning depends on context
f(T)	portion of Ξ due to internal molecular modes, vibrational, etc.
F(x)	The function used for the abscissa for the standard curve—$x = P/P_{vap}$
F_{1D}	1-dimensional packing factor
F_{2D}	2-dimensional packing factor
F_{3D}	3-dimensional packing factor
G	a parameter in mesopore fit (for χ it $= A_s/\overline{A}$))
g	factor to convert the volume to surface ratio to radius.
G	total surface area parameter in micro/mesopore analysis
$G(x, T, k_1...k_n)$	isotherm function with parameters $k_1...k_n$
G_b	mass velocity in kg m^{-2}s^{-1}
h	geometrical factor, $h = m_c$
H	parameter in the Adolphs prefilling equation
H	micropore parameter in the χ micro/mesopore analysis ($=pG$)
$\hat{H}'$	perturbation on the Hamiltonian QM operator
$\hat{H}$	the Hamiltonian QM operator
$\hbar$	Plank's constant ($/2\pi$)
H'$_{mn}$	the perturbation matrix
I_{BET}	intercept of the transformed BET equation
I_{hi}	ordinate intercept for high linear fit for a $\Delta\chi$ plot
I_i	ordinate intercept of the ith segment of the χ plot
$I_{int}(\Delta\chi, n_{ads})$	the intersection of the linear high and low fit for a $\Delta\chi$ plot
I_{lo}	ordinate intercept for low linear fit for a $\Delta\chi$ plot
J	mesopore parameter n micro/mesopore analysis
K	an equilibrium constant
k	used for node number for a segment of a sign function
k	fitting parameter for the Tóth equation
K	fitting parameter for the Tóth equation
k	Boltzmann's constant
k	the exponential constant in the Dubinin-Polanyi equations
K'	$1/K$ (equilibrium constant) in the Langmuir isotherm
k'	a constant which includes k_{FHH} and n_m
K_1, K_3	constants in the Bradley isotherm—can be written in terms of χ
k_a	slope in the linear relationship relating θ_1 to θ_2
k_{ads}	kinetic adsorption rate
k_b	intercept in the linear relationship relating θ_1 to θ_2
K_F	equilibrium constant in the Freundlich isotherm
k_{FHH}	pre-θ constant in the Frenkel, Halsey, Hill formulation

K_H	Henry's law constant
k_{HO}	harmonic oscillator "spring" constant
k_p	pre-θ constant in the exponent in the Polanyi and deBoer-Zwikker formulations
l	the total length of all the pores
L	column length
L	length of the potential box
l	length of adsorbed (1 D) molecule
l_p	the pore length of layers > the xth "layers"
m	slope of a linear fit
m	fitting parameter for the Tóth equation
m	particle mass of the xth layer
m_b	buoyancy mass gain ($m_b = bP$)
m_c	geometric factor for interface shape
$m_{c,a}$	geometric factor for g-l interface shape for adsorption
$m_{c,d}$	geometric factor for g-l interface shape for desorption
m_{mf}	molecular flow correction
$m_{mf}(P)$	correction for molecular flow (a function of pressure)
M_p	the molar mass of the buoyancy probe gas in $g\,mol^{-1}$
m_p	mass recording of the trial for mass flow correction
M_u	unit-less molar mass—relative molar mass compared to C-12
N	limiting layer number in the BDDT theory
N	(Gibbs' phase rule) degrees of freedom
N	number of adsorbed molecules
$\overline{A}_n$	the molar area for the nth adsorbate.
n	total moles, usually used for the adsorbate
n_1	the number of moles adsorbed in the localized layer
N_A	Avogadro's number
$n_{ads,I}$	amount adsorbed at inflection point
n_{ads}	amount adsorbed (mol, mmol or μmol per g)
$n_{ads,i}$	specific moles of adsorbate for ith adsorbate (per mass sample)
$n_{ads,n}$	amount admitted after the nth admission for the volumetic method
n_{ex}	amount adsorbed on external surface of particles
n_i	incremental number of moles adsorbed for one data point (volumetric)
N_i	number of adsorbate molecules of species i N_{BET} number of layers in the BDDT equation
n_L	amount adsorbed just before the onset of prefilling
n_m	fitting parameter for the Tóth equation
$n_{m,p}$	monolayer equivalence in pores
n_m	amount in a monolayer equivalence
$n_{m,1}$	multiplicative constant relating Δn_c to $\Delta \theta_c$, monolayer equivalence of adsorbate #1

n_{max}	the number of "allowed" layers in BDDT equation
$n_{m,\,ex}$	amount in a monolayer equivalence on the external surface of a porous material
n_p	amount adsorbed in the pores
N_p	the number of pores
n_S	moles/g of surface sites in the Langmuir isotherm equation
N_S	number of surface sites for the Langmuir model
n_{s+}	the virtual amount adsorbed if porosity were not a physical limitation.
n_{s-}	the amount in the pores to subtract from n_{s+}
P	(Gibbs' phase rule) phases
P	probability function
P	pressure
P(χ)	probability normal mass function (PMF) with χ as independent variable
$P^*_{vap,\,n}$	vapor pressure for the pure adsorptive.
P_1	~equilibrium pressure for a reading of a data point (BdB theory)
P_a	pressure of the hysteresis break point upon adsorption
P_c	the threshold pressure
P_{cr}	critical adsorptive pressure corresponding to t_{cr}
P_d	pressure of the hysteresis break point upon desorption
P_f	final pressure
P_G	P in the Langmuir isotherm
P_i	Initial pressure
$P_n(t)$	pressure as a function of time for the nth gas admission
P_p	probe gas pressure
$P_{r,\,e}$	fitting parameter for the Tóth equation
$P_{r,\,e}$	Tóth fitting parameter
$P_{R,\,n}$	vapor pressure of the nth component over a regular solution
P_{vap}	saturated vapor pressure
$p_{vap,\,x}$	partial pressure condition "x" versus under conditions "y" i.e., p_y
$P_{vap,\,x}$	pressure condition "x" versus under conditions "y" i.e., P_y
$P°$	P_{vap} often used in the BET theory
q	mass adsorbate in kg
q	heat transfer
Q	the partition function
Q	the integral energy of adsorption as defined by Hill
Q'	$(\overline{Q}')$ molar heat of adsorption (Morrison, Los and Drain)
Q'	integral heat of adsorption (Morrison, Los and Drain)
q_{1a}	heat of the transition liquid to solid
q_{st}	isosteric heat of adsorption
r	the distance between molecular centers in the plane of the surface

R	the gas constant
r_σ	Lennard-Jones radius to the LJ minimum
r^2	the r^2 from statistics
r_c	$r_c = r_p - t$ MKYO theory
r_{DP}	a parameter in the Dubinin-Polanyi class of isotherms, ex: $r_F = 1$ for the Freundlich
r_F	the pressure power term constant in the Freundlich isotherm
r_{FHH}	exponent to θ in the Frenkel, Halsey, Hill formulation
$R_{l:s}$	ratio of the density of the liquid to the density of the solid
r_{min}	potential minimum in the harmonic oscillator and LJ potential
r_p	the pore radius (cylindrical) or half distance across the pore (slit)
r_t	the radius of the immobile surface atom or ion
r_t	radius of the adsorbent surface atoms
$R_{2/3}$	ratio of liquid density to solid density raised to the $2/3$ power
s	a classical correction for molecular shape ~ 0.92
S_{BET}	slope of the BET transformed equation
S_{hi}	slope for high linear fit for a χ plot
S_i	slope of the ith segment of the χ plot
S_{lo}	slope for low linear fit for a χ plot
T	temperature/K
t	classical surface film thickness
t	time
$t_{\chi, p}$	equivalent flat surface layer thickness
$T_{1or2or3}$	temperature of volume designated with a subscript
t_{bt}	breakthrough time
t_{cr}	the critical film thickness for film collapse to prefill pores (BdB theory)
t_{mon}	a monolayer thickness
$t_{1/2}$	the "half life" constant for pressure decay
$t_{1/2}$	half life
$U(x)$	unit step function
v	velocity of the carrier gas
V	propagation velocity
V	volume of gas adsorbed at STP
v	velocity—of carrier front
v	effecting volume of a hard sphere molecule including unoccupied spaces
$v_{(c+a)}$	velocity of carrier gas including adsorptive
v_+	molecular volume including spaces between molecules
v_-	molecular volume excluding spaces between molecules
$V_{1or2or3}$	a calibrated volume 1 or 2 or 3
V_d	dead space
V_m	volume of gas at STP required to form a monolayer equivalency

Appendixes 397

V_p	total pore volume
V_s	the volume of the sample
W	the perturbation of the energy eigenvalues
W^0	the unperturbed energy eigenvalues
X	mole fraction of adsorptive unless otherwise stated
x	P/P_{vap} in the BET equations
$X_m(l)$	mole fraction of component m in bulk liquid state solution
X_n	mole fraction of the n component in the liquid phase
Y	mole fraction of the adsorbate
Y_n	mole fraction of the n component in the gas phase
Z_N	the configuration partition function for N particles
ε	energy of interaction between liquid molecules
$\varphi(X_2)$	function expressing deviation from ideal solution (usually Margules function)
$\langle \chi_c \rangle$	mean in the distribution of χ_c
$\langle \chi_p \rangle$	mean of a distribution for χ_p
$\overline{V}_{-F}$	occupied molar volume, liquid molar volume minus free space $= \overline{V}F_{3D}$
$\langle r \rangle$	average effective radius between adsorption "layers"
$\langle \rangle$	average of what is enclosed

Symbol list for Chapter 7

(The symbols used in Chapter 7 are from original publications and may be different from other sources, including IUPAC.)

β	$1/kT$
$\delta()$	Dirac delta function
$\delta F/\delta y(x)$	The form of the functional derivative
Δu_j	Potential energy for jth particle
Λ_{1or2}	A group of constants and T with m_{1or2}
μ_{1or2}	Chemical potential of component 1 or 2
Ξ	Grand partition function
$\rho\{x,y\}$	Number density
$\tau(\vec{r})$	A gathering functional—See Eq. (7.54) for definition
$a_{1\text{ or }2}$	Area of molecule 1 or 2
a_{vdW}	van der Waal area
C	$1/Q$
$C^{(1)}(\vec{r})$	Singlet direct correlation function
$C^{(2)}(\vec{r})$	Direct correlation function
E_i	Energy of the ith particle
E_{LJ}	Lennard-Jones potential
F	Helmholtz free energy function

$F(y(x))$	The form for the functional
F_{CS}	Carnahan-Starling hard sphere portion of F
F_I	Ideal gas portion of F
F_j	Force acting on jth particle
$g^{(X,Y)}$	Correlation function for probabilities
$H(\text{etc.})$	Classical Hamiltonian
k	Boltzmann constant
L	Length of 1D box
m_j	Mass of the ith particle
$\vec{r}_n$	Position vector of a particle
$\vec{v}_n$	Velocity vector of a particle
$N_{1 or 2}$	Number of molecules 1 or 2
$P\{x\}$	Probability of x
$P\{x,y\}$	Probability of x and y
Q	Partition function
T	Absolute temperature
w	Weighting functions to make agreement with bulk fluid
x	Position of particle from a wall
$y(x)$	**y** (bold) is a function of x
z	$(\pi a^3 \rho)/6$
Z_N	Configuration partition function
$\varphi(\vec{r})$	A functional resulting from an external potential
$\int G(y(x),x)dx$	The form of a functional integral

Superscripts

⊖   designates the quantity is the standard state with respect to 1 bar
*   designates vapor pressure over the liquid state

Embellishments

―   (bar over) indicates a molar (per mole) quantity; Ex: $\overline{V}$ is molar volume
bold   indicates the quantity is a function, Ex: $\mathbf{Q}(T)$
∧   (hat over) quantum mechanics operator
~   (tilde over) partial quantity
→   (arrow over) vector

Index lists

Authors
    Drain 395
    Frenkel 389

Halsey 389
Hill 389, 395
Los 395
Subject index
Configuration partition function 397
Dubinin-Polanyi 396
Partition function 395

Figure captions

Fig. A.1 The error in precision of the BET equation inside the BET
range (0.5–3.5)..379
Fig. A.2 Errors that the BET equation make with respect to the monolayer
equivalent determined by BET..379
Fig. A.3 Absolute Error of the BET method in terms of the ratio $\theta_{BET}/\theta_{actual}$ ...379
Fig. A.4 Detailed data table from the article by Gregg and Jacob 381

Table list

Perhaps the uninteresting history of χ

Some quick history about the χ theory: The QM derivation was derived in 1979 and first published in 1981 by myself and Dr. E. Loren Fuller in a form acceptable to reviewers. It has met considerable resistance then, primarily because the now discredit BET theory was considered the "gold standard" for determining the "surface area." Although QM perturbation theory was the original derivation, the first QM derivation that made it into the literature was a WKB approximation. Dr. Fuller was successful in publishing several papers dealing with the subject and left behind several publications that were unfinished at the time of his death in December 2007. You will be privy to some of his unpublished data in this book. After reading this book, it might be a good idea to read some of Dr. Fuller's publication in a new light, since he wrote them in such a way as to, frankly, trick the reviewers to accept them. This is thus the key to unlocking the deeper meaning in those papers.

References

[1] T.L. Loucks, Augmented Plane Wave Method, W.A. Benjamin, Inc, New York, 1967.
[2] S.J. Gregg, J. Jacobs, Trans. Faraday Soc. 44 (1948) 574–588.
[3] M. Thommes, K. Kaneko, A.V. Neimark, J.P. Olivier, F. Rodriguez-Reinoso, J. Rouquerol, K.S.W. Sing, Physisorption of gases, with special reference to the evaluation of surface area and pore size distribution (IUPAC Technical Report), Pure Appl. Chem. 87 (9–10) (2015) 1051–1069.
[4] IUPAC, Quantities, Units and Symbols in Physical Chemistry (The Green Book), third ed., RSC Publishing, Cambridge, 2007.

Acknowledgments

Special recognition for facilitating the possibility of the creation of this book goes to Prof. Tilman Schober, formerly of Forschungszentrum Jülich, who made the facilities available to me, and Dr. E. Loren Fuller, Jr., formerly with the Oak Ridge National Laboratory, who was instrumental in the development of χ theory. Thanks also go to Dr. Fuller and Dr. Kenneth Thompson for sharing large amounts of unmassaged data for my use, and to Prof. Qiao and Prof. Neimark for sharing their data and papers.

In memory of Dr. E. Loren Fuller a friend and mentor.

Author index

Note: Page numbers followed by *t* indicate tables and *np* indicate footnotes.

A

Abyzov, A.M., 242–243
Adamson, A.W., 97, 121–122, 146–147, 211
Adolphs, J., 36–37, 93–94, 169–171
Afonso, R., 30
Agron, P.A., 127, 225
Ahn, W.S., 317
Alolphs, J., 52
Aranovich, G.L., 20–21
Arnold, J.R., 9, 83–84, 200, 262
Astakhov, V.A., 247–248, 292–293

B

Barrande, M., 52
Barrett, E.P., 294–295
Basmadjian, D., 90
Beck, J.S., 324
Beebe, R.A., 260
Berg, W.T., 256, 258
Beurroles, I., 52
Bhambhani, M.R., 112, 122, 125
Bhanbhani, M.R., 221–222
Bhatia, S.K., 318, 324
Bradley, R.S., 228
Bray, W.C., 228
Broekhoff, J.C.P., 311–312
Bruaner, S., 27
Brunauer, S., 10, 43–44, 95–96, 101, 141, 206, 291, 335–336
Brunel, D., 21

C

Carnahan, N.F., 67–68, 351
Carpenter, F.G., 29, 66, 246–247
Chang, S., 317
Chmelka, B.F., 324
Chu, C.T.-W., 324
Churaev, N.V., 36–37, 169–171
Cohan, L.H., 49, 116, 309
Colina, C.M., 33, 48
Condon, J.B., 14, 36–37, 79, 93–94, 106, 110, 112, 147, 157, 171, 180–181, 209, 236, 241–242, 256, 261, 268, 293–295, 324, 331–332
Cordero, S., 20
Cranston, R.W., 96, 126, 218, 294–295
Cutting, R.A., 112, 122, 125, 221–222

D

Danner, R.P., 271, 293
Davis, H.N., 201
Davis, H.T., 2, 335
de Boer, J.H., 30
de la Puente, B., 237
deBoer, J.H., 96, 122, 125, 146–147, 218–219, 311–312
Debye, P., 171
Deitz, V.R., 66, 246–247
Deming, L.S., 10, 43–44, 95–96, 291
Deming, W.E., 95–96, 291
Deming, W.S., 10, 43–44
Denbigh, K., 268, 341–342
Dennis, K.S., 256, 260
Denoyel, B., 52
Derjaguin, B.V., 170–171
Di Renzo, F., 21
Dietz, V.R., 29
Do, D.D., 239–240, 316
Dodge, B.F., 201
Domingues, A., 20
Donohue, M.D., 20–21
Drain, L.E., 181–182
Draper, H.D., 228
Dubinin, M.M., 95–96, 180, 247–248, 268, 292–293, 331–332

E

Eager, M.H., 236
Eckstrom, H.C., 102
Emmett, P.H., 27, 79, 95–96, 101–102
Esparza, J.M., 20
Euckon, A., 142
Evans, R., 353
Everett, D.H., 3, 4*t*, 19, 97, 125, 178
Ewing, G.E., 237

F

Fajula, F., 21
Felipe, C., 20
Fenelonov, V.B., 96, 126
Feng, J., 324

Fischer, J., 357
Foster, A.G., 309
Fowler, R., 30
Fraesier, B.C., 353
Fubini, B., 21
Fuller, E.L., 36–37, 43, 93–94, 126–127, 147, 156, 178, 223, 225, 236, 241–242

G

Galareau, A., 21
Gales, L., 30
Gammage, R.B., 126–127, 178, 223, 225
Garrido- Segovia, J., 126
Garrone, E., 21
Gavrilov, V.Y., 96, 126
Gennady, Y.G., 336
Gil, A., 237
Giona, M., 120–121
Giustiniani, M., 120–121
Glasson, D.R., 127, 225
Goldman, F., 109
Goldmann, F., 301–302
Grange, P., 237
Gregg, S.J., 10, 33–34, 112, 125, 179–180
Grillet, Y., 82
Grün, M., 39–41
Guggenheim, E.A., 30
Guo, X., 237

H

Hachicha, M.A., 138
Halenda, P.H., 294–295
Haller, G.L., 2, 94, 336, 357
Halsey, G., 331
Harkins, W.D., 36–37, 79, 81, 107–108, 256
Harris, R.M., 21
Haul, R.A.W., 19
Henderson, D., 2
Hiemenz, P.C., 97, 120–121
Higgins, J.B., 324
Hill, T.L., 30, 181–182
Holmes, H.F., 126–127, 178, 223, 225
Holyst, R., 20–21
Hoover, W.G., 351
Huckel, E., 171
Huo, Q., 324

I

Inkley, F.A., 96, 126, 218, 294–295
IUPAC Geen Book, 3
IUPAC Technical Report, 3

J

Jaroniec, M., 95, 105, 112, 316, 327
Jhon, M.S., 317
Johnson, M., 353
Jones, L.L., 236
Joyner, L.G., 79, 294–295
Jura, G.J., 36–37, 79, 81, 107–108, 256

K

Kaganer, M.G., 36–37, 95–96, 106
Kanda, H., 316
Kaneka, K., 316
Kaneko, K., 3, 20–21
Karnaukhov, A.P., 96, 126
Khalfaoui, M., 138
Kington, G.L., 260
Knani, S., 138
Koch, S., 105
Korhause, K., 20
Kowalczyk, P., 20–21, 316
Krege, C.T., 324
Krug, M., 112, 327
Kruk, M., 95, 105
Kupgan, G., 33, 48

L

Lamine, A. B., 138
Landers, J., 336
Langmuir, I., 30, 43
Leonowicz, M.E., 324
Lide, D. R., 67–68
Linsen, B.G., 96, 125, 218
Lippens, B.C., 125, 218
Liyana-Arachi, T.P., 33, 48
Llewellyn, P., 2, 28, 83
Llewellyn, P.L., 46, 63, 75–76, 79, 244, 285, 309np, 311np
Lopez-Gonzalez, J.D., 29, 246–247
Lopez-Gonzolez, J., 66
Los, J.M., 181–182

M

Martin-Martinez, J.M., 126, 233–234
Masters, K.J., 126
Maurin, G., 2, 28, 83
McCullen, S.B., 324
McEnaney, B., 126, 233–234
McGavack, J., 228, 232
Melnitchenko, A., 34–35
Mendes, A., 30
Miyahara, M., 316

Molina-Sabio, M., 126
Monson, P.A., 5–6
Morrison, J.A., 181–182
Moscou, L., 19
Moyes, R.B., 6–7
Murdmaa, K.O., 268, 331–332
Myers, M.L., 5–6

N

Neimark, A.V., 2–3, 20–21, 94, 336, 357
Neimark, V., 336
Nguyen, C., 239–240
Nicolan, G.A., 227
Nordholm, S., 353

O

Okazaki, M., 316
Olivier, J.P., 3, 105, 356–357
Olson, D.H., 324
Ortiga, J., 34–35
Osinga, T.J., 96, 218

P

Pace, E.I., 256
Pak, H., 317
Palethorpe, S.R., 34–35
Parfitt, G.D., 125, 178
Partyka, S., 82
Patrick, W.A., 228, 232
Payne, D.A., 125, 275–276
Percus, J.K., 349–350, 353
Peters, S.J., 237
Pickering, H.L., 102
Pierotti, R., 19
Polanyi, M., 109, 121, 142, 301–302
Polley, M.H., 260
Prado-Burguete, C., 126, 233–234

Q

Qiao, S.Z., 318, 324
Qiu, S., 237

R

Radushkevich, L.V., 95–96, 247–248, 292–293
Rammler, E., 293
Ravikoritch, N.A.V., 39–41
Ravikovitch, P.I., 2, 20–21, 94, 336, 357
Ree, F.H., 351
Riccardo, J., 20
Rodrígues-Reinoso, F.J., 3, 46, 63, 75–76, 79, 126, 233–234, 244, 285, 309np, 311np

Roja, F., 20
Rosin, P., 293
Roth, A., 66
Roth, W.J., 324
Rouquerol, F., 2, 28, 82–83
Rouquerol, J., 2–3, 19, 28, 82–83
Rouquerol, R., 82
Rudzinski, W., 97
Ruthven, D.M., 91

S

Schlenker, J.L., 324
Schmitt, K.D., 324
Schober, T., 171
Schüth, F., 39–41
Semchinova, O.K., 242–243
Setzer, M.J., 93–94, 171
Sheppard, E.W., 324
Sherwin, C.W., 150
Shull, C.G., 294–295
Siemienwska, T., 19
Silvestre-Albero, A.M., 46, 63, 75–76, 79, 244, 285, 309np, 311np
Silvestre-Albero, J., 46, 63, 75–76, 79, 244, 285, 309np, 311np
Sing, K.S.W., 2–3, 10, 19, 28, 33–34, 83, 96, 98, 112, 122, 124–125, 169, 178–180, 221–222, 275–276
Smirnov, E.P., 242–243
Smith, W.R., 260
Sokolowski, S., 357
Solarz, L., 20–21
Starke, G., 36–37, 169–170
Starling, K.E., 67–68, 351
Steele, W.A., 355–356
Stucky, G.D., 324

T

Tanaka, H., 20–21
Tanchoux, N., 21
Tarazona, P., 353
Teichner, S.J., 227
Teller, E.J., 10, 27, 43–44, 95–96, 291
Terzyk, A.P., 20–21, 316
Thiele, E., 351
Thommes, M., 3
Thompson, J.G., 34–35
Thompson, K.A., 69–71, 236, 240–242
Tolman, R.C., 317
Torkia, Y.B., 138

Torregrosa, R., 126
Tóth, J., 107, 250
Trens, P., 21
Turk, D.H., 112, 122, 125, 221–222, 275–276

U
Uffmann, D., 242–243
Unger, K.K., 39–41

V
Vartuli, J.C., 324
Vishayakov, A., 357
Volmer, M., 30
Volsone, C., 34–35

W
Ward, R.J., 82
Wells, P.B., 6–7
Wenzel, L.A., 271, 293
Wheeler, A., 294–295
Wilson, R., 125

Wisneiwski, K.E., 305
Wojsz, R., 305

X
Xiao, F.-S., 237

Y
Yahia, M.B., 138
Yang, R.T., 91
Yevick, G.J., 349–350, 353
Yoshioka, T., 316
Yu, H., 237
Yu, J., 237

Z
Zaverina, E.D., 95–96, 247–248, 292–293
Zhao, D., 324
Zhao, X.S., 318, 324
Zhukovskaya, E.G., 268, 331–332
Zou, Y., 237
Zwikker, C.Z., 122, 146–147, 219

Subject index

Note: Page numbers followed by *f* indicate figures, *t* indicate tables, and *np* indicate footnotes.

A

Absolute adsorption, 5–6
Absolute method, 35–36. *See also* Harkins-Jura absolute method
Additive principle, 177–178
Adiabatic calorimetry, 77–79
Adsorbate
　area parameters, 40*t*
　definition, 4*t*
Adsorbate-adsorbent surface area, 81
Adsorbent, 4*t*
Adsorption, 4*t*
　Ar
　　on H_2 cleaned diamond, 243, 243*f*
　　on polytetrafluoroethylene (Teflon®), 240–241, 241*t*, 242*f*
　　on thoria, 223, 224*f*
　　on tin II oxide, 219, 220*f*
　binary, 267–275
　of CS_2 on carbon carbon, 302–303, 303*f*
　definition, 5, 10
　DRK-transformed plot, 98, 99*f*
　error analysis, 53–54
　of ethylene chloride on carbon by GP, 302–303, 302*f*
　of ethyl ether on carbon by GP, 302–303, 302*f*
　of gases on 25°C out-gassed thoria, 223, 225*t*
　of gases on lunar soil, 225, 227*t*
　heats of, 255–261
　of I_2 on CaF_2, 219, 219*f*
　multiplane, 246–255
　N_2 adsorption on thoria, 223, 224*f*
　for non-porous materials, 96–97, 96*f*
　on non-porous surfaces, 262–267
　of n-pentane, 302–303, 303*f*
　of SO_2 on SiO_2 gel, 232, 232*f*
　types, 10–14
　of water at 25°C after adsorption cycles, 223, 225*f*
Adsorption potential, 181
Adsorptive, 4*t*
Advantage of gravimetric method, 73–74
Alpha-α_s-curve standard, 96, 124–125, 128–131*t*, 233–234, 234*f*
Alpha-s standard plots, 221–222

Anatase, 256, 258
　Kr adsorption on, 258–260, 259–260*f*
　oxygen and nitrogen adsorption on, 260–262, 261–266*f*
Arbitrary function, 282, 284, 320
　porosity determination, 320–321
Argon (Ar) adsorption
　on $Al_2(SO_4)_3$, Bradley isotherms, 228–230, 229*f*
　on $CuSO_4$, Bradley isotherms, 228–230, 230*f*, 231*t*
　on H_2 cleaned diamond, 243, 243*f*
　on lunar soil, 137*t*, 225, 226*f*
　on polytetrafluoroethylene (Teflon®), 240–241, 241*t*, 242*f*
　on SiO_2 by Nicolan and Teichner, 227, 228*f*
　on thoria, 136*t*, 223, 224*f*
　on tin II oxide, 220*f*
　on SiO_2 data α-s plot, 221–222, 222*f*
Arnold flow method, 9, 83–84
　operations, 84–86, 85*f*
　system description, 78*f*, 83–84
Auto shielding physisorption (ASP) theory, 93–94, 142, 147

B

Barium sulfate, 219–220
Barrett Joyner and Halenda (BJH) method, 294–295
Bayard-Alpert type gauge, 69–71
BDDT equation, 291–292, 292*f*
Bed porosity, 43*np*, 77
Berg isotherm, Kr adsorption on anatase, 258–260, 259–260*f*
Berg isotherms and calorimetry, 258
BET, 31
BET analysis, 100–102
BET method, 253
BET single point method, 16, 18
BET theory equation, 32
Binary adsorbate method, 195–208
Binary adsorption, 195–208, 270–271
　by chi theory, 261
　in micropores, 267–274
　　comparison to experiments, 271–274
　　constant pressure, 270–271

409

Binary adsorption *(Continued)*
  Lewis' rule assumption, 268–270
  on nonporous adsorbents, 267
Boundary errors, 65
B point analysis, 10, 15–16
Bradley-derived van-Hoff plot, 230, 231*f*
Breakthrough time, 90
Broekhoff-deBoer (BdB) theory, 311–317, 328–329
  for adsorption in cylindrical pore, 313, 313*f*
  isotherms for adsorption on porous SiO_2, 314–316, 315–316*f*
  vs. Kelvin-Cohan calculation, 316, 317*f*
  of pore filling and pore emptying, 314, 315*f*
Bruauer, Emmett, and Teller (BET) theory, 16, 30–32, 48–49, 54, 93–94, 100–102, 104
  assumptions, 32
  BET constant, 32
  bottom line for, 35
  isotherms, 34
  linearization, 33
  problems and restrictions, 33–34
  surface area, 260–261
  surface dependent reactions, 34–35
Buoyancy, 71

C

Calorimetry, 5, 8–10, 77. *See also* Differential scanning calorimetry (DSC)
  absolute method, 79–81
  adiabatic, 77–79
  isosteric heat measurement, 79
Capillary filled, 47
Carbon, 233–234
  IUPAC standard isotherm, 125, 133*t*
  KFG standard curve, 126, 134*t*
  RMBM standard isotherm, 126, 134*t*
Carnahan-Starling (CS) approximation, 345, 351, 352*f*, 354–355
Carnahan-Starling-van der waal approximation, 351
Carrier gas flow method
  advantages, 86–87
  engineering aspects, 87–91
    Basmadjian symbolism, 90
    diffusion-sink situation, 88*f*
    Henry constant, 91
  signal, 86*f*
Carrier gas flow system, 9
Catalyst, 2
CaY-zeolite, 305
Chemical bonding, 2–3

Chemisorption, 6–7, 223
  definition, 3, 4*t*
Chi depth profile, 184–185
Chi(χ) plot, 1, 4*t*, 36–37, 93–94, 142
  additive principle, 177–178
  adsorption isotherms, 22–23, 22–24*f*, 25–26, 27*f*
  advantages, 41–42
  analysis, 102–104
  binary adsorption, 195–208
  depth profiles, 184–193
  disjoining pressure derivation, 169–174
  Dubinin isotherm, 208–210
  energy implications, 236–237
  features, 25*t*
  Freundlich isotherm, 208–210
  hard sphere geometry, 167–168
  heats of adsorption, 180–184
  heterogeneous surfaces, 177–180
  isotherm interpretation, 102–104
  Lennard-Jones correction, 167–168
  methane on MgO, 138, 139*f*
  n-layer calculations, 184–193
  vs. NLDFT, 358–359
  for non-porous single energy adsorbent, 38
  observation of, 235–246
  perturbation theory for adsorption, 150–156
  quantum mechanical derivation, 147–150
  spreading pressure, thermodynamic relationships for, 210–213
  temperature changes, 161–162
  threshold pressure, 237–246
  vapor pressure changes, 160–161
Chi(χ) theory
  distribution fit, interpretation of, 298–308
  hysteresis, 328–329
  n-layer simulation, 329–330, 330*f*
  vs. NLDFT, 358–359
  prefilling, 318–319
  QM derivation, 147–150
Chi theory equation, 158
Classifications of isotherms, 10–14
Clausius-Clapeyron, 21
CO adsorption
  on lunar soil, 225, 226*f*
  on 5A zeolite, 271*f*, 272
CO-N_2 adsorption, on 5A zeolite, 271*f*, 272
Configuration partition function, 343, 345–346
Configuration probability distribution function, 343
Constant pressure, binary adsorption, 270–271
Copper II oxide, 228
Copper II sulfate, 228–230

Correlation functions, 340–341, 344
Cranston and Inkley standard, 96, 126, 218, 294–295
CuO, water on, 233
Cylindrical pores, 300–301

D

Danner and Wenzel (DW), 293
 CO and O_2 adsorption
  on 5A zeolite, 297, 298f
  on 10X zeolite, 297, 297f
DD52 carbon, N_2 isotherm on, 285, 286f
Dead space, 62
Dead space calculating, 63
Dead volume, 71
deBoer-Zwikker formulation, 122
deBoer-Zwikker polarization theory, 122, 142, 146–147, 230
deBroglie wave equation, 154
Debrolie wave length, 352–353
Density functional theory (DFT), 5, 35, 93–94, 96, 104, 140, 274–275, 281, 335
 Carnahan-Starling approximation, 351
 χ theory vs. NLDFT, 358–359
 correlation functions, 340–341
 direct correlation functions, 344–345
 functional derivative, 339–340
 hard rod approximations, 345–346
 hard rods between two walls, 347–349
 Helmholtz free energy, 352–353
 modeling with presence of surface, 355–357
 Monte Carlo technique, 357–358
 non-local density functional theory, 353–355
 partition functions, 341–343
 Percus-Yevick solution expansion for hard spheres, 349–350
 Thiele analytical approximation, 351
Design error, 65
Desorption, 4t
DFT. See Density functional theory (DFT)
Diamond powder, 242–243
Differential scanning calorimetry (DSC), 82–83
Direct correlation functions, 344–345, 355
Disadvantage of gravimetric method, 73–74
Disadvantage of volumetric method, 68
Disjoining pressure, 165, 167, 169–170
Disjoining pressure isotherm, 255
Disjoining pressure theory, 27–29, 35–37, 42–43, 55, 93–94, 96, 141, 169–174, 252–255, 261, 310, 316, 358

Distribution of E_a values, 160, 179–180
Dog-leg, 238–240, 239f
DP. See Dubinin-Polanyi (DP) isotherm
DR. See Dubinin-Radushkevich (DR) transform
DRK equation, 106
DRK-transformed plot, 98, 99f, 106
Dubinin-Astakhov (DA) equations, 248, 292–293, 294f, 294t
Dubinin isotherm, 208–210
Dubinin-Kaganer (DK) isotherm, 95–96
Dubinin-Polanyi (DP) equation, 248, 293
Dubinin-Polanyi (DP) isotherm, 146, 180, 208–210, 210f, 247–251, 248–249f, 283–284, 293
Dubinin-Radushkevich (DR) isotherms, 247–248, 249f
Dubinin-Radushkevich (DR) equations, 106, 248, 292–293
Dubinin thermodynamic criterion, 159, 169, 181, 183–184, 253, 293
DW. See Danner and Wenzel (DW)
Depth profiles, 184–193

E

Effective surface area, 34
Energy distribution, 235, 238–239, 247–248, 251, 255, 285, 295, 305, 310–311, 318, 359
Energy distribution for chi, 239, 248
Energy of adsorption (E_a), 26, 32, 37–38
Equipment capabilities, 69–71
Equipment cost, 69–71
Equipment gravimetric description, 69–71
Equipment requirements
 gravimetric, 69–71
 volumetric, 60
Equipment volumetric description, 60–62
Error analysis, 53–55
 bed porosity, 77
 general, 74
 for gravimetric method, 73
 kinetics, 76
 pressure, 74–76
 sample density, 77
 temperature, 74–76
 for volumetric method, 65–68
Euler-LaGrange, 340
Excess adsorption, 5–6
Excess surface work (ESW) theory, 42, 93–94, 105–106, 214, 281
 for mesopores, 52
 monolayer equivalence, 36–37
External monolayer equivalence, 45

F

Field emission microscopy, 6–7
Film, 308
Film tension, 308
Fowler-Gugenheim isotherm, 30t
Free energy, 335, 339, 353
Frenkel-Halsey-Hill (FHH) isotherm, 122–123, 142
Freudlich-Dubinin-Polanyi equation, 247
Freundlich isotherm, 28, 30t, 32, 94, 120–121, 180, 208–210, 235, 247–251
Functional, 337–339, 338t
Functional derivative, 339–340
Functional derivative chain rule, 344

G

Gas burette, 9
Gas-like adsorption, 7, 7f
Gas-liquid interface, 309, 313, 318
Gibbs' free energy, 311
Gibbs adsorption equation, 28
Gibbs-Duhem equation, 120–121
Gibbs' phase rule, 211–212, 235
Gil, de laPuente, and Grange (GPG) data, 237–239
Goldmann and Polanyi (GP) data, 301–302, 304t
    adsorption
        of CS_2 on carbon, 302–303, 303f
        of ethylene chloride on carbon, 302–303, 302f
        of ethyl ether on carbon, 302–303, 302f
        of n-pentane, 302–305, 303f
Grand canonical ensemble method, 156, 196–205
Grand canonical partition function, 173, 195–196, 208, 213
Grand partition function, 343
Graphon 1 carbon, 219–220
Graphon 2 carbon, 219–220
Gravimetric method, 69
    advantages and disadvantages, 73–74
    determination method, 71–73
    error analysis, 73
    kinetic problems, 76–77
    pressure and temperature measurement error, 74–76
    sample density problems, 77
    system description, 69–71, 71f
    $vs.$volumetric method, 59
Gurvich rule, 109

H

Hard rod approximation, 345–346, 347f
Hard sphere approximation, 162–166, 175
Hard sphere model, 167–168, 174–175
Harkin and Jura (HJ) calorimeter, 79–81, 80f
Harkins and Jura (HJ)
    absolute method, 79–81, 94, 107–108, 256–258
    H_2O adsorption isotherm, 256, 257f
    parameters, 257, 258t
Harmonic oscillator approximation, 355, 356f
Heats of adsorption, 17, 36, 38, 38np, 180–184, 214, 255–261
Helium, 74–79, 86–87
Helmholtz free energy, 352–353
Henry's law, 23–24, 120, 180–181, 211, 235, 247–248
    assumption, 28–29
    disproof, 30–32
    isotherms, 31–32
Henry constant, 91
Heterogeneous surface, 41, 146, 177–180, 186, 214
Hill-deBoer isotherm, 30, 30t
Hydrogen bonding, 2, 2np
Hysteresis, 4t, 310, 328–329
Hysteresis loops, 19–21, 21t, 26, 47, 49
HY-zeolite, 305

I

Inert gas, 60, 63, 69–71, 74–76
Inflection point method, 10, 14, 16, 17f
    correction factor, 17, 18t
    simulation, 17, 17f
In-register adsorption, 6–7, 6f
Integral heats of adsorption, 181–183
Intermolecular force, 2–3
Ionic bond, 2, 6–7
Isosteric heat, 41, 79
Isosteric heat of adsorption, 79, 180–181
Isotherms, 95–96
    BET analysis, 100–102
    by Bradley, 228–232
    χ plot analysis, 102–104
    Dubinin methods, 106–107
    Dubinin-Polanyi, 247–251
    ESW theory, 105–106
    Freundlich, 247–251
    Harkins-Jura absolute method, 107–108
    mesoporosity analysis, 112–117
        geometrical values, 113–114, 114t
        Kelvin equation, 115–117
        parameters, 112, 113f, 113t
    micropore analysis, 109–112
    for multiplane adsorption, 251–255
    by Nicolan and Teichner, 227–228
    non-local DFT theory, 104–105
    objectives, 95–100
    overview, 93

Subject index

porosity determinations, 108–109
specific/molar parameters, 93, 93t
Tóth T-equation, 107, 247–251
type VI, 138–140
Ith data point, 117
IUPAC convention, 39, 40t
IUPAC pore classifications, 108
IUPAC standards, 125

K

Karnaukhov, Fenelonov, Gavrilov (KFG) standard fit, 96, 126
Kelvin-Cohan (KC) equation, 49, 51, 309–311, 331
Kelvin-Cohan formulation, 309
Kelvin equation, 49, 115–117, 309
Kernel, 105, 140
Knudsen number, 66
Kr adsorption, on anatase, 258–260, 259–260f
Krug and Jaroniec(KJ) data, 112

L

Langmuir isotherm, 15, 25–26, 30t, 96–97, 118–120, 141, 235, 252–255
Layer, definition of, 159
Lennard-Jones potential, 167–168, 186–187, 355
Lewis' rule, 268–270
Limit of detection, 68
Liquid flow system, 88
Liquid-like adsorption, 7, 7f
Liquid nitrogen, 60–61, 71, 74–78, 78f, 90–91, 90np
LMA233 carbon, N_2 isotherm on, 285, 286f
Log law, 45–46, 47f, 289–290, 289–290f
Log-normal energy distribution, 209
Lunar soils, 224–227
Lunar soil standard, 127
Lunar soil, standard isotherms, 127, 224–227
Ar adsorption, 137t, 225, 226f
CO adsorption, 138t, 225, 226f
gasses, adsorption of, 225, 227t
N_2 adsorption, 136t, 225, 226f
O_2 adsorption, 137t, 225, 226f

M

Macropores, 108
definition, 4t
standard curve pore analysis, 52–53
Magnesium oxide, 220f
Mass spectrometer analysis, 84
MCM-41 materials, 324
χ plot and pore distribution of Ar, 327, 327f
N_2 adsorption on, 324, 325f

Meso-micropore boundary, 319
Mesopore/micropore equation χ, 319–320
Mesopores, 41–43, 47, 108, 322, 322f
analysis, 308–319, 325t
geometrical values, 113–114, 114t
Kelvin equation, 115–117
parameters, 112, 113f, 113t
equation, 321–322
ESW theory, 52
Kelvin-Cohen value, 51
pore radius, 49–51
standard plot, type IV/V isotherm, 47, 48f
Mesoporosity analysis, 112–117
MgY-zeolite, 305
Micropores, 41–43, 108, 322, 322f
analysis, 109–112
BDDT equation, 291–292, 292f
χ theory interpretation of distribution fit, 298–308
DR and DA equations, 292–293
standard curve analysis, 293–298
binary adsorption in, 267–274
definition, 45
equation, 319–320
Langmuir analysis, 43–44
layering, 45–46
N-layer BET formulation, 44
standard plot, type I isotherm, 44–45
Microporosity, 291
Microporous carbon, Pc observation from data by GPG, 238, 238f
Microporous material, 290, 321, 331–332
Molar area, 165t, 298–300
Molar quantity, 3
Molecular cross section, 295
Monolayer, 4t
Monolayer equivalence, 4t, 36–38
Monte Carlos technique, 104, 255, 357–358
Multiplane adsorption, 246–247
Multiplane adsorption, isotherms for, 246–255
BET method, 253–254
χ and disjoining pressure equation, 252–255
Henry's law isotherms, 255
multiple energy, 255
Multiple energy, 255

N

Nermai direction profile, 185–186
Nickel antigorite, 219–220
Nickel oxide, 227

Nitrogen adsorption, 221–222, 233–234, 236–237, 243, 250–251, 260–261, 262–263f, 266
  on anatase, 260–262, 262–266f
  on DD52 activated carbon, 285, 286f
  on graphitized carbon black, 218, 219f
  on H_2 cleaned alumina, 243, 244f
  high-resolution isotherms, 244, 245f
  on LMA233 carbon, 285, 286f
  on lunar soil, 136t, 225, 226f
  on MgO aerosil, 219–220, 220f, 221t
  on Ni antigorite, 219–220, 221f, 221t
  on NiO by Nicolan and Teichner, 228, 229f, 229t
  of out-gassed thoria, 135t
  on SBA-15 silica, 286, 287f
  on SiO_2 by Nicolan and Teichner, 227, 228f
  on SiO_2 data α-s plot, 221–222, 222f
  on Sterling FT carbon, 246–247, 247f
  on thoria, 223, 224f
  on ultramicroporous carbon, Nguyen and Do data, 239–240, 239f
n-layer calculation, 184–194, 205–208
Non-local density functional theory (NLDFT), 353–355, 358–359
chemical potential, 356
  mismatch between standard curve and, 336, 337f
  non-porous portion of isotherm, 336, 337f
Nonporous adsorbents, binary adsorption on, 267
Non-porous surfaces, adsorption on, 262–267
Normal direction distribution, 189–193
Number density, 340–341, 346–348, 347f, 349f

O

O_2 adsorption
  on anatase, 260–262, 261–266f
  on lunar soil, 225, 227f
Olivier monolayer criterion, 146, 213
Omstein-Zernicke equation, 345
One-dimensional pressure, for hard rods, 346, 348f
OPALSPA, 320

P

Packing factor, 38–41, 102–103, 113–114
Packing problem uncertainty, 188–189
Partition functions, 341–343
Percus-Yevick solution, 349–350
Perturbation theory, 150–156
p formulation, 347–348
Physical adsorption isotherm, 14t, 15
Physisorption, 233–234, 246–247, 252–255
  definition, 2–3, 4t
  gas-like and liquid-like, 7, 7f
  in-register adsorption, 6–7, 6f

macropore analysis, 52–53
mesopore analysis, 47–52
micropore analysis, 43–46
surface area measurement
  gravimetric method, 8
  volumetric method, 9
ultramicropore analysis, 46–47
PMF. See Probability mass function (PMF)
Polanyi formulation, 121–122, 142
Pore analysis methods, 109–117
Pore filling, 21, 46
Pore size, 284–285, 291, 293–296, 300, 308, 319, 322, 324, 326–330, 358–359
Pore surface, 42
Porosity calculations
  arbitrary function, 320–321
  combined mesopore/micropore equation χ, 319–320
  mesopore equation, interpretation of, 321–322
  mesoporosity analysis
    Broekhoff-deBoer (BdB) theory, 311–317
    Kelvin-Cohan equation, 310–311
  micropore analysis, 291–308
    BDDT equation, 291–292
    χ theory interpretation of distribution fit, 298–308
    DR and DA equations, 292–293
    standard curve analysis, 293–298
  ultramicropore analysis, 284–290
Pre-filled, 47
Pre-filling, 45–46, 281–282
Probability mass function (PMF), 113, 116, 282, 295
Psudo-Henry's law, 90–91

Q

Qiao, Bhatia, and Zhao (QBZ)
  C-10, adsorption on, 326, 326f
  N_2 adsorption on MCM-41, 324, 325f
Quantum mechanics, 16, 37, 55
Quantum mechanical classification, 22–26
Quantum mechanical derivation, 147–150, 207
Quantum mechanical harmonic oscillator (QMHO), 146, 186–187, 289
Quantum mechanical perturbation theory, 176–177, 213
Quantum mechanical theory, 93–94, 122
Quenched solid density functional theory (QSDFT), 53, 93–94, 285, 336

R

Radiative sample heating, 75
Raoult's law, 262

Subject index

Reference Helmholtz, 353
Reversible pore filling, 21
RMBM carbon standard, 126, 134*t*, 233–234

S

Scaling factor, 106
Scanning calorimetry, 5
Semiempirical theory, 106
Sievert's equilibrium, 118
Silica aerogel, 222
Silicon oxide, 222*f*
Silvestre-Albero (SSLR) data, 285
Simultaneous measurements, 258–260
Single point methods, 14–16, 18
Smoothed density function, 353
SO_2
    adsorption on SiO_2 gel, McGavack and Patrick data, 232, 232*f*
    water contamination of, 232, 233*f*
Specific surface area, 3, 37–38
Spreading pressure, 210
Standard curve method, 293–298
    α_s-curve, 124–125, 128–131*t*
    Cranston and Inkley standard isotherm, 126
    IUPAC isotherms
        carbon samples, 125, 133*t*
        silica, 125, 133*t*
    KFG coefficients, carbons, 126, 134*t*
    for lunar soil, 127, 136–137*t*
    problems in, 124
    RMBM carbon standard, 126, 134*t*
    steps in, 123
    t-curve, 125, 132*t*
    thoria, 126–127, 135–136*t*
Standard isotherms, 94, 124, 217–234
Standard plot, 4*t*, 42–43
    macropore analysis, 52–53
    mesopore analysis, 47, 48*f*
    micropore analysis, 43–46
Standard *t*-curve
    Cranston and Inkley, 218, 218*f*
    deBoer, 218–221, 219*f*
Standard thoria plots, 223
Statistical comparisons, of isotherms to χ-plot, 275–277, 275*f*, 276*t*
Sterling FT carbon, 125, 178, 219–220
    N_2 adsorption, 246–247, 247*f*
Structural constant, 106
Surface area determination, 100–108
Surface areas, calculations, 298–300
Surface reaction, 3, 3*np*

T

t-curve, 96, 125, 132*t*
T-equation, 249*f*
Thermodynamic criterion. *See* Dubinin thermodynamic criterion
Thermodynamic disjoining pressure theory, 214
Thiele analytical approximation, 351
Thoria adsorption isotherm, 223
Thoria, 178
Thoria plots, 223
Thoria standard curve
    for argon adsorption, 126–127, 136*t*
    nitrogen isotherms, 126–127, 135*t*
    water adsorption, 126–127, 135*t*
Threshold pressure, 30, 157
    direct observation, 237–246
    energy indications, 236
    N_2 adsorption on microporous carbon, GPG data, 237–239, 238*f*
    N_2 adsorption on ultramicroporous carbon, Nguyen and Do data, 239–240, 239*f*
    observation by Thompson, 240–244
Tin II oxide, argon adsorption on, 219, 220*f*
Titration, 119
Tóth isotherm, 208–209, 247–251, 249–250*f*
Tóth T-equation isotherm, 94, 107, 141
Transformed BET plot
    organic material, 97, 98*f*
    silica material, 97, 98*f*
Two plane adsorption, 246–247
Type I isotherm, 10, 11*f*
Type II isotherm, 10, 11*f*, 15, 22*f*, 89–90, 96–97, 138
Type III isotherm, 12*f*, 23*f*, 26, 89, 89*f*
Type IV isotherm, 12*f*, 23*f*, 25, 34, 36
Type V isotherm, 13*f*, 24*f*, 34
Type VI isotherm, 11, 13*f*, 23–24, 26, 26*f*, 34, 138–140

U

Ultramicropores, 45, 109
    analysis, 46–47, 284–290
    definition, 43, 46, 52–53
Ultramicroporous carbon, N_2 on, 239–240, 239*f*
Ultramicroporous, definition, 45

V

Vacuum flow method, 9
Vacuum flow system, Arnold, 83–84, 83*f*
Valence bond theory, 29
Van der Waal equation, 66–68, 353
Van der Waal radius, 39, 166, 316–317

Virial equation, 28
Volmer isotherm, 30*t*
Volumetric measurement, 60
   advantages, 59, 68
   determination method, 62–65
   disadvantages, 68
   error analysis
      boundary errors, 65
      design errors, 65
      equation of state errors, 67–68
      limit of detection, 68
      molecular flow *vs.* viscous flow, 66–67
      poor calibration, 66
      sample temperature control, 65
      temperature error reading, 68
   vs. gravimetric method, 59
   kinetic problems, 76–77
   pressure and temperature measurements, 74–76
   sample density problems, 77
   system description, 60–62, 61*f*

VTMOP, 34–35, 35*t*
Vulcan carbon, 178

W

Water
   on CuO, Bray and Draper data, 233
   on Y-zeolites, Wisniewcki and Wojsz data, 305, 306*t*, 307*f*
Water adsorption isotherm, 223
Weibull distribution, 293
Weighting factor, 33, 53–54
Wisniewcki and Wojsz (WW) data, 305

Z

5A Zeolite
   CO isotherm on, 271*f*, 272
   CO-N_2 adsorption on, 271*f*, 272–273, 274*f*
   CO-O_2 mix on, 272–273, 273–274*f*
Zirconium oxide, 219–220